THE COAL NATION

The Coal Nation
Histories, Ecologies and Politics of Coal in India

EDITED BY

KUNTALA LAHIRI-DUTT
Australian National University, Australia

Routledge
Taylor & Francis Group

LONDON AND NEW YORK

First published 2014 by Ashgate Publishing

Published 2016 by Routledge
2 Park Square, Milton Park, Abingdon, Oxfordshire OX14 4RN
711 Third Avenue, New York, NY 10017, USA

First issued in paperback 2016

Routledge is an imprint of the Taylor & Francis Group, an informa business

British Library Cataloguing in Publication Data
A catalogue record for this book is available from the British Library

The Library of Congress has cataloged the printed edition as follows:
Lahiri-Dutt, Kuntala, 1956-
  The coal nation : histories, ecologies and politics of coal in India / by Kuntala Lahiri-Dutt.
    pages cm
  Includes bibliographical references and index.
  ISBN 978-1-4724-2470-9 (hardback)
  1. Coal trade--India. 2. Coal mines and mining--India. I. Title.
  HD9556.I42L34 2014
  338.27240954--dc23

                                                                    2013033196

ISBN 13: 978-1-138-27203-3 (pbk)
ISBN 13: 978-1-4724-2470-9 (hbk)

Contents

PART III   SOCIAL PERSPECTIVES TO INFORM MINING POLICY

List of Figures

List of Tables

The Contributors

Dr Nesar Ahmad is an expert on social development in mining areas of eastern India. He is currently working as a researcher in Budget Analysis Rajasthan Centre (BARC), India. He has worked on the issues of mining-induced displacement and its impact on women; the impact of economic liberalization on the poor, women and children; poverty and child rights; and liberal analysis of government budgets. His publications include (Jointly with Kuntala Lahiri-Dutt) 'Missing concerns: Engendering mining displacement in Jharkhand', *Gender, Technology, Development*, 2007, 10(3): 313–39. Nesar has undertaken doctoral research on 'Gender, Technology and Institutions in Indian Coal Mining Industry' at the Jawaharlal Nehru University, New Delhi. He is the author of *Women, Mining and Displacement*, published by the Indian Social Institute, New Delhi in 2002.

Dr Gita Bharali is a social researcher currently based in Indianapolis, USA. Until recently, she was the director of Research at North Eastern Social Research Centre (NESRC) located in Guwahati. Gita has undertaken research on development-induced displacement, cost-benefit analysis of people-displacing projects and gender issues at the NESRC, and has published several professional articles and five books. She was also involved in the drafting of State Water Policy 2007, Government of Assam, and was a regular contributor to the daily newspapers in Assam.

Dr Ananth Chikkatur holds Physics degrees from the University of Rochester and the Massachusetts Institute of Technology, and has extensive experience in natural gas market analysis, power market analysis, and carbon capture and storage technologies. Currently, he is a Manager at ICF International where his work includes energy and technology policy, with a focus on power and fuel market analysis. He has worked on broader energy policies in India, energy efficiency, and issues related to global climate change. His interests include technology innovation, cleaner coal power, energy efficiency, small-scale/rural energy systems, and the politics of, and responses to, climate change. Prior to joining ICFI, Dr Chikkatur was a Research Associate and Fellow at the Energy Technology Innovation Policy project at Harvard Kennedy School's Belfer Center for Science and International Affairs.

Dr Amarendra Das is currently an Assistant Professor in the Postgraduate Department of Analytical and Applied Economics of Utkal University in Bhubaneswar, Odisha. He has obtained his PhD from the Centre for Development Studies, Thiruvananthapuram, Kerala under Jawaharlal Nehru University, New

Delhi. His PhD dissertation examined the performance of public and private mining firms in India in productivity, environmental and social dimensions. He won first prize in the 2008 Global Development Awards and Medals Competition under the theme Natural Resource Management and Development. His teaching and research interests lie in environmental and natural resources economics and public economics.

Dr Debojyoti Das is a European Research Commission (ERC) Post-Doctoral Research Associate in the Department of History, Classics and Archeology at Birkbeck, University of London, working on a project titled 'Coastal Frontiers: Water, Power and Boundaries in South Asia.' His PhD in Anthropology at SOAS focused on the moral economy of highland farming in the northeastern part of India (Nagaland). He has been working on issues of political economy of highland agriculture, the anthropology of natural hazards and environment, and the anthropology of aid and development in the context of South Asia. He has contributed papers to *Economic and Political Weekly*, *Water History Journal*, *Journal of Borderland Studies*, *ORF Energy Monitor*, and several edited book volumes.

Dr Walter Fernandes, former director of the Indian Social Institute, New Delhi, and Editor of the journal *Social Action*, was the founder-director of the North Eastern Social Research Centre, Guwahati, Assam from 2000 to 2011. He has been working on development-induced displacement in India from 1947 to 2010. Currently, he is facilitating a large study of displacement, covering 20 of India's 28 States. Dr Fernandes' primary research focus has been on tribal and gender issues, and livelihoods and conflict resolution in order to promote peace. He has written about 35 books, 170 professional articles and 100 newspaper articles on these issues.

Tony Herbert is a Jesuit priest who has spent the greater part of his life in Jharkhand, India, living and working closely with the indigenous people of that area. He has focused on social awareness training programs, and for several years was director of Prerana Resource Centre, Hazaribag. He has played a significant role in various social projects: in the Panos Institute's oral testimony project among mining-displaced tribal communities; in Woodstock (Georgetown University) Center's Global Economy and Cultures Project, regarding the impact of globalisation on indigenous people; and also in assisting with the advocacy of various indigenous communities faced with the social and environmental impact of mining. He has authored several articles and a study titled 'A Cultural Journey Among the Bhuinyas of Jharkhand' (Published by Indian Social Institute, New Delhi in 2010). He can be contacted at atherbert@gmail.com.

Ms Radhika Krishnan, an electrical engineer by training, is currently undertaking her doctoral studies on environmental policy at the Centre for Studies in Science

Policy (CSSP) in Jawaharlal Nehru University, New Delhi. Ms Krishnan has a broad experience in research, particularly on coal but also on the mining of other minerals, and was part of the team at the Delhi-based research and advocacy group Centre for Science and Environment (CSE) that prepared the Citizen's Report (no 6, State of India's Environment), *Rich Lands, Poor People.*

Dr Kuntala Lahiri-Dutt is a Senior Fellow at the Crawford School of Public Policy, the College of Asia and the Pacific, at the Australian National University (ANU). Before joining the ANU, Kuntala was a Reader of Geography at the University of Burdwan, where her research on the feminist political economy of the eastern Indian coalbelt was funded by the Ministry of Environment and Forests (Government of India), the University Grants Commission, and other bodies. Since 1994, Kuntala has extensively researched the interface of mining, gender and community, accomplished with funding from a number of research grants, including the Australian Research Council. Her primary research in eastern Indian collieries has extended into the informal quarries in India to examine the grey area between legal and illegal mining, and more recently into the implicartions of mineral dependence in rural and remote parts of Indonesia and Lao PDR. Her body of work on Indian mining has contributed to the understanding of the political ecology of coal and mining-induced social changes in the Indian colliery belt. Her staff page lists her recent publications: http://crawford.anu.edu.au/crawford_people/content/staff/rmap/klahiridutt.php.

Dr Prajna Paramita Mishra holds a doctoral degree in Economics from the University of Hyderabad. Currently, she is Assistant Professor at the School of Economics, the University of Hyderabad, India. Dr Mishra's main research interest is in the area of mining, its economic valuation, and its impacts on livelihoods and the environment. Currently she is working on coal mining and livelihoods issues in the Indian state of Jharkhand and Chhattisgarh. She teaches Microeconomics theory and Environmental Economics. Prior to joining the University of Hyderabad, she worked as Assistant Professor in the Research Unit for Livelihoods and Natural Resources (RULNR), the Centre for Economic and Social Studies (CESS), Hyderabad.

Dr Dhiraj Kumar Nite obtained his Doctorate degree on the topic 'Work and Culture on the Mines: Jharia Coalfields 1890s–1970' from the Centre for Historical Research of the Jawaharlal Nehru University in 2010. He researches and writes on the daily lives of ordinary people in the mining communities of India and South Africa. Presently, he teaches courses on labour relations and wellbeing, race and capitalism in South Africa, and the modern world at Ambedkar University, Delhi. From 2010 to 2012 he was a post-doctoral fellow, and since 2012 a Research Associate with University of Johannesburg.

Dr Patrik Oskarsson holds a PhD from the University of East Anglia, United Kingdom, and is a researcher based in Gothenburg, Sweden. Dr Oskarsson's work relates to the political economy of industrialisation and economic development in India. Of special interest are the various ways in which poor and marginalised groups in rural India are affected in terms of livelihoods, environmental quality and natural resource base by economic development projects.

Ambuj Sagar was trained at the Indian Institute of Technology, Delhi, the University of Michigan, and the Massachusetts Institute of Technology, and is the Vipula and Mahesh Chaturvedi Professor of Policy Studies and the Dean of Alumni Affairs and International Programs at the Indian Institute of Technology, Delhi. Ambuj's interests lie in the policy of science and technology, environment, and development, with particular focus on the interactions between technology and society. He has worked with various agencies of the Indian Government, with numerous international organisations, as well as with other private and public-sector organisations in the United States, including as staff researcher for a major study on energy Research and Development for the White House. Ambuj is a member of the US–India Track II Dialogue on Climate and Energy, and a Board member of the US–India Educational Foundation. He was a member of the Indian Planning Commission's Expert Committee on a Low-Carbon Strategy for Inclusive Growth.

Dr Arupjyoti Saikia is the Associate Professor of History at the Department of Humanities and Social Sciences, Indian Institute of Technology, Guwahati. He received a Ph.D from the University of Delhi. He has authored *A Forests and Ecological History of Assam* (OUP, 2011); and has contributed articles to edited volumes and journals such as *Indian Economic and Social History Review, Studies in History, Indian Historical Review,* and *Journal of Peasant Studies.* His book, *A Century of Protests: Peasant Politics in Assam since 1900* is being published from Routledge (2014). A regular writer of prose in Assamese, Saikia was the recipient of the Yale University post-doctoral fellowship for the Agrarian Studies Program during 2011–12.

Preface

(Re)considering the Black Diamond of India

As McDivitt notes in the statement below, mining has created human civilization, and the hunger of human societies for minerals shows no sign of reduction.

> Since before the dawn of recorded history man has looked to the earth for materials from which to build his shelters and to make his tools and utensils. Over the centuries, this gouging and scratching at the surface to find clay, flint, bright stones or occasional pieces of native copper evolved into a burrowing beneath the surface in the broadening search for mineral materials. As history has progressed, the need for minerals has increased, and the search has gone on. ... The gargantuan appetite for raw materials and the ability of the earth to satisfy this appetite is the starting point for any consideration of minerals.

(McDivitt 1965,[1] 3–5)

This has been true for coal like no other mineral. Studying coal, scholars from a number of disciplinary backgrounds have shown that the relationship between its mining and human society has neither been straightforward, nor simple. Coal reserves, seen as endowments of nature, liberated human societies from the insuperable constraints imposed by the limitations of physical labour, but also rendered them utterly dependent on coal supplies over which they had little control. Coal formed the basis of the accumulation of industrial capital, industries and national wealth, but also led to murky industrial gloom and environmental wastelands, and intense struggles between capitalists and mineworkers.[2] Coal stands at the heart of the intertwined social, industrial and environmental histories, interlinking them with geology and human geographies. Politics is ingrained in coal mining; conventional modes of administration become difficult for coal reserves which often cut across multiple administrative units when spreading over large tracts of land.

Like most other countries, the state of India claims complete ownership of the valuable coal resources, viewing coal mining as the key to further the wellbeing

---

1 McDivitt James F. (1965) *Minerals and Man: An Exploration of the World's Minerals and its Effects on the World We Live in*, Resources for the Future Inc., Johns Hopkins Press, Baltimore.

2 Andrews, Thomas G. (2008) *Killing for Coal: America's Deadliest Labor War*, Cambridge, Ma: Harvard University Press.

of India's people. And indeed, when one is confronted with the fact that about 500 million people live without access to electricity – the official average allocation being one electric light bulb per dwelling – then one might begin to feel there is need to expand the coal mining industry.

It is surprising that one is then confronted with utter disregard for the communities within which new mines are established, and the corporate and policy environment within which they operate. Since the nationalisation of coal mining in India, the state-owned enterprise Coal India Limited (CIL) has been the primary coal extraction operator. One might expect that the coal mining industry would be poised for major changes with liberalisation and privatisation of the Indian economy. Indeed, the *Coal Mines (Nationalisation) Amendment Bill* was proposed in 2000 to fully liberalise the coal industry. The Bill has, however, been waiting to be considered by the Indian Parliament since then. Although the Bill has not been sanctioned yet, the *Coal Mines (Nationalisation) Act, 1973*, was amended in June 1993 to allow 'captive mining' by the private sector for generation of power and for washing of coal obtained from a mine, and for any other end uses notified by Government. These companies were granted mining leases primarily in order to produce coal for thermal power plants. With its monopolistic position, CIL still accounts for 85 per cent of coal production of India, the rest being supplied by SCCL (8.5 per cent), and other private producers (6.5 per cent). The last is the 'captive' mines that are owned in most cases by Indian entrepreneurs in search of capital gain from the current global resource boom, but gaining mining leases in order to produce therman power. These captive mining companies began to operate in an incongruous situation; in this state of affairs coal is accorded iconic and special status as a key agent of nation-building and national prosperity. The status is clearly embodied in major legal provisions such as the *Coal Bearing Areas Act*, which take precedence over other legislative measures that are meant to protect the interests of poor indigenous communities and the environment. Consequently, since the liberalisation of the Indian economy, the nation has witnessed on the one hand the establishment of large coal mines for power generation and on the other, disastrous performance, by both state-owned and privately-owned companies, in social areas, performance that is neither helpful for the poor in the coal-bearing tracts of India nor deemed appropriate by contemporary global or international standards. One then begins to see the tremendous environmental, social, political and economic costs hiding behind simplistic explanations of the need for coal. For example, the last few decades of aggressive expansion in coal mining have led to a gigantic increase in the number of internally displaced people evicted from their homes. Expanding coal mines have exacerbated ongoing conflicts over resources, in certain instances giving rise to protracted armed struggles. The exploitation of coal resources for the apparent benefit of the nation-state has hidden enormous personal and illicit gains made by corrupt individuals with access to state power and decision making. No reliable numbers are available either on the total number of people who lost their homes and livelihoods to advancing mines, or on the illicit profits made by the various actors involved in granting leases to private companies (see report by Singh and Watts 2011[3]).

The politics of coal in India unexpectedly took an interesting turn in March 2012, when the Comptroller and Auditor General of India (CAG) office accused the Government of India of allocating coal blocks in an inefficient manner during the period 2004–9, resulting in considerable loss of revenue for the nation. His main accusation was that since 2004, instead of auctioning coal blocks – enormous unexploited reserves of coal deposits – the government allocated them to mostly privately-owned mining companies for 'captive'[4] use of coal for power generation. The CAG calculated conservatively that such allocations led to a gain of over USD 194 billion to these companies. A number of high-level political and business figures were implicated in the scam that came to be known popularly as India's 'coalgate.' The CAG dramatically revised this figure when presenting the final version of the report to the Parliament in August 2012. Nonetheless, the report achieved far more than many earlier efforts of civil society activists in drawing media attention to the social and environmental costs of coal mining. The question remains as to whether it was able to expose the unfettered, damaging nature of the coal-based resource nationalism of the Indian state. The state monopoly over coal created by nationalisation of the industry in the early 1970s clearly needs to be redressed with comprehensive reforms in pricing and marketing policy, but will indiscriminare devolution of responsibilities to private companies be the solution?

Until the recent exposure by the CAG of the illicit gains made by state- and non-state actors in allocating coal blocks, coal mining received relatively little attention from scholars compared to either big dam projects or the forest management; the reasons for this were the project-to-project nature of mining; the scattered locations of mines in remote and rural areas; and comparatively short establishment periods – beginning from exploratory surveys to project construction and operation. The public outrage following the high-profile media coverage was matched by opposition party political demands that the government should resign. The question that arises then is: How should one respond to coal mining in contemporary India? Is corruption excusable by reference to India's status as a developing country? Are claims that India needs to ensure economic growth at any cost sustainable? Similarly, is there merit in relying on the claim that India should refuse western standards imposed on it by more developed countries which conspire against India's development? Or should academics and

---

3 Sarita Singh and Himangshu Watts, 'Corruption, inefficiency eat 25 per cent of CIL output: Sriprakash Jaiswal.' *Economic Times* (ET Bureau) Oct 19, 2011, 10.42am IST: http://articles.economictimes.indiatimes.com/2011-10-19/news/30297602_1_coal-shortage-coal-output-coal-india. Accessed on 25 March, 2013.

4 The term has come to be widely used in post-liberalisation India. It implies that all the production from these collieries are meant for either power or steel production and no coal can be sold in the open market. Thus, these coal mines are in a manner of speaking 'captive' to these power and steel plants.

professionals contemplate whether – and why – a sovereign nation must trample upon the interests of its poorest and weakest citizens in order to meet numerical targets in energy production? A positive approach to the findings of the CAG report would be to open the space for new ways of thinking about coal.

The 'Black Chief'

One might ask, why should coal in India be studied and what lessons might be in these studies for the rest of the world? The contested pasts of coal mining had attracted the attention of historians for much of the last century. Currently, however, coal has lost much of its shine over Western Europe; a logical extension of global scholarly interest on coal will now turn towards Asia, Africa and Australia.[5] With its billion-plus people, deep environmental sustainability issues and poverty contrasting with rapid economic growth, India offers an excellent case. The rich history of social science research emerging from the country, methodologically implies that one may expect new research approaches on coal and coal mining to emerge from India.

Globally, although coal has reached its peak in terms of dominating the energy mix, it continues to be a major player. This is reflected in the steadily rising prices of coal in the world; in 1987 the rough price of one ton of coal in northwest Europe was about USD 31, in 2012, the price was USD 122.[6] It is well-known that, although limited amounts of coal were used in rudimentary foundries since the early days, wind, water, and animals provided the only alternatives to human exertion until the industrial revolution in the United Kingdom. Burning of biomass provided the bulk of stationary energy supply for heating homes and cooking. The conversion of coal to coke, which eventually replaced charcoal in the iron and steel-making industry, enabled the modern technological world. However, even by 1850, by which time the railways were reasonably well-established, traditional renewables provided 90 per cent of the world's energy supply. The balance came from coal. Since then, the rise of coal has been spectacular; at the turn of the nineteenth century coal provided nearly half of the total energy supply, with biomass, wind and water comprising the remainder. This dominance continued until around 1950, when it was superseded by the inexorable rise of oil, in conjunction with use of the internal combustion engine.

Obsession with the limitation of coal reserves began early; the anxiety that technological inventions cannot prolong the life of mineral resources such as coal

---

5   Phimister, Ian (2004) 'coal mining and its recent pasts in comparative perspective', *Journal of Australasian Mining History*, 2: 115–130.

6   This is the actual price, adjusted for the then value of the USD. However, Zellou and Cuddington (2012) argue that in real terms since 1800, coal prices have actually *slowly decreased* from $120 to ~$60 in 2005 dollars with periods of ups and downs (see http://econbus.mines.edu/working-papers/wp201210.pdf).

is reflected in the work of the British economist W. Stanley Jevons (1906), who wrote *The Coal Question* to investigate whether the 'progress of the nation' might be affected by the exhaustion of coal mines.

Jevons underestimated the importance of coal substitutes such as petroleum, coal-seam gas, and hydroelectric power. Since the advent of oil, popularised by the motorcar and other rapid transportation such as the aeroplane, the dominance of coal in the global energy market has begun to decrease markedly. With regard to nuclear energy, a Bloomberg New Energy Finance report showed that, of all the countries in the world, China possesses the largest number of reactors (26), under construction in January, 2013; Russia was building 11, whereas India was the third, with 7 nuclear reactors under construction. Still, nuclear energy has a long way to go. The fall in the extraction of coal has been a remarkable feature of the twentieth century; its gains were by and large made in oil and gas. By 2000 the proportion of coal had fallen to 25 per cent of mined energy, whereas that of oil had risen by almost 40 per cent, with natural gas contributing 23 per cent of Total Primary Energy Supply (TPES).

Nonetheless, in its December 2012 Press Release, the International Energy Agency (IEA) predicted that the share of coal in global energy will increase, and by 2017 may once again surpass oil as the world's primary energy resource. The IEA expects coal demand to increase in every region of the world, excepting the United States, where it is being replaced by natural gas. Focusing on non-transport energy, the OPEC[7] report *World Oil Outlook* (2011) noted that coal satisfies one-third of present global demand. Although some energy pattern changes may also occur in India, coal remains important for the country, constituting 58 per cent of its total energy supply. It is likely that coal will continue to dominate energy use in India; the Working Group of the Planning Commission of India on coal and lignite has projected that the demand for coal for the year 2021/2 could be twice that of 2011/12 – almost 1,400 million tons – as compared to almost 700 tonnes currently. Coal is also of strategic importance for uninterrupted energy supplies to India that houses 17 per cent of the world's population but just 0.8 per cent of its known oil and gas reserves. Among the factors that make many less developed countries such as India more dependent than richer countries on coal than oil, four are of consequence. These are the lower price of coal, as compared to oil; the comparatively lower level of technology used to extract, process and transport it; the wider geographical distribution of coal resources; and the ease of its use. Currently, coal comprises 70 per cent of the TPES of China, as compared to India's 58 per cent. In comparison, for the more developed countries such as the United States, UK and Germany, the proportions are 22, 15 and 25 per cent respectively. One must remember that embodied energy – the sum of energy required to produce goods (calculated by reference to either tonnes of carbon dioxide created by the energy needed to make a kilogram of product, or by megajoules of energy

---

7 OPEC (2011) World Oil Outlook, Vienna. Available from www.opec.org (accessed on 28 March, 2013).

needed to make a kilogram of product) – is often not accounted for in energy extraction and transfer statistics. These figures can therefore be misleading, as the trade in embodied energy, for example in imported steel, is not considered. The other elephant present in these data arises from the potential of climate policy backlash; living in the Anthropocene, we are experiencing unprecedented changes of climate brought upon by the combustion of fossil fuels; it is likely that future policy measures will curb further unfettered growth of coal mining. Coal, therefore, continues be significant globally, and holds great importance for contemporary India.

Large consumers of coal are usually big miners, and vice versa. China is the largest coal miner, followed closely by the United States, with India in the third place after a large gap. However, somewhat anomalously, Indonesia is fourth on the list, followed by Australia, both countries using relatively little of their output domestically. Anomalies are also reflected in the broad distribution in the geography of coal mining; although conventional geologists and geographers would like to suggest that the global distribution of coal mining reflects the natural geological dispersal of coal deposits, the close relation between heavy industrial production such as iron and steel, and urban-industrial consumption needs and standards of living with coal mining (and power generation) are all too evident. Until 1943, for example, the power belt extending from the Mississippi Valley to Russia accounted for 90 per cent of the world's consumption of energy. As the heart of heavy industrial production becomes geographically more dispersed, one begins to recognise more clearly the political and economic factors of coal production. As coal mining expands into new areas of production in response to growing demands for environmental care, social license and transparency in revenue distribution, the global bodies constituted by large mining companies also establish new rules of access and distribution. Against the claim of pure economic benefit from mining, issues are raised regarding environment and social goods that have not been dealt with by the mining industry in the past. To what extent will the new states that now mine coal feel obligated to obey international guidelines and principles is a question presently lacking response.

The expansion of coal mining into new terrain is reflected in the polarisation of domestic consumption that emerges when population is taken into account. Australia is the highest per capita consumer of coal – 3.3 kilograms per person – as nuclear power is not used and the country has only limited hydropower production capacity. It is followed by South Africa (2.8 kg), which generates more than 90 per cent of its electricity from coal; the United States (2.4 kg); China (2.1 kg); Germany (1.4 kg); Russia (1 kg); with India far behind at 0.4 kg per capita. In other words, the average Indian uses only 400 grams of coal annually, as compared to the average Australian, who uses 3.3 kgs.

Such data, however, need to be interpreted in a new way; one may note that as compared to other coal producing nations, the damage from coal mining in India would probably affect millions more people than would mining coal in Australia. The key point is to find ways to deal with the impacts of large-scale coal mining projects that have caused local and regional conflicts by initiating complex social and cultural change that usually disadvantages local populations. In coal-bearing tracts in India, conflicts have unfolded in ways that are contingent on factors that not even the most rigorous development plans could anticipate. Throughout India, coal mining projects have been associated with varying degrees of environmental degradation and social conflict, resulting in public criticism of regional and central governments and, of course, politicians; in the escalating discontent of local populations; and in physical conflict that includes a state of quasi civil war in some coal-rich regions. The conspicuous power differentials engendered by large-scale coal mining projects give rise to these conflicts and inequities, which only intensify over time. Today India's coal-bearing tracts are cursed by this natural endowment, the extraction of which is leading to growing inequities, the decay of agriculture, and the displacement of the indigenous and poor classes. The all-consuming need for double-digit economic growth figures makes the ravages of the coal tracts well-known, but ill-understood.

The Dialogue

Mining is not a significant contributor to the Indian economy; the sector contributes less than 3 per cent of GDP. However, it is critical for economic growth – should mining cease, the impact on GDP would be substantial because of the flow-on effects on energy-intensive industries and activities. According to the Census data, in 2001 the mining and quarrying sector employed 2.2 million people who comprise 0.6 per cent of all workers,[8] of whom 6 per cent were women.[9] But these data hide the diversity and extent of mining as a human enterprise; there are millions of people, described by the Census as 'marginal' workers who are also involved in various kinds of smaller scale mines and quarries. One inference that can be drawn is that one segment of informal sector workers in India are involved in mining. Mining is also important because it directly supports a variety of local industries, such as building and construction. A huge amount of mining is conducted informally in sand and stone quarries and is seldom included in any official data; this indicates that the mining economy in India is probably neither fully appreciated nor comprehended by either the environmental activists or

8   But 1.4 per cent of all 'main' workers.

9   Indian census data classify the workers into two main categories: the 'main' workers are those who had worked for the major part of the reference period (that is, 6 months or more), and the 'marginal' workers who did not work for at least 183 days in the preceding 12 months to the census taking. Main workers comprise the larger segment of workers, that is, 78 per cent. The data are sex- and rural-urban segregated, giving the numbers of male and female workers in these categories. These definitions are clarified in the Census of India's homepage: see http://censusindia.gov.in/Metadata/Metada.htm#2k accessed on 28th March, 2013.

the researchers. In India, mining has generally come to be seen as synonymous with large, corporatised, formal extractive operations. One must take note of the existence of non-corporatised forms of mining, which have a long and complex history and associated problems: besides the more apparent ones relating to the environment, there are important issues around non-formal labour and production organisation and relations in these mines.

Although a growing body of literature from civil society movements complements the significant body of research outputs generated by social science academics, the two groups refuse to converse with each other. The growing body of literature remains partisan and does not adequately represent the multiple voices and approaches on the subject of mining's effect on people and the environment from interdisciplinary, holistic perspectives. The authenticity of research outcomes are contested, depending on who makes the claim. Many academic researchers tend to engage exclusively with those within their disciplines who are well-established and whose voices are heard by a wider public: an expression, in some cases, of status. Above all, with few exceptions researchers prefer to work within the academic establishment and to reject the ongoing work of activists, organisers and trade unionists, who maintain engagement with the social and instrumental spheres that academics only write of. Silos therefore exist between disparate professions; not only do academics refrain from appreciating the work accomplished by civil society activists and trade unionists; within the broad fold of social science, those with economics or geographical training tend to ignore the valuable contributions made by historians. The rigid disciplinarity that marks social science research in India not only prevents productive dialogue between academics, but also between academics and activists. Consequently, not only does knowledge of mining remain fractured, the mainstream media finds it difficult to represent these realities in a context of strongly influenced opinion. There is growing yet scattered evidence that besides research outputs from academia, exemplary research efforts are being undertaken by civil society organisations on a range of complicated social, political, cultural, historical and gender issues concerning mining in India.

In this dismal context of mutual isolation and wilful cecity, a workshop appropriately titled 'Dialogue across Boundaries: People and Mining in India', was held in August 2008 by a number of grassroots civil society groups. The aim was to initiate a productive conversation among the various stakeholders regarding mining. Hosted at the Centre of Science and Environment, the workshop was organised by the Mine Labour Protection Campaign, MM&P (Mines, Minerals & People) and other activist groups such as Santulan and Prasaar, and was assisted by the Resource Management in Asia Pacific Program of The Australian National University. The timing of the Dialogue was crucial; it occurred not long after the CSE had published its Sixth Citizen's Report on State of the Environment in India: *Rich Lands, Poor People*, questioning whether sustainable mining is possible. Consequently, the Dialogue aimed to take stock of existing social science research on the range of social issues that define mining in India; to introduce scholarly work to grassroots-based activist groups that are protesting against social injustices;

and to build effective dialogue and trust between the two. Representatives from a number of civil society groups conversed with young scholars studying the broader political, historical, and economic factors describing the environmental impacts of mining; the outcomes demonstrated that a close scrutiny of theoretical, methodological and standpoint challenges relating to a specific social problem can illuminate both research and action. In a small way, the meeting was a step towards addressing and closing the gap between the separate worlds inhabited by researchers and practitioners. One can hope that it will leave its imprint on future efforts by social scientists researching mining in India.

One of the goals of the Dialogue was to explore the possibility of bringing into the public domain a publication that showcases the diversity of issues, interests, and emergent areas in research and practice that cut across disciplinary and thematic boundaries. One point that was discussed concerned the diverse nature of the minerals being mined; how different minerals give rise to varying production processes, leading to divergent production and labour relations. For example, diamonds and gem stone mining remains more speculative in nature than coal mining; hence, mining of these minerals gives rise to more individually-oriented production structures. Gold ore, like copper, is present only in small proportions, necessitating processing close to mining operations. Similarly, the end use of marble and sandstone as dimension stone in the construction industry requires block-cutting, which induces quarry-owners to flout existing mining stability rules and to cut the stone in near-vertical pits on-site. In contrast to these minerals, coal and iron ore comprise quantities that can be exported for processing or use in their entirety.

Those present at the Dialogue agreed that coal was supreme among minerals in India, encouraging participants to focus on this mineral: its mining and its past, present and future in its entirety. By ushering in the machine age elsewhere in the world, coal as a mineral became all-powerful. As coal enabled the smelting of metal for machines that transformed human society in the twentieth century, it was described by T. S. Lovering in 1944 as 'the black chief of the mineral jinn' who has gifted to humans seemingly mythical power in producing energy. For India, coal attracts the iconography of the 'black diamond', the exploitation of which symbolically equates with the nation-state. Desperate attempts by the Indian state to extract increasing quantities of coal in order to generate more power are intended to sustain the high rates of economic growth that the country has experienced in recent years.

The symbolic value and absolute economic importance of coal in India is difficult to understand because of the relative dearth of interdisciplinary research into the coal mining industry as a whole. First and foremost is the knowledge that 75 per cent of all coal is used for electricity generation, primarily for urban-industrial India: this statistic is by far the most visible and drives public opinion. One cannot help but acknowledge a sense of national pride, a resource nationalism regarding coal that is only equaled by the nineteenth century nationalist pride of the United Kingdom in its expansionist industrial enterprise. India's coal nationalism exceeds the resource nationalism of Latin America countries, which are now said to be experiencing a post-extractive resurgence marked by the increasing control of the state over the extraction of mineral and other energy resources. In seeking to advance the debates on social, political, and environmental consequences and conceptualisations of coal, this book brings together a collection of chapters that consider coal in its entirety. These chapters comprise not just the selected papers from the Dialogue workshop but also invited papers and topical contributions scattered in other journals. They draw upon the conceptual and methodological frameworks of a number of disciplinary traditions in the social sciences. They show that coal in India is much more than just the material that it is thought to be, and illustrate how the nation has laid its trust and built its ambitions on this commodity and how in return coal has assumed an unprecedented social and political primacy in the country. If there is a singular policy message, it represents the combined voices of researchers, trade unions, activists and community groups, conveying that the ways in which the Indian state operates at all levels of government with regard to coal mining must change. A sovereign state's primary responsibility is toward its peoples, not the commercial business houses and corporations. In order to reflect that responsibility mining enterprises must first seek approval from local communities so that they do not merely obtain a business license, but also the social license to operate. To achieve that, the way coal-mining related business is conducted has to change. Many of those with power, money and influence act as feudal lords in their quest for wealth, exhibiting a lack of concern with the rest of society. The advanced extent of corruption and unaccountability involved in coal mining has been obvious for years, but it is only now, after its gargantuan proportions have been exposed officially, that a sense of urgency to change the situation has become palpable. The contributions in this book define the relevance of this debate.

Acknowledgements

As in any ongoing conversation, the beginning of this book is difficult to trace. It likely began in 1993, when I received the first tranche of research funding from the Ministry of Environment and Forests to map the changes that had taken place in the local environments of Raniganj-Jharia colliery areas in eastern India: the fires burning under them, the social and urban changes and the perceptions of women and men living in the villages and towns. Or, it could have begun much earlier, when, as a young girl, I first entered a colliery village and felt an immediate need to understand how and why the social fabric of that place seemed so very different from what I was familiar with. Possibly my interest was triggered by my father's longtime association with the intricate politics of the coalbelt, and the fact that several of his associates had given up their worldly comforts in order to live and work with coal mine labourers, to help them organise themselves in order to be able to claim healthier working conditions and higher wages. One may also say that it could have begun when, in 2002, the Panos Institute invited me to coordinate their Oral Testimony Project among mining-displaced indigenous people in the newly established state of Jharkhand, where the destruction caused by coal had already begun. Attended by a large number of activists and academics, the Roundtable held in Ranchi in July 2002 was a milestone. Among those who participated in that Roundtable, Smitu Kothari is no longer with us, but his words and ideas have persisted through the minds and actions of those fortunate enough to hear him speak. Another contribution was probably made in 2007 when an international conference on the 'Social and Environmental Consequences of Coal' was organised jointly by the Indian School of Mines University, the University of New South Wales, the Australian National University, and the India International Centre, New Delhi. Each of these interventions were different in nature involving different partners, and addressed different issues and audiences; the common feature was the escalating need to establish a comprehensive understanding of coal. On retrospect, it now seems difficult to pinpoint which one of these most accurately represents the nascence of the conversations that comprise this volume.

Throughout my working life I have tried to add to my understanding of the importance of coal not only to the Indian economy, but also to the politics of coal and its symbolic value in the national psyche. Over all these years my debt of knowledge has accumulated. In editing *The Coal Nation*, I incurred the debts of many people, among whom only a few can be mentioned in the limited space of this book. To begin with, I thank all the participants at the Dialogue, over 55 individuals who gave their time generously, attended, presented, raised issues and actively commented on the issues at hand. These included not only academic presenters but also activists and NGO workers, representing wide regional interests from across India. In particular and among others, I thank Professor Vineeta Damodaran (of the Centre of World Environment History, Sussex University), Mr Xavier Dias and Ms Ajitha (mines, minerals & People), Sri Bhanumati Kalluri (Samatha), Mr Sreedhar R. (Environics Trust), Ms Manju Menon and Kanchi Kohli (Kalpvrikhsh), Dr Kavita Arora (Rajiv Gandhi Institute of Contemporary Studies), Mr Chandra Bhushan, Ms Monali Jeya Hazra and Nivit Kr Jadav of Centre for Science and Environment, Dr Rohan D'Souza of the Centre for Science Policy and Dr Md, Zakaria Siddiqui (of JNU), Dr Nitish Priyadarshi and Professor Ramesh Sharan of Ranchi University, Ms Mridula Gupte (Directorate General of Mines Safety, Rajasthan), Ms Heather Bedi Plumridge (Cambridge University), Dr Apurba Ratan Ghosh (Burdwan University), Mr Xavier Dias, Dr Naveen Ramesh (Division of Work Environment), Late Prodipto Roy (formerly of Council for Social Development), Late Sri Haradhon Roy (of Colliery Majdoor Sabha India), Mr Ashim Roy (INTUC, Trade Unionist), Sri Philen Horo (of Prerana Resource Centre, Hazaribagh), Mr Bastu Rege (of Santulan), D. S Paliwal (Workers' Solidarity), K.S. Nikhil Kumar (Symbiosis, Pune), Ravi Rebapraggada and Pravin Mote (of m,m&P), Rana Sengupta (MLPC), Himanshu Upadhyaya and Nishant Alag (Environics Trust), Yousuf Beg (Patthar Khadan Mazdoor Sangh, Panna), Mr S. A Azad (Prasar), Emily Bilo (Haq Centre for Child Rights), Mr Paranjoy Guhathakurta (media personality), Swati Birla, Saraswati (Independent Film-maker), S. Madhavan (independent photographer), Sri Prakash (independent filmmaker), Ana Gabriela (El Pais, Spanish Newspaper) and Bipin Kumar (The Other Media), and most importantly, my academic colleagues, Professor Gurdeep Singh (of Centre for Mining Environment, Indian School of Mines University), Dr David Laurence (Mining Engineering, University of New South Wales), Mr Rahul Guha (Deputy Director General of Mines Safetry), Mr Joydeb Bannerji (formerly of Coal India Limited), and Mr Amol Haldar (historian of Raniganj coalbelt) of Burdwan.

Another group of people I gratefully thank includes a number of generous friends and colleagues from academic institutions around the world who offered their time to review the papers contained in this book.

Earlier versions of some of the papers were published in *Economic and Political Weekly* of India; I thank Mr Rammanohar Reddy, the editor of the journal, for his generosity in allowing wider dissemination of the papers which have now benefitted from revisiting and revisioning.

Small amounts of research subsidies from the Resource, Environment & Development (RE&D) group, the Australian National University (ANU), allowed me to travel to the fieldsites, to co-organise the Dialogue, and to develop this book into its present form. I thank RE&D, in particular the Convenor Dr Colin Filer, for his kindness in awarding these grants and for providing an enabling work environment for pursuing academic endeavours such as this book. In general, the collegial atmosphere of RE&D and the University, and the rich resources of ANU libraries have played key roles in enhancing my understanding of the social issues related to mining in less developed countries. I am also thankful to the CartoGIS

Unit of the College of Asia and the Pacific at the ANU, in particular Ms Kay Dancey and Ms Jennifer Sheehan, for their brilliant map-making skills from which this book has benefitted.

I also gratefully acknowledge the competent assistance of Ms Madhula Banerji in the initial phases of editorial work, and of Dr Andrew Watts for adding the final editorial touches to the book. Completing the editorial tasks of *The Coal Nation* would not have been possible without encouragement, love and inspiration from my husband, Dr David Williams, formerly of the Division of Energy Technology at the Commonwealth Scientific and Industrial Research Organisation (CSIRO) of Australia, and an expert on the technical aspects of coal. For every piece of scientific information, David was a not merely a source of data but also one of explanation; he acted as a sounding board for all my ideas. Last but not the least, I thank Ms Katy Crossan, the commissioning editor for the publisher, Ashgate, and her efficient team, for publishing this book.

Reviews for
The Coal Nation

Combining the insights of social history, political economy and political ecology, The Coal Nation *is able to expose not only the contested and often violent history of India's exploitation of coal, but also charts the larger contours of India's colonial and post-colonial development. A vibrant and wide-ranging collection of pathbreaking contributions to our understanding of how coal came, and continues, to energize India.*

Michael Watts, University of California, Berkeley, USA

Civilisation as we know it is grounded in the dirty labour practices of dirty fossil carbon. But this fact is as unfamiliar to most students of civilisation as it is to students of development and economy. In this important multidisciplinary book, Lahiri-Dutt's team has prospected Indian coal to expose rich seams. A significant achievement requiring global attention.

Barbara Harriss-White, Emeritus Professor, Oxford University, UK

The great strength of this book is its breadth. Drawing on history, geography, economics, engineering and policy studies it demonstrates the value of interdisciplinary co-operation. This approach provides a unique understanding of an industry that played a critical role in Indian history and will grow in importance with the doubling of production that is predicted for the next decade. Alive to the industry's potential for expanding socio-economic development, the book explores problems of safety, dispossession, ownership, ecological degradation and climate change. The chapters provide specialist knowledge of different periods, places and problems, yet converse with each other to offer an integrated picture of the industry as a whole. In so-doing it poses pertinent questions for policy makers and for public discussion.

Peter Alexander, University of Johannesburg, South Africa

Chapter 1

Introduction to Coal in India: Energising the Nation

Kuntala Lahiri-Dutt

'Coal is a national asset.' – Coal Fields Committee, quoted in Read (1931: 105)

Coal, the Material

As an organic material and a mineral resource, coal is a carbon-rich substance produced through natural processes over millions of years. From material resource to commodity, coal traverses a long path, and is created and recreated by its users; it acquires meaning and assumes utility through its interactions with human society. Not only the material itself, the images that surround coal – as dirty, old-fashioned, cheap industrial raw material – are also created by those who use it. They are as important as the material itself. It is interesting that coal does not make us think of the rich, but of the poor. It brings to mind a bleak imagery of soot-covered men, the miners, trudging back from the pithead to their grim living quarters. But it is also true that the lowly lump of coal was once live organic matter – forest vegetation, gigantic ferns that existed as long as 400 million years ago. Transformed into a commodity, coal gave unimaginable power to humans through the capacity to fundamentally change – as well as challenge – their relationship with nature, all of which has come at a very high price. The industrial age emerged from the use of coal, and to the present day has detained us in a haze of coal smoke; it is in that smoke that we can discern the history and the future of the modern world. Coal's impact is far from over; this is the reason why the social historian of British coal, Barbara Freese (2003, 2), commented that coal gives us 'a disturbing glimpse of our future.'

This statement is possibly more apt for India than for any other country. With its separate Ministry,[1] coal occupies pride of place in contemporary India, the energy future of which it shapes and the economic and political milieu of which it influences. Coal mining was one of the pivotal modern industries shaping India's colonial trajectory but, unlike tea plantations or jute mills, the coal mining industry assumed iconic status as a national symbol after Independence. At the same time, the coal mining industry created a unique working class comprising people who came from villages to work in the collieries, and who formed unions to protest against worker exploitation. In recent years, the supremacy given to

---

1   Having a separate Ministry for coal is one indication of coal's important to India's economy. No other major coal producing or mining nation can boast of a Ministry of Coal.

coal mining in forest-covered frontier areas traditionally used by tribal and rural communities has dispossessed and pauperised many such people. Combustion of coal to produce electricity constitutes the compelling need that has prevented the Indian state in engaging with the impending realities of a climate-changed future. Given its current role in meeting India's energy demands, one can predict that the prime position of coal in the country will remain, at least for the foreseeable future.

This chapter comprises the backdrop to the chapters that follow: it illuminates the intricate and dynamic relationships between people and coal, and situates coal within the context of India's history and current economic growth trajectory. Within this context it is possible to consider coal as something which is as much produced by society as it is by nature: it aims to illustrate, alongside the historical growth of the coal mining industry, how coal as a mineral commodity has shaped and continues to influence India's identity as a nation. It also sketches out the broad contours of emerging challenges raised by each chapter. Let us first look at the material itself before tracing how coal shaped social and political life in India, making it a 'coal nation.'

As a fossil-fuel, coal was produced from biomass that had grown millions of years ago when the world was warmer and more humid. This warmth and humidity encouraged the prolific growth of vegetation which, mainly in shallow lakes, lagoons or deltaic basins, generated and accumulated a layer of dead plant matter. The basins were subject to constant accretion and subsidence, the dead vegetation becoming buried beneath inorganic sediment of sand and silt. This cycle would be repeated numerous times, resulting in multiple alternating layers of organic and inorganic material. Initially the deposited plant matter partly decomposed in the manner of landfill waste to leave behind the more resilient lignin component of the vegetation – what is known as peat. Geologists (see e.g., James 1982, 2) tell us that gradually the residual plant matter accumulated to the thickness of many metres in some places and would be buried to depths of many hundreds of metres. They would consequently experience elevated temperatures due to the geothermal gradient. Over an extended geological period this heating would further degrade the organic material, which would first lose its oxygen by loss of water and carbon dioxide, then more hydrogen through loss of methane, leaving behind carbonaceous sedimentary rock. This is the 'coalification' process: how coal came into existence.

There are regional variations to this general story. The coalification process may transform the peat into lignite, then into bituminous or soft coal, and finally into hard anthracite coal, a form that approaches pure carbon. Anthracite coal was put to use in America for domestic and industrial purposes, but in most other countries in the world, it is bituminous coal, easier to extract due to its softer nature and wider occurrence, that led to coal becoming the paramount among the minerals. Generally, the deeper the burial, the higher is the rank of the coal: if burial was only shallow, deposits end up as lignite or brown coal.

Warm temperatures and excessive humidity during the Carboniferous and Permian geological periods gave rise to large forests of which Glossopteris and

Gangamopteris are the most well-known species, the trunks and branches of which were washed by rivers into narrow, elongated and shallow lakes to give rise to today's coal deposits. Most of the coal in India was formed by such shallow lacustrine deposits. At that time, the Indian subcontinent comprised part of a larger landmass that is known as Gondwanaland. Whereas the coal deposits in Europe and North America are from 350 million years ago (the Carboniferous), those in the landmasses that formed Gondwanaland are from the Permian Period (250 million). Thus, coal deposits in Australia, South Africa and Mozambique are similar in age to the deposits located in India and Pakistan, and also Bangladesh. However, there are significant deposits of more recent coal, for example in China, Indonesia and even in Australia, where some brown coal in the state of Victoria is only a few million years old. Similarly, in rising from the depths, the Himalayas crushed the younger-age coal lying in its beds. This series of coal beds were laid down over many hundreds of millions of years in the past and the process is still being initiated in certain parts of the world, such as in the shallower seas around the Indonesian islands.

The coal-bearing beds of India are designated by the geologists as 'Gondwana coal measures' and are generally found in river beds such as the Damodar, Mahanadi, Godavari and Wardha in West Bengal, Jharkhand, Orissa (now spelled Odisha), Madhya Pradesh and Andhra Pradesh. The other coal beds are known as 'Tertiary coal measures,' which occur in Assam and elsewhere in the northeast, but because they were crushed by the rising Himalayas, they are of somewhat less value than the Gondwana deposits. Besides these, patches of lignite coal are found in Bikaner in Rajasthan and South Arcot in Tamil Nadu. Of all Indian coal reserves the most commercially important are the 'Damuda' series, particularly the Barakar and Raniganj stages. Some of these coal beds contain coking coal, considered to be of the highest grade because of its suitability of use in steel making, but the largest segment is non-coking coal, which is generally comprised of contents of a higher volatility than other coal types.

Because the geological character of India varies, coal deposits are not evenly spread across the country; there is a clear concentration of coal in its east and east central parts. Such uneven distribution common for mineral resources; this inequity gives what is known as a natural advantage to some geographical locations while depriving others of such benefit. Political power has sought to either enhance such advantage or curb it. As the Pittsburgh Plus system demonstrates in the USA, the price of coal at one point of time was controlled in such a way as to offer additional benefit to the coal-producing areas of Pittsburgh. In postcolonial India, however, centralised economic control by and large stripped potential benefits from the natural coal endowments in these locations; as a consequence the local economies of these coal-producing regions did not make significant economic gains, while having to bear the environmental and social costs of coal mining. I will elaborate on this aspect in the latter sections. For the present purpose, the location of coal deposits in India (Figures 1.1 and 1.2) provides a sufficient overview.

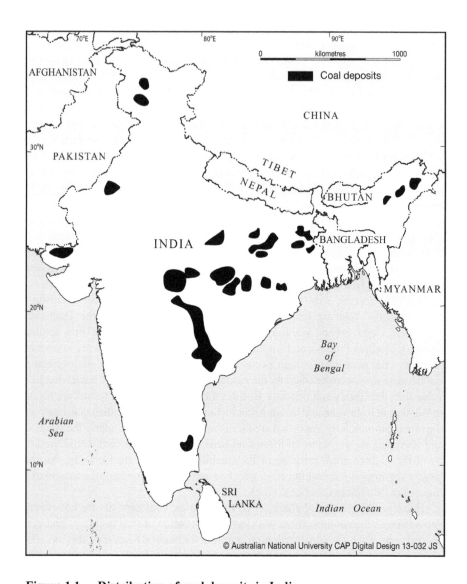

Figure 1.1   Distribution of coal deposits in India

Source: The Australian National University College of Asia and the Pacific

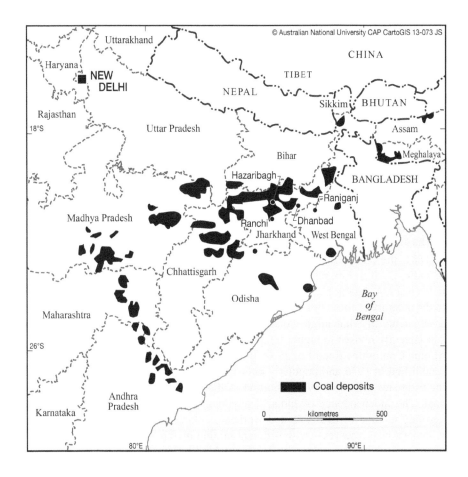

Figure 1.2   Detail of important coal deposits

Source: The Australian National University College of Asia and the Pacific

It is not likely that all of the coal resources of India have been accurately mapped,[2] as the exploration boreholes seldom extend deeper than 400 metres. Currently, only about 45 per cent of the potential coal-bearing area in India has been covered by regional surveys, leaving over half of the potential area for further exploration. One can thus say that much of Indian coal is still being discovered. Indeed, the data required to allow a precise estimation of the amount of coal at a significant depth can be difficult to ascertain, if not impossible. Hence, different terminologies such as 'indicated', 'potential', 'proven', and 'measured' present a wide range of alternative assessments with regard to how much coal resource awaits exploitation.[3] A number of geological surveys since colonial times have attempted to provide a more comprehensive picture of the resource situation of coal in India. Ghosh (1997, 11–12) outlines these surveys; the earliest systematic assay was conducted by D. H. Williams, the Geological Surveyor of the East India Company, in 1845. This was followed by W. T. Blanford's detailed study during 1858–60. In 1881, the colonial government published a report on the Economic Geology of India. Cyril S. Fox resurveyed the Raniganj–Jharia and Madhya Pradesh coal fields in 1924. Again, each of these reports presents a different picture with regard to the quantity of the estimated coal that is workable, and is informed by differing assessments of what might constitute good quality coal. Fox's estimates were subsequently revised to higher figures by Sir Lewis Fermer in 1935 and the Coal Mining Committee Report of 1937. Report of the Indian Coalfields Committee – established in 1946 and popularly known as the Mahindra Committee – updated the estimates primarily by redefinition of the quality and the workability of the coal. The independence of India, if anything, multiplied such inquiries on the absolute volume of available and workable coking and non-coking coal – that is, India's coal reserves – with different mining methods, resulting in a series of investigations being conducted after the Mahindra Committee: the Committee on Conservation of Metallurgical Coal, under the Chairmanship of Dr M. S. Krishnan (1950); the Working Party for the Coal Industry, under the Chairmanship of Shri B. Das (1951); the Energy Survey of India Committee (1965); and the Fuel Policy Committee (1974). The general consensus arrived at was that the landmass of India contains a large enough reserve of coal, in particular coking coal, that it will not be exhausted in the immediately foreseeable future; for example, the Energy Survey Report noted (1965, 263): 'On any basis of probable consumption they do

---

2   About 61 per cent of total coal resources occur within 300 metres below the surface, and 26 per cent within the depth range of 300–600 metres (CCO 2006).

3   As noted by Chand (2008: 19) the key difference is between 'resource' and 'reserve'. The term 'resource' refers to the amount of mineral that is potentially exploitable from under the ground. The term 'reserve' means the amount of that resource which has been shown to be technically and economically feasible to exploit. Changes in extraction technology may well alter the estimation of the reserve. However, terms such as 'potential', 'proven', 'indicated' are still found in texts. Consequently, estimates of the amount of extractable mineral need to be viewed with caution.

not present grounds for fear over the next century or more...to the fears that have been held regarding the adequacy of coking coal.' The point of debate, then, was the extractability of this coal, which occurred in very thick seams, and as such was difficult to mine; and the previous custom of leaving large amounts of coal in pillars which allowed only about 40 per cent of the coal to be mined. Considering the pros and cons, the Report observed that (257–8): 'it will clearly be necessary to make careful studies of possible ways in which, in the future, the rate of recovery may be increased.' Similar anxieties can be detected in many other studies of coal (see, e.g., Prasad 1986). Chand (2008, 23) suggests that the extractable coal in India is far less than existing assessments have determined.

The anxiety that presents itself through these initiatives to know accurately the size of the resource is similar to nineteenth-century England's preoccupation with its dependence on coal and its interrelatedness with conceptions of the sovereignty of the country, best expressed in the works of William Stanley Jevons' (1865) book, *The Coal Question: An Inquiry Concerning the Progress of Our Nation*. Jevons calculated how much coal would remain for use after years of extraction. In India, the anxiety is less about the supply of coal under the landmass, and more about the recoverability of this resource, as expressed in analyses such as that by Vasudevachary (1994). Again, one can interpret this concern as being linked to two Indian characteristics: passion and ingrained respect for the use of sophisticated technology; and a reinvigorated sense of nationalism around coal. The newly independent country embarked on projects of nation-building, within in which coal was seen as the mineral that would drive development. Regional disparities in coal quality and accessibility were considered to be one of the major obstacles to planned national economic development; much of the high-quality coking coal is concentrated in eastern India. Locational considerations led to policy prescriptions: coal prices were equalised throughout the country. Moreover, to mobilise coal quickly and efficiently within the regions, it was suggested (Report of the Fuel Policy Committee 1974, 48) that India will need 'an efficient and adequate transport system which would ensure the flow of available fuel resources from the points of availability to points of requirement.' Calculations of gross reserves, proven reserves, indicated reserves, and inferred reserves were quantitative, supporting assessments that were geological in nature and thereby creating a subtext of the plentiful availability of coal and the ability of technology to extend its useability. In this regard, the anxiety over coal in India is not related to its polluting capacity – its contribution of fly ash, sulphur dioxides and carbon dioxide to the air, and the general pollution that the use of coal causes. At times, references to 'clean' coal technologies occur, implying that technological improvement would reduce the ill effects of coal on the environment, thereby relying upon the reassurance of future technology as a panacea for existing problems. Such confidence is expressed in the replacement of less efficient, older power plants with new, technologically advanced plants intended to reduce emissions. Thus, although the superstructure of the Indian coal industry is based upon the physical deposits of coal, the senses of plenty, compulsion and need that have accompanied debate surrounding these resources are central to our understanding of the social, political and economic significance of coal in India. With this observation in mind, let us briefly look at the social life of coal.

The Kingdom of Coal

As real as the material fact of coal is the reality of coal as a social object. Part of the social nature of coal derives from its use; writings of early philosophers such as Aristotle and Theophrastus show that the combustion of coal has been a part of human life for a very long time. Romans may have used coal for heating during their occupation of Britain. The earliest existent charter to mine coal was given in 1210 to the monks of Newbattle Abbey near Preston, Scotland (Gregory 2001, 55). By 1233 coal was being mined in Newcastle and elsewhere in England, and in the fourteenth century British coal was being exported to Northern Europe. What might have heralded an early start to an industrial age was brought to a halt by the Black Death, which devastated Europe. In the Ruhr, coal mining began in 1590 and slightly later in Belgium; but it was not until the 1600s that human civilisation entered what can be called 'the coal age.' The coal age ushered in multiple uses of coal in diverse industrial processes – such as lime burning, brick-making, glass manufacture, brewing, and the evaporation of seawater for the production of salt – that began to replace wood charcoal. A shortage of firewood for charcoal production due to deforestation was the main driving force in the adoption of coal (Gregory 2001, 55). By this time, coal had begun to displace wood for heating large buildings such as monasteries. Thus, coal was a 'salvation from a prolonged energy crisis' (Hatcher 1993, 7). In comparison with wood, the higher energy content and the ease with which it remained dry made coal more attractive. Added to this was the realisation that coal, once coked, can be used in the manufacture of iron. By 1760, 17 blast furnaces had made their appearance in Britain, and the figure grew by leaps and bounds as pig-iron production grew. The world by now had entered the coal age; this heralded massive transformations of society.

The impact of coal on the environmental and social landscapes was spectacular. This constituted the 'kingdom of coal' – an unforeseen scarred landscape devoid of shades of green, housing black-faced workers toiling in the depths of the earth and living in slums. The use of this fuel gives rise to blighted landscapes no matter where it occurs. Miller (1985, xix) describes this kingdom in Pennsylvania as one that comprises:

> gigantic pits and craters, the work of bulldozers and dynamite and of power shovels...heaps of clay, crumbled rock, and slate...oily-black brackish ponds [that] give off the smell of sulphur. In places, acrid white smoke rises from the ground, eerie evidence of fires raging far below in the coal seams...fires that burn on and on, even through days of drenching rain.

This description would be true of any part of the world, at any point of industrial history. The classic example is that of the eighteenth-century Black Country in the English Midlands, around the collection of villages and small towns in the coal and iron-producing region that earned its name from coal soot. The region was the site of an intense metal-working industry founded on the close proximity of limestone, iron ore, and the South Staffordshire coal seams. The clouds of black smoke and slag heaps of the Black Country transformed the landscape. At the peak of the region's industrial success there were about 1,500 collieries and foundries in the area, with a smoke-belching chimney every 200–300 metres (Jones 2008).

In the older coal mining areas of eastern India, such as Raniganj–Jharia, smoke oozes not just from visible seams that have unintentionally caught fire in open-cut mines, but from the cracks in nearby ground beneath one's feet, leaving the charred and ash-like ground spongy. The more than 70 mine fires that Saxena (2002) identified as being spread over only one portion of the Jharia coalfield were in most cases started by humans accidentally due to lack of care in mining. Some of these fires have been burning for decades, and these older fires have consequently rendered the ground underneath many villages and towns unstable (Lahiri-Dutt and Gangopadhyaya 2007). Underground voids were left by older mining operations when backfilling was not standard practice, and hence the ground can – and indeed does – frequently subside. The water table throughout the colliery region has fallen, drastically reducing the availability of water. The mining of coal over centuries has not only completely altered local flora and wiped out local animal species, it continues to undermine the rural agricultural economy, creating an unbroken urban belt running for two hundred kilometres. Further west, the North and South Karanpura fields in Jharkhand, is frontier territory, in the sense that coal mining is relatively new and closely associated with both the liberalisation of India's economy, and the newly created Jharkhand state's need to earn revenue from mineral resources.

Coal mining gave rise to coalfield societies and coalfield towns whose life rhythms were in tune with 'the nature of the work processes, the character of the social structure of the communities, the topography of coalfields,' creating the 'often turbulent histories' of these places (Berger et al. 2005, 4). Describing how 45 'coal pioneers' opened coal mines and organised production in southern West Virginia in the late nineteenth century, Sullivan (1989, 5) commented that 'the coal town was an integral part of the production apparatus,' presenting a 'symbiotic relationship' between the mine, the men and the communities. Such a symbiotic relationship between labourers and mines exists also in India, where coal mining has given rise to extensive urbanised regions, but where urban growth has not necessarily benefited local communities. From the nascence of the industry, the influx of labour from other regions into coal producing areas of Raniganj–Jharia created a broad urban belt, what has been termed (by Rothermund and Wadhwa 1978) a 'secondary' urban enclave. During the Raj, it acted to service the primary colonial enclave of Calcutta metropolis, but did not stimulate the social and economic development of the immediate hinterlands, which performed the role of labour suppliers, building a coolie linkage between village and mine. Remarkably, this pattern has persisted until the present, and in the regions of more recent colliery expansion. Writing about the phenomenon over three decades ago, Rothermund and Wadhwa observe (1978, 226):

> the area is characterized by a fragmentation of markets for agricultural produce, the isolated local market remaining unrelated to the urban market.... Accordingly, there is not much of a response to innovations in agricultural production, and capital formation in terms of land reclamation...remains at a low level.

Later on, Rothermund, Kropp and Dienemann (1979) explained that the urban exclave which emerged under colonial rule was controlled by entrepreneurs who had no interest in ploughing the capital back into the region: there was no free market, local entrepreneurs had no chance to organise production and supply of coal on their own, and coal mining was undertaken only in order to satisfy the limited demand of the export-oriented industries. What is interesting is that the 'demand-centred' organisation of the coal mining industry observed in 1978 by Rothermund and Wadhwa (226) has remained until now; in other words, the entirety of the coal bearing tracts of India is seen as resource hinterland that exists in order to serve the primary metropolitan demands created by the urban-industrial nexus. In the next section, we consider the colonial history of coal mining in India, followed by a discussion of its postcolonial context.

Colonial Past

The kingdom of coal as we know it in India is primarily a gift of the Raj, a by-product of colonial rule. Nonetheless, evidence exists substantiating the early use of coal in India: the Sanskrit word *angaar* stands for coal and the ancient texts of Ŕgveda; Taittirīya upaniṣad and Kautilya's *Arthashastra* all mention the burning of coal for smelting iron and other ores. Evidence of sophisticated iron-making – for example in the skilled manufacture of wrought iron pillars of the Kutab Minar in New Delhi – also suggests that coal might have been in use during the Mughal period. The remains of old slag heaps can still be found scattered around the hillocks of Jharkhand that contain coal deposits. Throughout this region, coal-related place-names – such as 'kalipahari', meaning black hill – also indicate the possibility of early coal use. However, unlike Britain, heating was not necessary over most of the country, and the main source of fuel remained wood, biomass and charcoal; even though the use of coal might have been known, it was not widespread. Therefore, the two covenanted employees of the East India Company who were travelling along the Damodar River on a winter night shortly after the acquisition of Bengal by the British East India Company in 1765 must have been surprised when they came across a group of local tribespeople burning a black rock to warm themselves. Coming from a country that had just begun using the steam engine, the two recognised this rock as coal.

The credit for 'discovering' coal in modern India thus goes to two Englishmen, John Sumner and Suetonius Grant Heatly.[4] On 11th August, 1774, they applied to Governor General of Bengal Lord Warren Hastings to produce and sell coal in Bengal and its dependencies. In exchange, they also offered to supply to the British East India Company ten thousand *maunds* (a Bangla term, no longer in popular use, roughly equivalent of 40 kgs) of pit coal every year for a period of five years (Ghosh 1997, 34). They subsequently began to work on six mines, with exclusive rights granted by Warren Hastings to work and sell coal in Bengal for a period of eighteen years. In these initial years the Company felt it more appropriate for its commercial interests to import coal from England, and the Bengal coal was seen as less effective and poorer in quality than its imported counterpart. Consequently, further exploration and mining of Bengal coal was discouraged. However, in his book on the early history of coal mining in Bengal, H. D. G. Humphreys (1956, 147) explains that in 1809 the high cost of bringing coal from England encouraged Governor General Lord Minto to order further inquiries into the availability and quality of local coal. The search team, headed by Thomas Mariott, found a substantially thick bed of coal occurring next to the Damodar River near the village of Mudjea (now spelled as Mejia). Humphreys quoted the report given by Lt Delamain to the Military Board of this inquiry: 'It seemed, however, of a slaty structure, harder than the common coal, soiling the fingers a little when rubbed…and the colour approaching to grey.' This was later discovered to have derived from the same location as that used by Sumner and his colleagues, and was again discarded as of low quality. From 1813 onwards, the Governor General Lord Hastings began to take serious interest in exploiting local coal, and again initiated a series of inquiries. By this time, private entrepreneurs such as shipwrights of Calcutta Port had begun to use Bengal coal for their forges (Ghosh 1997, 36).

From this chequered history and the consistent efforts by the British to find coal locally, one gains an understanding of the high-esteem in which coal was held by the British rulers of India. Entering the coal age themselves, coal, and the power generated by it, was seen by the colonial British as the great engine of moral improvement, the greatest instrument of civilisation for the people, and the factories and mines dependent upon it were schools of instruction' for the colonised. It is therefore not surprising that coal mines were the most prominent amongst the various modern industrial enterprises established in India under colonial rule. Well after coal mining commenced, the first pound of tea from Indian

---

4   The phrase denotes an anachronistic European viewpoint. A similar 'discovery' took place in the United States when the French explorer Louis Jolliet and his Jesuit companion, Jacques Marquette stopped at a native Indian encampment on the Illinois River in 1673 and sketched the black banks of *charbon de terre* on their maps. By 1682, Cavalier de La Salle had claimed the Mississippi River Valley had recognised the potential wealth of minerals, particularly the coal banks (Biggers 2010: 49–50).

plantations was brought to Calcutta in 1838; the first jute mill was established in 1873; and the first iron foundry began production in 1874. It is through coal mining that India internalised many of the modernist values and ideas that colonialism introduced. The idea of coal as the driving force of the modern industrial economy was one of these ideas. It is no wonder that for contemporary India, armed with 20 nuclear reactors and a significant renewable capacity, coal remains of paramount importance for the sovereignty of the nation.

Most of the early developments of Indian coal mining took place in Bengal. The earliest Indian coal company, named Alexander and Co., was established in 1815 when the colonial government advanced a large loan at a low rate of interest[5] to the celebrated entrepreneur William Jones, a mining engineer who is described as the 'Father of coal mining in India,' to start a coal mine in Burdwan. Jones obtained a lease for 99 bighas of land from the Maharani of Burdwan and operated the mine at a loss until his death in 1821. Alexander and Co. paid off Jones' debt and sold the mine on mortgage to one Captain Stewart, who worked it for a small number of years without commercial success. Alexander and Co. foreclosed on Stewart in 1822 and began to run the Raniganj mine, but also failed in 1824. Still, the creditors of the firm continued to operate it and production increased from 4,000 bazaar maunds (market weight) in 1824 to 400,000 in 1832. This went on until 1836, when Prince Dwarakanath Tagore purchased the Raniganj mine for Rs 70,000 and established the famous Carr, Tagore and Co. At that time, Raniganj mine was the oldest, largest and the richest coal mine in India. Tagore was more than a merchant; he was a dreamer and realised that the main difficulty lay in the transportation of coal from Raniganj to Calcutta through the flood-prone paddy fields of the lower Damodar valley. Since smooth and year-round transportation was the key to success, he invested in a 120-horsepower Forbes steamer and the Lower Howrah Dock on the Hooghly River near to Calcutta. He hired local tribal communities, Santhals, who used crowbars and wedges and were originally trained by William Jones in the Raniganj mine. Members of another local, lower caste community, Bauris, were generally employed in the Chinakuri mine; trained by Betts, the former owner of Chinakuri, they generally used picks. The equipment of picks and hammers for coal mining were supplied by the factory of Jessop and Co., located near Calcutta. The political significance of Carr, Tagore and Co. is apparent from the fact that they effectively built a monopoly over coal production, transportation and distribution. Tagore's extraordinary story of success is shows that Indian entrepreneurs were acting as partners in establishing the coal mines. However, Tagore became more than a partner: according to King (1976, 94), the purchase of the Raniganj coal mine gave him 'virtual control over the supply of fuel in the Bengal presidency.' In spite of investments in steamships, the transport of coal from Raniganj to Calcutta continued to remain a problem. All the shipping was to be undertaken during a ten-day period during the rainy season when the

---

5   Of Rs 40,000 at 6 per cent interest, see King (1976: 94).

level of Damodar River was high enough to permit the passage of coal barges. Barraclough (1951, 98) has described how hundreds of boats and barges used to be moored in Calcutta's 'Koilaghat' (literally, the coal landing dock) and the Amtaghat depot on the Western bank of the Hooghly River in Howrah. Each of these boats carried 200 to 600 *maunds* of coal, costing Rs 10 per *maund*, and managed three to four loads per season. The colonial government was the sole customer of this coal, using the resource to power its steamships.

In 1843, Carr, Tagore and Co. merged with Jeremiah Homfray's Gilmore, Homfray and Co. to form a joint stock association, the Bengal Coal Company. King considers this as Tagore's 'final contribution to the coal industry' (115). Gradually, this company became the largest in India. Coal was still transported in large tug boats powered by steam to the capital in Calcutta, located less than 200 kilometres southeast of the colliery belt. Transportation was expensive: in 1842, over 57 per cent of the annual expenditure of Carr, Tagore and Co. was incurred for boat hire and storage charges at depots.[6] There is no doubt why the suggestion – by Lord Auckland's 1837 Committee for Investigating Coal and Mineral Resources of India – to use Assamese coal did not gain momentum.

Although India's rulers were attempting to wean themselves from coal transported from faraway England, the colonial government nonetheless continued to remain heavily dependent on the Bengal Coal Company. The turning-point in the early history of coal in India was when the railways were established in 1857, connecting Calcutta to the Raniganj coal fields. By 1860, fifty collieries were at work in Raniganj, producing 99 per cent of Indian coal (Buchanan 1934). The ease of transportation raised the demand for coal by leaps and bounds, opening new markets in the jute mills and other factories that began to be established in the Hooghly industrial belt.

As mentioned earlier, a significant point to note is the prevalence of indigenous entrepreneurs in early coal mining. Thirteen of the 17 existing concerns were owned by Indians. The low levels of technology and capital investment ensured that Indian landlords (zamindars) could easily start a coal mine. Baboo Gobinda Pandit, the main rival of Bengal Coal Co., was the *Raja* of Siarsol, a zamindar under the Burdwan Maharajas, who contributed land for the Ondal railway station, which is still the largest coal-handling and transporting facility in India. Throughout the nineteenth century the Raniganj fields, with their extended sections in Jharia–Bokaro–Kargali, remained the key coal producer.

It would, however, be incorrect to consider the early coal mines as an open playing field for indigenous capital. In Simeon's (1996, 85) view, three main groups of actors eventually came to control the economic regime of the colonial

---

6   According to King (1976, 106) the total annual expenditure of Carr, Tagore & Co. in that year was Rs 175,000 of which Indian labour costed 50,000, rents and land taxes were 10,000, European salaries at mines was 6,000 and legal expenses 4,000, with miscellaneous expenses claiming 5,000. A hefty amount of 100,000 was spent on boat hire, storage at depots and other transport related expenditure.

coal mines, operating in informal modes of regulation that developed out of a preindustrial social context supported by legal structures that through inertia and laxity accommodated the system. These were, namely, the landlords or the rentiers and mine owners, the managing agency houses, and the Railway Board, the latter two firmly in British hands. While the first group, the zamindar mine-owners, was arguably replicating the production relations known to them from agrarian feudal structures in order to ensure a steady supply of labour, the managing agency houses and the Railway Board were efficiently generating surplus by controlling trade and the market. The British managing agency system was created to serve the urban-industrial enclaves, and it led to demand-centred organisation of the mining industry. In this organisation, the managing agency houses, located usually in Calcutta, had evolved out of the earlier commercial concerns that had traded in agricultural commodities and functioned as 'primordial banks' (85). The proclivity of these managing agency houses – such as Andrew Yule, which was one of the largest in coal trading – for speculation led to a spectacular commercial collapse during late 1820s and again in 1847–8 (Tripathy 1979; Singh 1966). By the end of the nineteenth century, both their form and functions had changed, but they remained agents of Limited Liability Companies in charge of the daily function of productive concerns, and the agents received, in exchange, a percentage return from the profits (Commander 1981, 92). By 1908, all public companies working in coal in Bengal were controlled by such enterprises. According to Papandieck (1978, 167–8), control by these agents – whose core interests lay outside the region, and indeed, the country – had a number of serious implications for the coal mining industry, including a lack of large-scale investment in the concerns controlled by them, which made them little more than market predators. The Railway Board purchased the produced coal, but set the selling price of coal at its convenience, its absolute control derived from its position as the principal consumer and transporter and its ownership of captive collieries.

Coal heralded the advent of modern mining in India and created a working class of its own. In his 1896 Report, the Chief Inspector of Mines James Grundy noted that the coal-mining industry was the largest employer of mineworkers in India, hiring over 67 per cent of all mineworkers, and coal mines were the most significant employer of women. Captain Stonier's 1901 *Chief Inspector of Mines Report* noted that at the turn of the century, India headed the 'British colonies, dependencies and possessions' in coal production. He also recorded a surge in the number of workers between 1880 and 1900. Many of these workers were from local tribal communities and lower castes. The coal mines were where India's tribal groups and oppressed castes began their transformation into a proletarian working class, moving from the fields to the factories, exchanging their ploughs for picks (Read 1931, 106).

Nonetheless, coal mines of India truly remained a resource periphery; well into the early decades of the twentieth century, the entire production of Raniganj coal was being transported to Calcutta for use in steamships that plied the coastline;

iron foundries and other industries surrounding the Calcutta metropolis. This made the coal mining industry a secondary enclave that served the primary colonial industrial enclave closer to the metropolis. The turn of the century, however, saw the development of coal-based industries such as iron-smelting furnaces and paper mills in the coal producing region.

By the time of the First World War, the Indian coal industry had come to acquire all of the characteristics that caused its chronic economic-social ailments in later years. The growth of coal mining had brought in its wake the exploitation of labour, displacement of agriculture, lease disputes over land, problems of transportation methods, the overall technological and administrative inefficiency of most of the mines, incompatible mechanisation and, above all, the haphazard extraction of coal and incidental mining hazards. The period between 1942 and 1945 witnessed the occurrence of a coal famine, partly due to increased demand, but mainly caused by the sudden fall in production caused by wartime labour scarcity and lack of expedient transportation facilities. To resolve these circumstances the *Colliery Control Order* of 1944 was passed, assuming control over prices, production and distribution of coal. The Coalfield Recruiting Organisation was introduced to maintain – often forcibly – uninterrupted supply of labour to the mines. The resultant supply of immigrant labourers not just significantly altered the social character of the coal-bearing tracts but also shaped part of the national debates over labour safety.

Postcolonial Anxieties over the 'Death Pits'

If coal was symbolic of colonialism before 1947, it started to become closely associated with nationalism after India's independence. Most of this sense of nationalism arose from early trade union activities; leaders such as S. A. Dange (1945, 3), a frontrunner of the Communist Party of India, described the coal-pits of India as 'nothing but death pits for the miner.' More than safety, his primary concerns were around poor wages relative to production efficiency and lower coal prices (p. 4): 'The Indian miner is getting a much smaller share in the value produced by him than what the miners of other countries are getting.' Indeed, even during the Second World War when India's coal production rose sharply to meet the wartime demand, the wages earned by the colliery workers had not increased.

Both demand and production of coal rapidly had increased until India became independent in 1947. In that year there were 902 coal mines employing 322,000 workers and producing about 27 million tonnes of coal in India. During the 1950s, the Indian government began its centralised five-year planning system, following the vision of the new Prime Minister of India, Jawaharlal Nehru. Nehru imagined a new India, a growing power that produced its own steel and iron, and the coal mines played a key role in fleshing out this dream. Under the five-year plans, state policies favouring industrial development further encouraged coal mining.

The newly constructed steel plants needed coal not only as a raw material for manufacturing, but also as a source of energy. Coal was rapidly becoming the driver of the Indian industrial sector, and many households in urban areas made the transition from biomass fuels to coal-fired cooking ovens, and to the convenience of having electricity at home. Thus, by 1960, coal came to play an important role for the domestic sector, particularly for middle-class urban consumers. At the same time, trade unions were growing stronger in the collieries, some of which were owned by local landlords and run along the lines of feudal labour and production relations (Kumar 1996). These leaders united the workers to demand better and regular wages, increased safety, and more secure jobs. Non-payment of bonuses and low wages, lack of social security, and the establishment of a Labour Welfare Fund, as well as management malpractice and breaches in regulation were major issues around which the trade unions mobilised until mid-1960s. Pramanik (1993, 88) gives an account of the deplorable working and living conditions of miners during this time:

> Malpractice, corruption, mismanagement were common under private ownership and "law of the jungle" prevailed....Workers were exploited ruthlessly and... were denied payments....The "musclemen" (*lathiyals*) of the owners silenced the outspoken workers...Manipulation of accounts, depriving workers' [of their] dues, evasion of government royalties and taxes were the policies of the private colliery owners.

Consequently, debates in the Indian Parliament regarding the future of the coal mining industry became heated. Around this time prominent members of the Indian Parliament, particularly those from the Communist Party of India (e.g., Indrajit Gupta) pressed for nationalisation of the coal mining industry. In a question placed to the Minister of Steel and Mines on 27 August, 1965, Gupta argued: 'As small coal mines are a drag on the whole coal industry, even from the point of view of cost of production and modernisation, isn't nationalisation the only way out instead of amalgamation?' Leftists and trade union leaders in the Parliament maintained that state-ownership would provide better working conditions for the workers. Findings from the 1946 report of the Mahindra Committee were reasserted by key figures such as S. Mohan Kumarmangalam (1973, 54), who wrote:

> Over much of the industry, the conditions of the labour are still in a shocking state; living accommodation is inadequate and deplorable; educational and medical facilities are scanty and few amenities exist to relieve the strain and tedium of work underground.

The poor record of labour safety in collieries and the frequency of accidents, however, formed the other major argument in favour of nationalisation was. A number of serious accidents took place in various coal mines in India during 1962,

1963 and 1964, resulting in injuries and deaths of workers. D. Sanjivayya, the Minister of Labour and Employment, presented information in the Parliament on 12 April 1965 that showed that in 1962 there were as many as 229 fatal and 3,125 serious accidents in 80 coal mines in the country.

Further fatal colliery accidents came to light in the Parliament during this period. The May 1965 accident in New Kenda Colliery was caused by a sudden roof collapse, inflicting serious injuries on the workers present there. Two mineworkers were killed in Darula colliery in Pandaveshwar in West Bengal in November 1965, also due to a roof collapse. Another accident occurred in May in Dhori colliery, caused by an explosion that initiated 'a series of coal-dust explosions.' The rate of accidents remained high throughout 1960s: in July, 1967, an accident in Dhemo Main Colliery, Asansol, killed two workers. All these reports drew attention to the need to tighten the state's control over the coal mines in order to improve the safety of workers. At the same time, production of coal was increasing in India: in 1967, the country was annually exporting about 200,000 tonnes of coal to Burma (now Myanmar), and 150,000 tonnes of coal to Ceylon (now Sri Lanka). The possibilities of selling surplus coking and non-coking coal to other Asian countries were improving, and the state realised that in order to gain a secure footing in these export markets, it was imperative to establish greater control over the coal mining industry. Looking back at the times, one might say that the precocious language of coal nationalism was created during this time through these debates, in order to present independent India's response to earlier coal colonialism. Through these extensive debates in the parliament, coal was elevated to the status of a national treasure: a treasure that dispenses death to its toiling masses but that also deserves the governance of the state. The misery of the mineworkers through each tragedy described in Delhi wove a narrative that extols not just individual sacrifice but national glory and power.

By the time Nehru's daughter, Indira Gandhi, became Prime Minister, the ideological wave was such that bringing all the collieries under state control by creating a major state enterprise appeared to be the logical option to ensure multiple objectives were met: the provision of a steady flow of coal at a regulated price to the new steel plants and other ancillary industries; control of the production and supplies and hence the price of coal by a central body; removal of feudal interests; improvement of working conditions of workers; and building consolidated knowledge of coal reserves and their planned development. State ownership or take-over of mining companies was favoured at the time, particularly among the less-developed countries, as is evident in the nationalisation of Indonesian tin and the Venezuelan iron ore mining industries (Radetzki 1985). At that time, there were several reasons for governments to pursue mineral activities: first, the belief that a country's mineral wealth is the nation's patrimony; secondly, to prevent private interests from profiting by exploitation of such assets; thirdly, the inability of fiscal tools such as taxation with which to appropriate adequate and regular mineral rents; and last, the strategic importance of extraction and processing of minerals to assure domestic manufacturing supply, including the defence industry. Radetzki (1985, 14) comments that 'the ultimate and most far-reaching intervention measure was for the government to become the owner of the industry.'

Coal mining was brought under public ownership in phases between 1971 and 1973, followed by the *Coal Mines (Nationalisation) Act*, passed by the Indian Parliament in 1973. The Coal Mines Authority Ltd. (CMAL) was also established in this year to operate the nationalised non-coking coal mines. In September 1975 the nationalised coal industry was restructured with the establishment of Coal India Limited (CIL), a holding company with its headquarters at Calcutta. CIL now controls eight subsidiary companies. Seven of these are coal-producing entities directly engaged in the extraction and distribution of coal. The eighth, Central Mine Planning and Design Institute Limited (CMPDIL), is solely engaged in mine planning and designing in the coal sector. In spite of state ownership of CIL, many of the old relations of production have remained, giving rise to new difficulties. Some of the pre-modern labour structures, such as labour replacement (*badli*), were so entrenched that they continue even today. Typical of a public sector enterprise, CIL became an inefficient and corrupt corporate organisation that was neither answerable to the state nor to the poorest of the people who live on the richest coal-bearing tracts of the country. By late 1980s, it became clear that the expectations of a socially and environmentally responsible organisation – such as due diligence in land acquisition, resettlement of displaced people, rehabilitation of the environment, and financial viability – were not being met. The economic environment also began to change: in response to the liberalisation of the Indian economy in 1992, there was a proposal to liberalise the coal mining industry. This proposal has been languishing in the Indian parliament for several years; the extent of liberalisation has been limited to the partial deregulation of the market pricing of coal.

Today, CIL is the world's largest coal mining organisation. Coal is regarded as equivalent to national wellbeing because it supplies two-thirds of India's total electricity. Due to the large profits it delivers – because of the high demand of an energy-hungry country where nearly half the population is still without access to electricity – CIL is regarded as a 'Maharatna' (a 'major gem') company, a blue-chip public sector enterprise that produces nearly all of India's coal, with the exception of a few captive coal mines that feed the thermal power plants.[7] Commercial coal sales can legally only be conducted by and through the public sector coal companies and their subsidiaries; coal produced from captive mines by the private sector cannot be sold on the open market. The restriction of captive mining does not apply to state-owned coal development undertakings such as SCCL, Neyveli Lignite Corporation (NLC), and the Mineral Development Corporations of the State Governments. One can expect the private sector to play a limited role in the Indian coal industry in the foreseeable future.

---

7   In March 1996, the Government also allowed captive mining of coal for production of cement.

Generating the Crisis of Coal

The liberalisation of the coal mining industry since the early 1990s has followed the rising sense of crisis in the availability of energy, framed as part of the perceived military and strategic need to maintain sovereign political power in the face of the rapidly growing Indian economy (see the Brookings Institution studies series by Madan 2009). The sense of crisis in policy circles is most acutely felt – and conveyed – in thinking about the finiteness of resources in relation to the growing population of India. The compelling reasons for increasing coal production lie in ensuring energy security; the Planning Commission (2006, xxiv) elaborates upon the need for energy security by stating that:

> India's energy security, at its broadest level, is primarily about ensuring the continuous availability of commercial energy at competitive prices to support its economic growth and meet the lifeline energy needs of its households with safe, clean and convenient forms of energy even if that entails directed subsidies. Reducing energy requirements and increasing efficiency are two very important measures to increase energy security. However, it is also necessary to recognise that India's growing dependence on energy imports exposes its energy needs to external price shocks. Hence, domestic energy resources must be expanded. For India it is not a question of choosing among alternate domestic energy resources but exploiting all available domestic energy resources to the maximum as long as they are competitive.

Given that coal accounts for around 55 per cent of India's current commercial energy consumption, and about 75 per cent of the total coal consumed in the country is used for power generation, coal is therefore inextricably linked to the goal of ensuring India's energy security. The same document notes that at the current levels of consumption, India's proven coal reserves can last for about 80 years. If the inferred reserves can also be proven, then the total coal reserves could last for 140 years, again assuming current levels of extraction and consumption. However, if domestic coal production continues to grow at 5 per cent per year, the total extractable coal reserves will be depleted in around 45 years.

The Planning Commission (2006) predicts that the next decade will bring pronounced coal shortages to India. In a mid-term appraisal of the tenth Plan, it projected a shortfall of 11 mt in the terminal year (2006–7) of the Plan. This shortfall was predicted to rise dramatically by the end of the 11th Plan (2011–12). The magnitude of the shortfall has been estimated differently, depending on where the coal would be sourced from to eliminate the shortfall. A Ministry of Coal expert sub-group estimated a shortfall in 2011–12 of 503 million mt if the entire coal requirements were to be met from coal produced within the country, and of 472 million mt if the shortfall were to be met from the import of coal. Coal shortages, the report notes, are likely to become acute in the first two years of the

11th Plan and are likely to rise rapidly by the end of the 11th Plan unless Coal India's unprecedented capacity expansion plans materialise during this time.

Not only are resources constructed in these discourses by their utilitarian value in meeting human needs, they also assume wider cultural meanings associated with economic development, nationalism and nation-building. The procurement of a certain physical quantities of material is framed as a problem; this is essentially a problem of there being less supply than is required. In documents such as the report mentioned on energy resources, the availability of electricity is equated with development. A picture of insecurity and a bleak future is drawn to convey the sense of urgency in dealing with the problem (see, e.g., Malik 2002; Garg and Shukla 2009). The formulation of the problem in absolute terms leads to simplistic and predetermined solutions. These solutions result from a quantitative and positivist view of resources as able to be managed by technical experts, leading to the comparisons of absolute numbers of population, land and resource amounts, which are seen as necessary to reach certain levels of development, thereby developing a sense of scarcity. It also entrenches the belief that there is nothing untoward with the way we consume and use resources; that nothing is wrong with its ownership, allocation or distribution, and that no other alternatives exist to large-scale coal extraction. The concept of energy security is conceived on simplistic assumptions such as linkages between growing populations leading to environmental degradation; scarcity leading to decreased economic activity and migration; the weakening of states resulting in conflicts and violence. This conceptual position has equated energy security with strategic and military security: on the one hand, resources have become the environmental trigger of conflicts and crises; and on the other, the sense of urgency has helped the Indian state to shed ethical values and its responsibility to the poorest citizens.

Regulatory Framework

Let us now briefly examine the resource regulatory framework in order to understand the legal context or the rules within which coal mining operates in India. The management of mineral resources in India is the responsibility of both Central Government and State Governments as per the Constitution of India. The *Mines and Minerals (Regulation and Development Act 1957*[8] (MMRD) and the *Mines Act 1952*, together with the regulations framed under them, constitute the basic laws governing the mining in India. For coal mining, the MMRD is one of the three overarching pieces of legislation affecting the coal sector. The two others are more specific to the coal sector, and include the *Coal Mines (Nationalisation) Act 1973*, and the *Coal Bearing Areas Act* (CBAA). The MMRD

---

8   This Act was later revised in 20th of December, 1999, to Mines and Minerals (Development and Regulation Act, 1957, ('MMDR' or MMDRA). See http//mines.gov.in/policy/printmmrdnew.html (accessed on 17 December, 2012).

classifies all minerals into two categories – major and minor – according to their strategic and military importance. Coal, lignite, mineral oils, iron ore, copper, zinc, atomic minerals, etc. are major minerals listed in Schedule A, which is reserved exclusively for the public sector. Schedule B lists minor minerals, which the private sector is permitted to mine along with the public sector. According to MMRD, prior approval of the Government of India is necessary before the grant or renewal of mineral concessions for minerals specified in Schedule 1 of the Act. Presently, 23 minerals are included in Schedule 1; of these 11 are atomic minerals, one is a fuel mineral (coal and lignite), and the remaining ten are metallic ores and industrial minerals. Important supplementary rules in force under the MMDR are the *Mineral Concession Rules 1960* (MCR) and the *Mineral Conservation and Development Rules 1988* (MCDR). The MCR outline the procedures and conditions for obtaining a Prospecting Licence or Mining Lease. The MCDR prescribe guidelines for ensuring mining is undertaken according to specific environmentally responsible requirements. It is noteworthy, however, that the MCR and MCDR do not apply to coal, atomic minerals, and minor minerals. Amendment to the Act was effected in 1978, 1986, 1994, 1999, and most recently in 2012. While the first two amendments increased governmental control, the last two relaxed them. The changes in the regulatory dispensation in 1994 and 1999 envisaged considerable devolution of authority from the federal government to the states. Currently, MMRD is known as MMDR, the intention being the attribution of somewhat greater prominence to development than to regulation.

The *Coal Mines (Nationalisation) Act 1973* reinforces the spirit of the MMDRA, because by nationalising the mines it has firmly consigned coal to the purview of the public sector. The Act categorically states that 'no person, other than the central government or a government company or a corporation owned, managed or controlled by the central government shall carry on coal mining operation in India, in any form.'[9] The responsibility for overseeing mines and mineral development – and the implementation of legislation – is divided among the central and state governments; the former having exclusive power to make laws with respect to regulation of mines and major minerals development. The state governments are largely responsible for implementing the laws, but are constrained to act within the framework laid down by the central government. The states do, however, have the power to make rules in respect of minor minerals under Section 15 of the Act. They are able to devise specific supplementary legislation to promote investment within the state, and have, over time, begun to take initiatives to attract private players, including foreign investors, to the states.

I have noted that coal in India has been surrounded by legal provisions aimed at attributing primacy to coal mining over all other uses of land. This is acutely evident in the *Coal Bearing Areas (Acquisition and Development)*

---

9   See the full Act in see http://www.indiankanoon.org/doc/72652/ (accessed on 17 January, 2013).

Act 1957 (CBAA).[10] The CBAA was passed to 'establish greater public control over the coal mining industry and its development, [and] provided for the acquisition by the state of unworked land containing coal deposits or of rights in or over such land.' Thus, CBAA can overrule tribal ownership of land. Such a law goes against the grain of democracy. In coal mining, the CBAA and the *Land Acquisition Act 1884* (LAA) together give the state the extreme power of usurping any property belonging to any citizen or corporate entity if extractable amounts of coal occur under it. A High-Level Committee was established by the Planning Commission of the Government of India during the mid-term review of the Tenth Five Year Plan (2005) to increase the speed of the flow of private investment and to examine procedural delays and the absence of infrastructure. This committee reviewed the *National Mineral Policy 1993* and the *Mines, Minerals (Development and Regulation) Act 1957*, and suggested major changes in the legal framework around coal (High-Level Committee 2006). It noted that more robust and socially informed mineral resource laws are needed based on an agreed set of broad principles. The foremost of these principles would be respect for the rights and interests of all those involved. Moreover, current environmental laws are focused only on *mitigating the negative impacts* of mining. Instead, we need to frame policies and laws that can deliver sustainable developmental benefits for local, regional and even global communities. For local communities to enjoy the rights to continue their livelihoods – founded on, and in proximity to – mineral resources, to participate in decision-making, and to benefit in real terms from the economic benefits of coal mining, the laws must emphasise the need for a participatory and inclusive approach at the process level that mining companies currently do not or cannot adopt. Revisiting the laws surrounding mineral resource extraction in India would thus involve a fuller understanding of the role of the community in local economies; it would provide access to resources for local people, and integrate community interests in mine management plans. If the coal resources of India are truly vested in the national interest, then the state should ensure that opportunities benefit the nation in an equable way. India is now passing through a volatile time that foresees the eventual divestment of the nationalised coal mining industry and the end of the CIL monopoly over coal mining. How then will the social impacts that the coal mining industry causes be dealt with?

The Emerging Scenario

The economic liberalisation first initiated in 1991 was shortly followed in 1993 by a *National Mineral Policy*, which governed, among other things, public and private sector participation in the wider mining sector. The *Coal Mines (Nationalisation)*

---

10   The full text of CBAA is downloadable from http://www.coal.nic.in/cba-act.pdf (viewed on 17 March, 2013).

Act 1973 was amended in 1993 to allow captive mining in approved end-user industries, namely iron and steel, cement, power generation, and coal washing. It was clearly defined that companies engaged in captive mining must undertake to supply coal only to the approved end-user industries, and not sell any coal on the open market. This entailed several proscriptions, but still marked the start of the process of increasing liberalisation. In December 2000, Central and State Public Sector Units (PSUs) and State Government Undertakings were allowed to mine coal at par with the Central Coal Companies. Over a decade since then, the results have been disappointing at best. According to the Ministry of Coal (2005), efforts to promote captive mining only began in earnest from 2003, even though the policy itself was promulgated ten years earlier. When captive mining was first allowed, 148 blocks were identified for allotment under government or captive dispensation; of these, 89 are already allotted or earmarked for allocation. These 89 blocks contain total geological coal reserves of 13.5 billion tonnes – eight billion in the 'proven' category – and could theoretically yield a total production of about 100 mt of coal annually (Ministry of Coal 2005). However, as the subsequent analysis shows, the use of captive mining in increasing domestic coal output has not yielded the results hoped for by the Indian Government. The most striking feature is that of the 68 letters of allotment, only about ten mines have gone into production to date. This points to a situation in which the slow gestation period from allotment to coal production due to a corrupt, slow and inefficient regulatory process renders impossible the government's stated objective of significantly increasing production through captive mining. This appears to reflect serious neglect or lack of understanding of implementation requirements in the goal-setting process. It is possible that for this reason the major players in international mining have largely stayed away from the Indian coal sector. Since 2003 the government has increased the strictures by which it monitored the progress of captive mines, aiming to reserve the right to cancel an allotment if progress is found to be unusually slow. However, foreign investors blame the failure of captive mining on a multitude of factors, primary among which is the regulatory quagmire of discretionary rules and opaque procedure.

Within this scenario of poor foreign investment in coal, many Indian companies from both public and private sectors have proactively begun to acquire foreign coal assets abroad to augment their supplies and captive mining activities. A glance at such companies that have already made overseas acquisitions, or are trying to do so, yields several names, including CIL. Through its subsidiary for overseas ventures, Coal Videsh Limited (CVL), CIL has begun exploring for coal in South Africa, Indonesia, Mozambique, Zimbabwe and Russia, either through complete ownership or joint ventures. CVL is also actively exploring opportunities in Australia and Canada. The Steel Authority of India Limited (SAIL), India's largest steel producer, is also acquiring overseas mines as a means of obtaining high-quality coal that can be blended with Indian coal to produce steel. According to the Chairman of SAIL, currently about 65 per cent of SAIL's coal requirements are met through imports and the remaining 35 per cent through domestic sources.

India's largest independent metallurgical coke producer, Gujarat NRE Coke Ltd, became the first Indian company to acquire coking coal mines in Australia, in late 2004; the agreements were to acquire the coal mining leases in the whole of old Avondale Colliery and part of Huntley Colliery in the Southern Coalfields of New South Wales. Tata Power Company purchased a one-third share in Indonesia's Bumi Resources, and both Tata Power and Tata Steel as part of the Tata Group are actively seeking coal assets in Australia, New Zealand, Mozambique, and Indonesia in a bid to triple their steel production.

In India's coal-bearing tracts the sense of urgency that accompanies the framing of crisis has encouraged and allowed private and public companies to invest in resource extraction. Some of these mineralised lands are owned by indigenous, tribal communities (*adivasis*) where signs of development and amenities indicating an improved quality of life have yet to reach. In the Indian state of Jharkhand, much of the land that is being excavated by extractive industries is inalienable *adivasi* land. Some of these lands might have been used by these communities for generations, but without ever having been being recorded in revenue records. These are deedless lands, commons that play important roles for these communities. Mineral resource extraction in these areas attracts outsiders, changes the resource and environmental base of livelihoods, and fundamentally alters the lives of members of these communities. Adequate compensation is never forthcoming. Allowing companies based in metropolitan cities or owned by individuals as share-holders to mine in these lands implies a one-way drain of wealth out of these areas and an accumulation of surplus value at the cost of rural displacement and dispossession;[11] these are peripheral areas that produce resources and serve the urban-industrial economy.[12] At the same time, the land and mineral ownership laws have persisted from colonial times, as have laws to acquire land. The Indian coal mining industry needs to engage with debate in the global mining community, which has now come to accept the primacy of rights of communities over mineral resources. The situation in India is recognised as constrained, even by mainstream bodies such as the World Bank, which, in its 'Mining Sector Reforms and Investment—A Global Survey' carried out in 2001 (cited in Planning Commission 2006), identified India as one of the least desirable

---

11   Even mainstream economists agree that natural resources, particularly minerals, and conflicts are closely linked and can eventually impact on the development agenda (see Bannon and Collier, 2003).

12   In context of Canadian indigenous-owned land, Laforce (2010) argued that mining companies entered the arctic north of America at a time when state power did not reach the remote frontier areas where mining booms were taking off through 'gold rushes'. Consequently, mining entrepreneurs stepped in to fill the normative field that the state failed to establish, allowing them to define the rules of the game which the state eventually adopted. Even in Australia, where contract-making and negotiations between mining companies and aboriginal communities have increasingly become standard practice, raise major issues for aboriginal relations with other political actors and institutions, including government, environmental groups and the judicial system (O'Faircheallaigh, 2010).

countries compared to other resource countries such as Australia, Brazil, Chile, China, and Indonesia.[13] It is important to note that these countries have greatly liberalised their mining laws and made them extremely investor-friendly. So far, India's efforts have been concentrated in simplifying procedures to facilitate captive mining, and to some extent the introduction of transparency in the process of consultation; however, much is left to be desired. A quick entry into open-cut coal mining also requires that the Indian state adopts due measures to protect its citizens. Unless the social and environmental impacts of coal mining are adequately addressed by suitable experts and multi-stakeholder forums, it will continue to be fraught with difficulties. This was noted by the High Level Committee appointed by the Planning Commission of the Government of India in September 2005. While the Committee's tasks did not clearly include consideration of the social impacts of mining, it devoted a significant section to 'Local Communities and Mining Activities.' It lamented the poor social care in mining regions (2006, 69):

> Land is often used without the consent of the indigenous people. Mining companies should act as if consent to gain access to land is required even when the law of the land does not require this. In making decisions, the cultural circumstances of the local people and loss of access to common resources should be kept in mind. Where resettlement takes place, companies need to ensure that living standards are not diminished, that community and social ties are preserved, and that they provide fair compensation for loss of assets and economic opportunity. Responsibility for ensuring the long-term wellbeing of resettled communities needs to be defined and monitored.

A body of literature is emerging that demonstrates that the local poor, particularly indigenous community, receive the least benefits from such developments, and are among the most affected (Padel and Das 2010). Throughout India there is growing resentment and resistance to land grabs for mining–industrial–urban requirements; the Sixth Citizen's Report on the State of India's Environment by the New Delhi-based think-tank, Centre for Science and Environment, titled '*Rich Lands, poor Peoples*' suggests that social unrest and conflicts in India is closely associated with the mineral-rich belt that extends from Jharkhand through Odisha to Chhattishgarh and the Andhra Pradesh states of the country (Bhusan and Hazra 2008). The question of rights and access to the land and its natural resources is at the centre of debate over unauthorised coal mining. The protection of common pool resources to help poor communities survive in colliery tracts is crucially important, as is the need to find ways to vest the power to co-manage mineral extraction with the local communities. It is also important that a wide debate occurs between social scientists, planners, international agencies and civil

---

13   Mining is a risky venture to invest capital in, and has long gestation periods and uncertain returns on the invested capital; hence it is extremely sensitive to the regulatory framework operating in a country.

society, on the issue of justice in coal mining areas. It is understandable that the public sector would resist any attempts to implement regulatory and technological changes that reduce its monopoly power. The resistance is unlikely to be explicit, but most likely will be mounted through administrative delays and bureaucratic bottlenecks, long identified as the primary restrictions to the efficiency and expansion of coal mining in India.

The Chapters in the Book

The contributions in this volume engage with the key debates specified above. The book structures the contributions into three sections: the first, titled *Justice, Legality and History*, deals with laws and legality in coal mining. This is an important aspect, particularly for India, where adherence to the rule of law[14] is an important and robust correlate of development. In the field of mining, resource governance has come to assume the status of a panacea which, if applied properly, will be able to cure all problems. At the same time, a growing body of literature on mining, primarily from countries other than India, suggests that liberal legalism suffers from systemic problems that affect its integrity and effectiveness (Szablowski 2007, 117). When a resource or environment-related law is violated, either by the state or by citizens, it opens up an area of inquiry that illuminates the grey zone within which the meanings of both legality and legitimacy are contested by the actors. Chapters in this section also deal with non-legal spaces; these are areas that have remained beyond the purview of existing laws, which are not adequate to cover the full gamut of human activities. Because non-legal spaces remain beyond the legal framework, they are able to expose the fragility of law. Deep insight into the historical accounts of the establishment of the coal mining industry also helps us to understand how and in what context some of these laws came into existence.

Five chapters comprise this section. The first chapter is titled 'Between Legitimacy and Illegality: Informal Coal Mining at the Limits of Justice,' by Kuntala Lahiri-Dutt. It presents critical analysis of what constitutes illegal coal mining in eastern India, and considers mining within the moral economy of local tribal and other communities who have been displaced over the years by advancing mining activity. A common sight in eastern Indian coal tracts is that of hundreds of cycles, each loaded with 200–300 kilograms of coal packed in sacks, being pushed like beasts of burden, or packhorses. Questions arise as to who these people are and what they are doing. This chapter critiques popular notions of illegality in the context of coal mining in eastern India and shows that the social and environmental contexts of such mining differ depending on whether it is the older colliery belt, the Raniganj-Jharia fields, or the newer coalfields

---

14   The tradition of 'rule of law' relates to the principles of 'natural justice', first developed by courts in the UK in the nineteenth century as a code of administrative procedure. For more on rule of law and its social determinants, see Dawson (2013).

where mines are just beginning to expand, the Karanpura fields. Consequently, the causes of illegal mining are different, but the answers must be sought within the overall lack of social justice within the colliery belts. Thus, what is presented by the media as an illegal activity that is encroaching into the commons is seen by those who are engaged in it as claiming their basic moral rights to a precarious survival. If in the older regions the major coal mining companies, by ignoring their responsibilities of social and environmental care, have de-agrarianised the entire region and thus pauperised indigenous and lower caste populations, in the areas of new coal mining the physical displacement and innumerable and intangible secondary impacts have forced people who had previously depended on forest and water resources from the land to make a precarious living digging coal. Thus, to deal with what is commonly presented as illegal mining, one must understand what constitutes environmental and social justice for the local inhabitants.

The third chapter of the book, titled 'Coal in Colonial Assam: Exploration, Trade and Environmental Consequences' is by Arupjyoti Saikia, who explains that while economic imperatives are important, they were not the only factor in the rapid expansion of coal mining in Assam after the Anglo-Burmese War in 1825. Saikia shows how coal was more than just a subject of imperial economic investment: that it was also associated with advances in geological science in tracing the potential repositories of coal and other minerals across the province of Assam, leading eventually to the entry of imperial capital, assisted by the government in securing legal rights to land and natural resources. Thus, the key factor behind establishment of coal mines in Assam was the firm establishment of rights over land and other resources of Assam by the colonial state. The history of coal appears to be less contested than the history of tea or petroleum, the two other important imperial ventures of the colonial period in Assam. Coal mining arrived in Assam later than in the Raniganj–Jharia fields of Jharkhand; the earliest convincing suggestion that Assam would be an important source of coal in British India occurred during the Anglo-Burmese war in 1825. At this time coal was yet to be productively extracted from other parts of British India. The situation changed rapidly in the next couple of decades, and within three quarters of a century the British Empire oversaw a swath of coal mines in Assam. Undoubtedly, coal was an important source of its political and economic strength. This chapter deals with aspects of geological exploration, securing rights to extract and process minerals, and an examination of the political contest over coal as its economic value has increased.

The fourth chapter of the book, titled 'Border Mining: State Politics, Migrant Labour and Land Relations along the India-Bangladesh Borderlands' is by Debojyoti Das. This chapter explores the history of informal coal mining and its legal context in the North-eastern state of Meghalaya. Meghalaya occupies a special status among the Indian states; being a sixth schedule state, here the residents own the minerals that lie under their land. The chapter explores the issues that arise from the contradictory positioning of coal as a major mineral, to be used for the national good by major companies, and – unlike in the rest of the country – the strong rights of its residents not only to land but also to sub-surface resources, including coal. The chapter links India's colonial legacy with the cultural politics of resource use in contemporary Khasi-Jaintiya society, where ethnic identity is used as a cultural tool valorising indigeneity and the capitalist production of coal by native mine owners, and where the very idea of political borders still seems alien.

The fifth chapter, titled 'Slaughter Mining and the "Yielding Collier": The Politics of Safety in the Jharia Coalfields 1895–1950' is by Dr D. K. Nite, who enquires into how the idea of a 'conscientious miner', the *chalak* colliers who are endowed with practical expertise was created in opposition to the idea of unformed *burbak* colliers. The ideas were mobilised by colonial mine administrators to enhance safety, but, in so doing, colonial mining capital connected systems of labour subordination with safety, thus initiating what can be described as the politics of safety, which was a marked development in the coalfields. The chapter shows how miners negotiated the dangers of the mines and how they conceived and articulated their ideas of safety. The fatalistic *khadan-kali* tradition of colliers remained impermeable to the modern safety perspectives, whereas the informed, initiated colliers applied new safety behaviours and politics to the workplace. Contrary to the conventional ideals, safety was anything but a secular and apolitical matter: it was subject to socio-political and religious dimensions in mining life. The better-paid and higher-level mining personnel – the *babus* – remained largely blithe by the safety agenda of mineworkers. The industry remained an agent of appropriation, pushing the workers into perilous working situations. The chapter brings to the forefront the colliers' work beliefs, experiences and work patterns in order to grasp the dynamics of a coal mining community that lived a precarious existence. In particular, it elaborates upon how the cultural and political locus of the mineworkers helped them to adapt to changing work relations in colonial India.

The last chapter in this section is by Kuntala Lahiri-Dutt, titled 'Stranded Between the State and the Market: "Uneconomic" Mine Closure in the Raniganj Coal Belt.' This chapter discusses the social and environmental issues surrounding the coal mining operations of the Raniganj belt of West Bengal, and argues that poor environmental and social care leads to the misleading depiction of open cut mines as economically more efficient than underground mines. The state-owned Coal India Limited (CIL) has been beset by economic problems and has been responsible for causing serious social disruptions and environmental hazards in its areas of operation. The monopoly status that the public sector companies have enjoyed for over three decades has acted as a disincentive to improve the social and environmental performance of the industry, the major effort being directed at improving operational processes through the introduction of new technology. It closely scrutinises a decision by Coal India Limited and one of its subsidiaries, Eastern Coalfields Limited (ECL), to close down what are regarded as 64 'uneconomic' collieries located mainly in the Raniganj coal belt of West Bengal: the collieries that are depicted as uneconomic are those that depend upon underground mining technology. It also suggests that this agenda of closure attempts to hide the active efforts to have collieries run by private operators, an agenda of ECL since liberalisation of the coal sector.

The second section of the book, titled *Mining Displacement and Other Social Impacts*, comprises six chapters considering the various social issues around coal in India, with particular emphasis on the most critical of them: displacement of the poor. The data of coal mining displacement is staggering, as is evident from the detailed database created by Ekka and Asif (2000) for Jharkhand state alone. Resource extraction has further pauperised or marginalised the poor, treating them as the detritus of mining development and abusing their fundamental human rights. Human rights lawyers show that the direct or silent complicity of mining corporations in such abuses enhance resource conflicts. Official discourse justifying displacement is often framed in the name of mineral resource development, thus leading to what has come to be known as 'development-induced' displacement. People are treated as infinitely moveable and uprootable, making their displacement seem like an acceptable by-product in the production of wealth and the creation of various forms of public order (IDMC 2009). When development is seen as beneficial, as 'in the national interest', it renders the sufferings of displaced people invisible. The relationship between development and displacement has been explored from the 'reformist-managerial' school by Cernea (2003), who views development as necessary, and displacement as an unintended but inevitable outcome of the development process, the 'pathology of induced development' (2003, 37). For the most part, this approach is favoured by planners and managers of mining.

The alternative 'radical-movementist' approach to displacement often informs social movements opposing large development projects; it views the phenomenon of displacement as a symptom of a development model in crisis, the outcome of structural biases within the dominant development paradigm. As noted by Ganguly-Scrase and Lahiri-Dutt (2013), development-induced displacement is not merely an accidental feature of development gone wrong, but is an inherent part of neoliberalism, which in its drive for ever greater efficiency renders traditional livelihoods redundant and systematically devalued. In many instances, therefore, displacement has become intrinsic to the contemporary ideas of mining development in India.

The chapters in this section combine both of the above approaches, focusing on the consequences of displacement with the intention of developing procedural means by which the negative consequences of displacement can be avoided or ameliorated. The seventh chapter, 'World Bank, Coal and Indigenous Peoples Lessons from Parej East, Jharkhand,' is by Tony Herbert and Kuntala Lahiri-Dutt, and examines the effectiveness of the World Bank's grievance resolution mechanism, the Inspection Panel (IP). The Bank-funded Parej East mine was one of the massive colliery projects that were to be initiated in Jharkhand. This large open-cut colliery displaced a significant number of indigenous families and affected the livelihood of many others. The chapter analyses how the country-level management of the World Bank can bypass the judgment handed down by the IP. It recounts the saga of a local non-governmental organisation that complained to the IP regarding the practices of the Bank. It critiques the IP for its lack of action, showing that such complaints do not receive adequate attention from either the state or Bank Management.

Chapter 8, '"Captive" Coal Mining in Jharkhand: Taking Land from Indigenous Communities' is by Kuntala Lahiri-Dutt, Nesar Ahmad and Radhika Krishnan. Resource restructuring in post-liberalisation India involves securing land from farmers to allow for commercial and industrial uses, including mining. Coal dependence in overall energy supplies has led to acquisition of coal from overseas as well as the granting of mining blocks to private companies for captive coal mining. What is exceptional is that some of the lands that are being given away as mining leases are technically inalienable, being tribal-owned land. The chapter analyses how privately owned coal mining companies, mining their own captive power plants, displace indigenous people by circumventing protective legislation supposed to prevent the alienation of tribal land. It argues that the legal framework around coal mining needs to be revisited and co-management of coal mining introduced.

Chapter 9, 'Coal Mining in Northeastern India in the Age of Globalisation,' by Walter Fernandes and Gita Bharali, takes us to North-eastern India, where it locates coal mining-related displacement within the public and social context of the region. Here, mining is presented as the key to development for a region that has always been treated by the central government and its policymakers as a resource periphery. Ecologically unstable and socially complex, the North-eastern states have become yet another resource frontier where the expansion of mining has been rapid, leading to enormous social and environmental consequences for the region that are mostly negative.

Chapter 10, 'Marginalising People on Marginal Commons: The Political Ecology of Coal in Andhra Pradesh,' by Patrik Oskarsson, shows how coal plays a crucial role in the ongoing efforts to increase energy production dramatically in the south Indian state of Andhra Pradesh. A number of large-scale coal mining and thermal power projects are planned in the northern tribal areas and along the eastern coastline. This chapter examines the trade-offs in land use which are being made at the state-level politics of coal as part of this energy expansion. A prevailing characteristic is the dependence on common marginal land; that is, land with officially low, or even non-existent, displacement of people and their livelihoods since government records do not acknowledge local land and resource uses. The political ecology of coal in the state thus comes with significant distributional concerns about who should be able to benefit from scarce land resources. This paper shows that despite a wealth of supportive legislation and strong civil society efforts, the interests of the poorest of the poor are harmed by coal-fuelled power generation.

Chapter 11, titled 'Water Worries in a Coal Mining Community: Understanding the Problem from the Community Perspective,' by Prajna Paramita Mishra, uses a case study of the Ib Valley in Odisha to show how the mining of coal induces water scarcities in the areas in which it operates. This chapter examines water problems that are two-fold: scarcity and pollution. It presents findings from a very large study of 260 selected households from mining villages and 100 households from control villages. It demonstrates that mining villages face acute water problems not only during the summer, but throughout the year, whereas this problem is entirely absent in control villages. The study rejects the general view that underground mines are less vulnerable than open-cut mines, and argues that with regard to water, both types of mines are equally vulnerable.

The last chapter comprising this section is titled 'Gender in Coal Mining Induced Displacement and Rehabilitation in Jharkhand,' by Nesar Ahmad and Kuntala Lahiri-Dutt. This chapter focuses on the impacts of coal mining on women and men to point out that the impacts of mining are gendered because women and men play different roles in these communities and receive differential levels of state and social recognition and recompense. Differences in gender roles mean that women and girls bear a disproportionate share of the unpaid work in almost all households in rural India; the inequities are exacerbated by shifting burdens: of work; of procuring primary subsistence; and of providing for children. When mining projects lead to restructuring of agrarian relations, decline in household assets, and in differential entry into labour markets and differential burdens of care, the capacities of women and men to seize new opportunities and to cope with the risks and fall-outs from such investments are invariably different.

The last section of the book is titled *Social Perspectives to Inform Mining Policy* and deals with the critical policy considerations emerging from the previous chapters. Many scholars, particularly economists, believe that developing countries and states can prosper by extracting and exporting their mineral wealth. There is now ample evidence (Oxfam 2001, 5) to the contrary. Countries that depend on mineral exports can be among the most troubled states; they suffer from exceptionally slow rates of economic growth, their governments tend to be weak and undemocratic, and they more frequently suffer from civil wars than resource-poor states. Extractive industries that are capital-intensive produce social and economic problems that fall heavily on the poor; they follow a boom-and-bust cycle that creates insecurity for the poor, and they lead to high rates of corruption, repression and conflict. Policy issues therefore assume priority in considerations of the function and future of coal mining.

Chapter 13, titled 'Colonial Legislation in Postcolonial Times,' by Nesar Ahmad, analyses the laws and policies governing involuntary displacement caused by coal mining, and the consequent rehabilitation and resettlement. It notes the vintage of resettlement laws and the marginal or cosmetic changes that have been undertaken so far, before suggesting that they be reviewed in an open, just and fair manner. The chapter offers compelling reasons to suggest that the time has come for these laws to be changed. The author urges policymakers and readers to consider the provisions made by the Indian Constitution for protection of the interests of schedules caste and scheduled tribe (SC and ST) communities, to note the inherent contradictions between these provisions and coal legislation.

Chapter 14 is titled 'On the States' Ownership and Taxation Rights over Minerals in India,' by Amarendra Das. This chapter appraises the constitutional provisions for the ownership and regulation of minerals in India under a legal framework supported by economic logic. It argues that to ensure intra-generational and inter-generational equity, ownership of minerals should be vested with state governments and regulatory power with the central government. However, for the development of mining areas, state governments should be provided with adequate elbow-room with which to mobilise revenue from minerals. Therefore, the present system of uniform royalty rates determined by the central government should be discontinued and states should be made free to determine their royalty rates and other levies.

Chapter 15, titled 'Key Policy Issues for the Indian Coal Mining Industry,' is by Ananth Chikkatur and Ambuj Sagar. It argues that decisions relating to coal mining must be placed within the broader perspective of India's energy needs. The authors show that coal use currently accounts for more than 50 per cent of total primary commercial energy consumption in the country, and for about 70 per cent of total electricity generation. Furthermore, about 80 per cent of coal produced in India is used for power generation. Hence, the power and coal sectors are interlinked in India, and coal is likely to remain a key energy source for India for at least the next 30–40 years, especially since India has significant domestic coal resources, relative to other fossil fuels, and a large existing installed base of coal-based electricity capacity. In this context, the final chapter addresses the key policy issues and concerns with regard to coal in India.

References

Andrews, Thomas G. 2008. *Killing for Coal: America's Deadliest Labor War*, Cambridge, Ma: Harvard University Press.

Bannon, Ian and Paul Collier. 2003. *Natural Resources and Violent Conflict: Options and Action*, Washington DC: The World Bank.

Barraclough, L. J. 1951. 'A further Contribution to the History of the Development of the Coal Mining Industry in India', Presidential Address, Mining, Geological, and Metallurgical Institute of India. *Transactions*, Vol. 47 (April 1951): 2–22.

Berger, Stefan. 2005. 'Introduction' in Stefan Berger, Andy Croll and Normal LaPorte (eds) *Towards a Comparative History of Coalfield Societies*, Aldershot: Ashgate, 1–11.

Biggers, Jeff. 2010. *Reckoning at Eagle Creek: The Secret Legacy of Coal in the Heartland*, New York: Nation Books.

Brown, Carolyn. 2005. 'Nigerian coal miners, protest and gender, 1914–49: The Iva valley mining community' in Stefan Berger, Andy Croll and Normal

LaPorte (eds) *Towards a Comparative History of Coalfield Societies*, Aldershot: Ashgate, 127–45.

Buchanan, Daniel. 1934. *The Development of Capitalist Enterprise in India*, New York.

Cernea, M. 2003. 'For a new economics of resettlement: a sociological critique of the compensation principle.' *International Social Science Journal*, 55(175): 37–45.

Chand, S. K. 2008. *The Coal Dilemma*, New Delhi: The Energy and Resources Institute.

Coal Controllers Organization (CCO). 2006. *Coal Directory of India*, Kolkata: Ministry of Coal, Government of India.

Commander, Simon. 1981. 'Industrialization and sectoral imbalance: Coal mining and the theory of dualism in colonial and independent India.' *Journal of Peasant Studies*, 9(1): 86–96.

Dange, S. A. 1945. *Death Pits in Our Land: How 200,000 Indian Miners Live and Work*, Bombay: New Age Printing Press.

Dawson, Andrew. 2013. 'The social determinants of the rule of law: A comparison of Jamaica and Barbados.' *World Development*, 45: 314–24.

Ekka, Alexius and Mohammed Asif. 2000. *Development-Induced Displacement in Jharkhand, 1951–1995: A Database on its Extent and Nature*. New Delhi: Indian Social Institute.

Energy Survey of India Committee. 1965. *Energy Survey of India Committee Report*, New Delhi: Government of India.

Freese, Barbara. 2003. *Coal: A Human History*, London: William Henemann.

Fuel Policy Committee. 1974. *Fuel Policy Committee Report*, New Delhi: Government of India.

Ganguly-Scrase, Ruchira and Kuntala Lahiri-Dutt. 2013. 'Dispossession, Placelessness, Home and Belonging: An Outline of a Research Agenda.' in Ruchira Ganguly-Scrase and Kuntala Lahiri-Dutt (eds) *Rethinking Displacement: Asia-Pacific Perspectives*, Aldershot: Ashgate, 3–30.

Garg, Amit and P. R. Shukla. 2009. 'Coal and energy security for India: Role of carbon dioxide (CO_2) capture and storage (CCS).' *Energy*, 34(8): 1032–41.

Gregory, Cedric A. 2001. *A Concise History of Mining* (Revised edition of the 1980 publication by Pergamon Press of New York), Lisse: AA Balkema Publishers.

Hatcher, John. 1993. *The History of the British Coal Industry, Volume 1, Before 1700: Towards the Age of Coal*, Oxford: Clarendon Press.

Humphreys, H. D. G. 1956. 'The Early History of Coal Mining in Bengal', in Mining, Geological, and Metallurgical Institute of India. *Progress of the Mineral Industry of India, 1906–1955*. pp. 147–159.

Internal Displacement Monitoring Centre (IDMC). 2009. *Internal Displacement: Global Overview of Trends and Developments in 2008*, Chateleine: Internal Displacement Monitoring Centre and Norwegian Refugee Council.

James, Peter. 1982. *The Future of Coal*. McMillan.

Jones, Peter M. 2009. *Industrial Enlightenment: Science, Technology and Culture in Birmingham and the West Midlands, 1760–1820*, Manchester: Manchester University Press.

King, Blair B. 1976. *Partners in Empire: Dwarakanath Tagore and the Age of Enterprise in Eastern India*, Berkeley: University of California Press.

Kumar, Shobha Sadan. 1996. *Mining and the Raj: A Study of the Coal Industry of Bihar, 1900–1947*, Patna: Janaki Prakashan.

Kumarmangalam, S. Mohan. 1973. *Coal Industry in India: Tasks Ahead*, Oxford India Book House Publishing Co.: Calcutta.

Laforce, Myriam. 2010. 'L'évolution des régimes miniers an Canada: L'émergence de nouvelles formes de régulation et ses implications, *Canadian Journal of Development Studies, Special Issue on Rethinking Extractive Industry*, 30(1–2): 49–68.

Lahiri-Dutt, Kuntala and Prasun Kumar Gangopadhyaya. 2007. 'Subsurface coalfires in the Raniganj Coalbelt: Investigating their causes and assessing human impacts.' *Resources, Energy and Development*, 4(1): 71–87.

Madan, Tanvi. 2009. The Brookings Institution Energy Security Studies, India, The Brookings Foreign Policy Studies Energy Security Studies, Available from http://dspace.cigilibrary.org/jspui/handle/123456789/5661 accessed on 27 May, 2012.

Malik, R. P. S. 2002. 'Water-Energy Nexus in Resource-poor Economies: The Indian Experience.' *International Journal of Water Resources Development*, 18(1): 47–58.

Miller, Donald L. 1985. *The Kingdom of Coal: Work, Enterprise, and Ethnic Communities in the Mine Fields*, Philadelphia: University of Pennsylvania Press.

Ministry of Coal 2005. *The Expert Committee on Road Map for Coal Sector Reforms*, www.coal.nic.in, accessed on 10 January 2007.

Ministry of Coal 2006. *Annual Report 2005–06*, www.coal.nic.in, accessed on 10 January 2007.

O'Faircheallaigh, Ciaran. 2010. 'Aboriginal-mining company contractual agreements in Australia and Canada: Implications for political autonomy and community development.' *Canadian Journal of Development Studies, Special Issue on Rethinking Extractive Industry*, 30(1–2): 69–88.

Oxfam. 2001. *Extractive Sectors and the Poor, An Oxfam America Report*, Boston: Oxfam.

Padel, Felix and Samarendra Das. 2010. *Out of This Earth: East India Adivasis and the Aluminium Cartel*, Hyderabad: Orient Blackswan.

Papandiek, H. 1978. 'British managing agency houses in the Indian coalfield.' in D. Rothermund, and D. C. Wadhwa (eds) *Zamindars, Mines and Peasants: Studies in the History of an Indian Coalfield and its Rural Hinterland*, New Delhi: Manohar, 165–224.

Planning Commission of India. 2006. *National Mineral Policy: Report of the High Level Committee*, available online at m ines.nic.in/File_link_view. aspx%3Fltp%3D1%26lid%3D116, accessed on 11 January, 2013.

Planning Commission of India. *Midterm Appraisal of the Tenth Five Year Plan (2002–2007)*, http://planningcommission.nic.in/midterm/cont_eng1.htm, accessed on 15 January 2007.

Planning Commission. 2006. *Integrated Energy Policy: Report of the Expert Committee*, Planning Commission, Government of India. Available online at http://planningcommission.nic.in/reports/genrep/rep_intengy.pdf (accessed on 1 April 2008).

Pramanik, P. 1993. *Coal Miners in Private and Public Sector Collieries*, New Delhi: Reliance Publishing House.

Prasad, Anubhuti Ranjan. 1986. *Coal Industry of India*, New Delhi: Ashish Publishing House.

Radetzki, Marian. 1985. *State Mineral Enterprises: An Investigation into their Impact on International Mineral Markets*, Washington, D.C.: Resources for the Future.

Read, Margaret 1931. *The Indian Peasant Uprooted: A Study of the Human Machine*, London, New York and Toronto: Longmans, Green and Co.

Rothermund, D. 1978. 'introduction' in D. Rothermund, and D. C. Wadhwa (eds) *Zamindars, Mines and Peasants: Studies in the History of an Indian Coalfield and its Rural Hinterland*, New Delhi: Manohar, 1–20.

Rothermund, D., E. Kropp and G. Dienemann (eds). 1979. *Urban Growth and Rural Stagnation*, New Delhi.

Saxena, Naresh C. 2002. *Jharia Coalfield Today, Tomorrow and Thereafter*, Centre for Mining Environment, Indian School of Mines, Dhanbad.

Simeon, Dilip. 1996. 'Coal and colonialism: Production relations in an Indian colliery, c. 1895–1947.' *International Review of Social History*, 41, 83–108.

Singh, S. B. 1966. *European Agency Houses in Bengal, 1783–1833*, Calcutta: Saraswati Press.

Sullivan, Charles Kenneth. 1989. *Coal Men and Coal Towns: Development of the Smokeless Coalfields of Southern West Virginia*, New York & London: Garland Publishing, Inc.

Szablowski, David. 2007. *Transnational Law and Local Struggles: Mining, Communities and the World Bank*, Oxford and Portland, Oregon: Hart Publishing.

Tripathi, A. 1979. *Trade and Finance in the Bengal Presidency, 1789–1833*, Calcutta: Saraswati Press.

Vasudevacharry, A. K. 1994. *Problems and Prospects of Coal Industry in India: A Case Study*, New Delhi: Mittal Publications.

PART I
Justice, Legality and History

Chapter 2

Between Legitimacy and Illegality: Informal Coal Mining at the Limits of Justice[1]

Kuntala Lahiri-Dutt

Unintended Collieries

Illegal mining is commonly represented in the media as arising from poor policing and corruption in mining tracts. People who are involved in the illicit production and marketing of coal are depicted as large-scale thieves, raiders and destroyers of the environmental commons. This chapter suggests that such mining constitutes a significant aspect of everyday life in the coal-bearing tracts of eastern India. The representation of such mining as posing threat to the well-being of the rest of the community hides unpleasant realities of the coal mining tracts: poor environmental care by large mining projects of both the state-owned and privately-owned mines; social disruption and displacement caused by them through physical relocation and occupational changes of farming and forest-based communities; and general decay in traditional subsistence bases of peasant and indigenous communities. To unveil the relations between what are seen as legitimate and illegitimate and forms of mining, this chapter digs through the complex layers of mining laws and investigates into what is generally seen as illegal mining. What coal as a resource means to the ordinary people in everyday contexts of the rural areas of eastern India lets us query whether the mining laws, as they stand now, are robust enough to promote social justice and equity in order to protect the interests of the disadvantaged citizens. In doing so, this chapter offers a critique of what causes and constitutes illegality in coal mining, and offers an alternative perspective of community's moral rights over access to local resources in a context in which a large number of people's livelihoods depend on informal mining.

Illegal mining usually makes sensational news in the popular media of India. Such news comprise a mix of human-interest stories that are peppered with environmental concerns, highlighting how these business transactions occurring outside the boundaries of law-making make enormous profits and generally painting an appalling state of affairs. The media depict illegal coal miners as thieves, and indeed such is the stigma that in case of accidental deaths from mine collapse, even the closest relatives fear to claim the bodies of the dead. The

1   A version of this chapter first appeared in *Economic and Political Weekly*, on 8 December 2007. I thank the Editor of EPW for his permission to use this revised version.

media portrays local politicians preventing the police from enquiring into mine disasters; mafia involvement and the complicity of company and government officials in dishonest dealings; the threat to the environment from these mines and the imminent hazards to roads and rail tracks that these mines pose. Accidents in particular are newsworthy; often they make it to the front page, especially when large numbers of dead are involved or if the security of middle-class life appears to be threatened.

Illegal mining, however, is a part of everyday rural life in the coal-bearing tract stretching from Raniganj in West Bengal westward to Dhanbad–Ranchi–Hazaribagh, where the collieries extend into Jharia and the North Karanpura areas in Jharkhand. Three subsidiary companies[2] of Coal India Limited (CIL) – harbouring disparate histories and problems – are responsible for mining operations in India. Illegal mines are found in the older mining areas of Raniganj–Asansol–Dhanbad, which possess a considerable number of working and abandoned underground mines. Such mining has also expanded rapidly around the large open-cut coal mines that have opened in the last decade or so in the Ranchi–Hazaribagh area. Although this chapter deals only with eastern India, informed observers say that illegal mining is common throughout the coal tracts of India. These may be described as 'unintended' collieries, a tiny part of India's informal sector which, according to Harriss-White, comprises 83 per cent of India's economy.[3]

Once a furtive activity like the rice trading which occurred in Calcutta's suburban trains in the 1960s, illegal coal mining now openly inhabits the public space. It is now impossible to drive along any length of the highway from the Raniganj region to Ranchi or Hazaribagh towns without encountering evidence of illegal mining in the myriad ant-like processions of ragtag men – known as cyclewallahs – pushing bicycles laden with sacks of coal weighing over 150 kilograms.

To query the scale of unintended collieries, I jointly undertook a field survey in 2003–4 of the small-scale distribution of illegal coal in eastern India (Lahiri-Dutt and Williams 2005). The cyclewallahs often cover up to 25 km in a day of work and on an average day on the roads around the edge of the coalfields between Ranchi and Hazaribagh their numbers may well be 2,000 or more. The survey determined that even on a conservative estimate, about 2.5 million tons of coal was transported by cycles in 2003–4. We did another phase of survey in 2012 in order to update the data; we found that within less than ten years span of time, the amount of coal carried on cycles has gone up to ~3.7 million tonnes. This amount is equivalent to the production of a reasonably large colliery. More significantly, the accompanying social and economic survey found that whilst incomes from coal transportation have not increased notably, many of those pushing the cycles

---

2 These are Eastern Coalfields Limited, Bharat Coking Coal Limited (BCCL) and Central Coalfields Limited (CCL).

3 The term 'unintended collieries' is inspired by Jai Sen's article 'Unintended Cities' and is gratefully acknowledged.

have left their traditional agricultural occupations for full-time involvement in this trade. The cycles are now better reinforced to carry bigger loads than before, and we also found that coal transport on cycles now contributes a significant part to the livelihoods of the people who live in the local villages.

One might remember that the coal transported on cycles is only the visible tip of the iceberg of a 'black coal economy', only a minute proportion of the unseen quantities hauled by trucks all over the coal-bearing tracts of India, each one of which carries up to 70 cycles' weight of coal. A veteran from the coal industry suggested that around 70–80 million tons of coal is produced in India annually in addition to the official production figure of about 350 million tonnes.[4] This illegal coal – black, invisible and underground in every sense of these terms – forms part of an economy that has intricate networks and complex linkages deep within every aspect of life in the coal producing regions of India.

These unintended collieries pose a challenge to our understanding of the social changes engulfing coal mining – and probably all mineral-bearing – tracts. Since the economic reforms, these mineralised regions have seen a flurry of activities as a result of the enormous demand for energy, minerals and industrial and building materials. Local people have often been unable to take full advantage of the new economy, whereas environmental organisations have risen in unqualified critique of all kinds of mining, even calling for a moratorium on all mining (see Vagholikar et al. 2003). In my view, the causes are buried under layers of complexity comprising: outdated colonial laws of land acquisition and state-ownership of coal resources; lack of safeguards and protection of poor people; despicable social and environmental practices undertaken by formal mines; the disregard of mining engineers and technologists for social impacts; the perpetuation of a raj-like licensing regime in CIL; and the overall trend towards informalisation of the economy. Illegal coal mines are a local expression of ordinary people's protests against unjust national mineral laws enacted by legislatures that have failed to ask simple questions such as 'who owns the nation's mineral resources, since when, and why?'; 'who controls their use?' and 'who is profiteering and looting the Indian people of coal reserves, and under what circumstances?' They also speak volumes about the performance of CIL as a mining company that represents the state and its interests. Above all, illegal collieries reflect several inescapable global trends – in mining, in mineral prices – and indicate a complex future in view of increasing pressure to liberalise the coal mining sector. The possible resolution of such problems would depend on asking the right questions, and this chapter aims to draw attention to the possibility of rethinking India's mineral resource

---

4   For example, Meghalaya is a 'fifth schedule' state in the Indian Northeast, implying that mineral resources there belong to local land owners. See Chapter 4 by Debojyoti Das, this book which describes how the abundance of coal, classified as a 'major mineral' meaning that technically it can only be mined by the state or state-condoned corporate interests, has thrown the 30,000 or so workers in coal mines in Meghalaya into a non-legal vacuum.

management and its mining laws through questioning the composition and function of illegitimacy in mining.

In dealing with illegal mining, the invocation of macro-economic theories of resource dependency is inadequate to explain fully the phenomenon of illegality. Theories of 'resource curse' and 'resource war' tend to reduce the complexity of mining livelihoods to a singular factor without political and historical context (see Lahiri-Dutt 2006; Le Billion 2007; Omeje 2006).[5] Conventional understandings of mining-related social change or even the most sophisticated Environmental Impact Analysis techniques are inadequate for developing a socially sensitive, politically engaged, historically informed and locally embedded understanding of the phenomenon. Clearly, rethinking is urgently needed; this chapter involves challenging the picture of lawlessness repeatedly painted by the urban-based middle class, a picture that accepts laws as immutable, and state's interests as preceding those of local people. One can then revisit the commanding heights philosophy of coal mining laws and the monopoly provided by those laws to CIL. In this chapter I have attempted to use thick geographical and historical contextualisation, and avoided citing too many international comparisons.[6] Let us first take a brief look at the informalisation of the economy and illegitimacy in mining in other countries.

Informalisation and Illegitimacy in Mining

Illegal mining is prevalent throughout the mineral-bearing tracts of the developing world. Such mining can be traditional – as in the Philippines – but more commonly it is a part of the informal sector of the economy, and results from a range of pressures:

> the economic crisis, urban unemployment in the cities, poverty in the agricultural areas and the violence that prevailed in the 1980s gave rise to growing social phenomena – individual, family or collective migration to zones other than the place of origin, searching for safety and economic survival (Jennings 1999: 35).

---

5   For example, from January 2006 the Movement for the Emancipation of the Niger Delta (MEND) declared a war on the petroleum company Shell; it kidnapped and ransomed over 50 workers, destroyed pipelines, overran offshore rigs, and killed Nigerian soldiers. The movement's demands were restitution for the environmental damage wrought by the oil industry, greater control over oil revenues for local government, and development aid to improve living conditions. Watts (2004) has previously documented the means by which bureaucratic and government control over the oil resources of Niger delta have led to widespread impoverishment.

6   Diamonds in Africa are the best example; seen widely as a 'rebel's best friend' (Buhaug and Rod 2006), the issues of which are addressed dramatically in the Edward Zwick film *Blood Diamond* (2006).

The use and extraction of minerals by different means such as digging, panning, sorting and amalgamation comprise an integral part of the vast informal economy on which little or no official data exists. In terms of sheer numbers, these people are not insignificant; a recent World Bank estimate suggests that over 20 million people in the world depend on mineral resource extraction for their living, a figure that is immensely more than those employed by the large and formal mining industries (CASM 2005). Indeed, employment in the formal mining sector has been steadily declining, whereas the numbers in informal mining have increased considerably (ILO 2002). A significant amount of minerals is produced this way, and can often account for a greater segment of a country's mineral production. For example, informal mining generated up to 65 per cent of Peru's gold production in 2005–6. The representations of those engaged in informal mining vary: they are known as Garimpeiros in Brazilian Amazonia, Galampseys in Ghana, Barranquilas in Bolivia, Ninjas in Mongolia and Gurandils (literally, 'those who jump from cliff to cliff') or PETIs (acronym for 'those mining without license' in Bahasa Indonesian) in Indonesia. Only sparse data are available on China, but according to experts, the number of people engaged could reach 15–16 million if cheap industrial minerals such as sand, stone and gravels are included.[7]

Whereas some countries might have a long artisanal tradition of mining, in most contemporary cases, informalisation of mining can be related to increasing poverty in rural areas. International decision-making circles have now developed a nuanced understanding of this kind of mining as 'a poverty issue which must be addressed by a comprehensive approach' (CASM 2005, 22). People enter the informal mining sector as an alternative or supplement to subsistence agriculture, thereby gaining marginally improved access to cash incomes for the maintenance of family livelihoods. Hilson and Potter (2005) noted that the policies associated with the Structural Adjustment Program of Ghana have fuelled the uncontrolled growth of poverty driven gold mining and have further marginalised its impoverished participants. However, in almost all ex-colonial countries, the legal framework is such that minerals are owned only by the states. Consequently, throughout the Third World the phenomenon of illegal mining is increasing as greater numbers of people take up this profession. For example, in Mongolia, a semi-desert country and one of the last frontiers of human settlement, the number of Ninja miners increased from 10,000 to 100,000 between 2000 and 2004 (MBDA 2003), representing around 20 per cent of the rural workforce (ILO 2006). During the same time the Mongolian government had aggressively wooed foreign mining capital. As major mining companies entered the fray, Ninja miners were pushed into marginalised environments, panning for gold in harsh conditions often only at night to avoid apprehension by authorities (Appel 2004).

---

7   Personal communication, Professor Shen Li, an authority on ASM in China and also the head of CASM China Network. See http://www.casmsite.org/regional_CASM-China.htm

History has seen such periodic phases of mineral rushes in the Americas and Australia, though the early gold rushes in these countries of white settlement are now glorified as heroic elements of the colonial frontier economy. By contrast, in the Third World countries the lawless chaos envisioned in contemporary rushes has emphasised the illegality of such mining and suggested analogies of curse and war.[8] Sierra Leone is the most remarkable case in which the illegal mining of diamonds has funded warring rebel groups. However, attention to the economics of mineral revenues in isolation, leaving aside questions of justice and political ecology, can impart the impression that all conflicts over resources are caused by the minerals as such, making them the problem and eroding our historic understanding of resources as a natural endowment. For example, the Central Intelligence Agency (CIA) described leaching of petroleum from oil pipelines by 'militant impoverished ethnic groups' in Nigeria – an 'archetypal oil nation' – as violent expressions of conflict. According to Watts (2005), summoning notions of a resource war in this case detracts attention from a situation that is rooted in the colonial history of the country, whereby different ethnic groups have consistently sought to expand their access to and control over resource revenues occurring within their territories and have resisted governmental control over resources.

Of Small Mines and Major Minerals

To explore the question of legitimacy in Indian coal mining, let us first turn to the legal or regulatory framework of coal and other minerals in India, and its licensing and policing systems. We will then go on to examine how responsive Indian legal and political structures have been to the social and economic issues arising in its mining regions. This section shows how coal is categorised as a major mineral, and mines and mining are classified into different categories according to production size. In this classical and seemingly watertight classificatory mode, there is no space left for small mining of a major mineral such as coal. But first let us obtain a snapshot view of mines classification in India.

Besides the *Indian Mines Act* of 1952, which is primarily intended for labour welfare and safety and health issues, the *Mines and Minerals [Regulation and Development (MMRD)] Act* of 1957 is the principal legislation governing mineral prospecting, exploration and mining in India. According to MMRD, a 'mine' means any excavation where an operation for the purpose of searching for or obtaining minerals is conducted, and includes many other specific activities and operations. 'Minerals' means all substances which can be obtained from the earth by a variety of mining, digging, drilling, etc., and includes mineral oils, which in

---

8   Some of popular theories of resource conflict, resource curse and resource wars present them as typical features of the less developed world, are rooted in a positivist view of resources, prophesise doom in developing countries, demonise the rightful needs and livelihoods of poor people, and prescribe top-down measures such as conflict management.

turn include natural gas and petroleum. The MMRD Act and any other mining development plan are guided by the overall *National Mineral Policy* (NMP)[9] first outlined by the Government of India in 1993, and then revised in 2002. The objectives of the NMP are primarily 'mineral development' through explorations of 'mineral wealth' in terrestrial and off-shore areas, taking national and strategic considerations into account, and to ensure their adequate supply and best use. The NMP is meant to promote the mineral industry as well as research, training and development in minerals, subject to present needs and future requirements, but while seeking minimise adverse effects on forest, environment and ecology, and ensure the safety and health of all concerned.

These objectives raise a number of important concerns. First, if the ordering of the objectives reflect the priorities of the state, then where does one place the interests and well-being of ordinary people, who fail to be included specifically, but instead are subsumed within 'all concerned'? Secondly, the policy does consider the possibility and the need for undertaking assessments and mitigations of social impacts. Thirdly, it fails to deal comprehensively with the governance and regulation of mineral producers, including voluntary regulation. Finally, where do we situate informal mines in this policy, and how do we deal with the phenomenon of illegitimacy under the current laws? According to Chakravorty (2002), an expert on small mines, they constitute 88 per cent of reported mines in India, and produce approximately 10 per cent of the total value of mineral production of the country.

Informal mines are worked by the poorest people, toiling at the lowest wages in the country. The security, health and safety conditions are appalling. Here one must remember that whatever the size, all mines in India are subject to a plethora of government rules and regulations – the MMRD Act, *Mines Act*, *Forest Act*, *Environment Act*. The *Minerals Conservation and the Development Rules* (MCDR) of 1988 divides all minerals into 'major' and 'minor', and rests the responsibility of mining major minerals such as coal with the state. The Indian Bureau of Mines (IBM), working under the MMRD Act identified, according to Rule 42 of the MCDR 1988, two further categories: A and B category mines, determined on the basis of employment numbers and the standard of mechanical equipment used

Whereas 'minor' minerals are defined with relative precision by the MMRD Act as 'building stones, gravel, ordinary clay, ordinary sand other than sand used for prescribed purposes, and any other mineral which the Central Government may, by notification in the Official Gazette, declare to be a minor mineral', there is some confusion regarding the definition of B class mines.[10] The outstanding feature in this definition is scale: small production, small capital investment, labour intensiveness, shallow nature of deposits and low-technology deployment. Thus,

---

9   Available online at http://mines.nic.in/nmp.html accessed on 2 August 2007.

10   Sahu (1992) has described B class mines as 'those whose production, or excavation quantity is limited in tonnage and not very large, mostly manually operated and sometimes employing machines to small capacity. Such mining activities are usually confined to deposits which are shallow in depth and small in extent.'

some of the coal mining operations on privately-owned land would come under this category of mining. On the other hand, certain labour-intensive underground collieries of the Eastern Coal Fields (ECL) could also qualify as 'small' mines. Clearly, the current policies and regulations on the mining of minerals are not built to deal with the complex realities of informal or illegal mining, but rather attempt to simplify mines, mining and minerals.

It is worse when mines are classified only according to production amounts; the National Institute of Small Mines (NISM 1993, 1994) defined the categories of mines in India according to their production.[11] The size of operation determines the duration of a mine; however, for an economic activity such as mining, which has close social linkages, such legal definitions are not helpful because they give the impression that a large colliery is a scaled-up version of a small quarry. This reductionist concept relies on the popular language of scale classification and obscures the unity or diversity of mining practices and linkages across differing scales. As large formal extraction processes are regarded by government as the only acceptable forms of mineral extraction, processing and use, those practices that cannot be fitted within the categories tend to be rejected and delegitimised.

Clearly, the existing laws are not comprehensive, and nor are they adequate to handle the informal mining sub-sector, which is made up of a number of disparate elements, some licensed, others illegal. As a sector it claims a long artisanal tradition extending to pre-colonial days, while sub-groups of informal mining are an offshoot of recent mining developments. Illegal miners are precluded from lobbying for recognition, and the current laws offer very few practical possibilities for them to mine coal legally, a situation that has resulted in serious consequences for the well-being of local populations and the environment.

Coal: Corporate Monopoly

Under the amended *Coal Mines (Nationalisation) Act 1993*, only two groups are eligible to mine coal: a central or state government company or corporation, or 'a person to whom a sub-lease has been granted by the above mentioned Government company or corporation having a coal mining lease, subject to the conditions that the coal reserves covered by the sub-lease are in isolated small pockets or are not sufficient for scientific and economic development in a coordinated manner and that the coal produced by the sub-lessee will not be required to be transported by rail' (GoI n.d.). The rule is unambiguous: there is clearly very little space for individual operators to mine coal. CIL, established in the euphoric days of nationalisation of the early 1970s, retains the full legal and financial benefits of

---

11 Small-scale mines are those which produce up to 0.1 million tonnes per year (mtpy); medium are those mines from 0.1 to 0.5 mtpy, and large-scale produce greater than 0.5 mtpy.

coal mining, the ownership of which is vested in the state; over the years CIL has come become synonymous with notions of the greater common good.

Decision-making regarding the mineral resources of India has so far been characterised by a preponderance of engineers, geologists and bureaucrats, with politicians claiming they represent the entirety of people's interests.[12] The governance of coal resources is vested almost entirely with CIL. The *Coal Mines (Nationalisation) Amendment Bill 2000* allows state governments or other recognised organisations to mine coal from smaller deposits only if CIL provides a certificate of no intention to mine, which is unlikely as it would preclude CIL's future mining interests in that area. The Ministry of Coal (MoC) has awarded CIL a near monopoly, giving rise to highly institutionalised control over India's coal resources. An opening of the coal sector to competition – attempted by the World Bank and hinted at by the Planning Commission report of 2006 (see GoI 2006) – will not necessarily solve its problems. There is a need is for policymakers to connect to social realities in mining areas and explore how to retain wealth generated from mining within the region; how to benefit people from mining expansion; how to make laws that do not render people and their livelihoods illegitimate. Until 1993, mine employment was being allocated as compensation for land, but with increasing mechanisation and the preference for open-cut mining technologies, the number of available jobs has dwindled. While large projects have necessitated the *Resettlement and Rehabilitation Policy* of 1994 (revised 2000), 'income restoration' as envisaged in the policy has primarily amounted to insecure jobs with contractors and limited assistance towards non-land based self-employment.

Social Impacts

In spite of Ministry of Environment and Forest controls, CIL subsidiaries have vandalised the environment with little or no concern for the social implications this might have. Environmental degradation associated with new mining projects has had serious social consequences: decay in forest-based livelihoods; crumbling social order; declining farming and the shift of peasantry away from farm-based livelihoods; to say nothing of physical displacement by mining. Even without environmental degradation, large mining projects are acknowledged internationally to have caused significant social changes, with serious implications

---

12 A statement in the March 1993 document of National Mineral Policy of the Government of India typifies the gap between policy and reality, particularly regarding indigenous access to legal mining, which prescribes that "In grant of mineral concessions for small deposits in scheduled areas, preference shall be given to the scheduled tribes.' Excepting rare cases, a scoping study on small mines and quarries did not identify small quarries owned by indigenous people, although many such mines operate on tribal lands, especially throughout the Jharkhand–Chotanagpur belt.

for the livelihoods of local communities (see Lahiri-Dutt 2001). Yet, this is an area that seems to have remained beyond the periphery of mineral governance in India. Social impacts in India are particularly associated with new mining expansion. Rao (2005) noted that displacement from traditional occupations has forced people into scavenging in Jharkhand. The neglect of social and cultural issues around minerals and mining has created a space for extreme leftist or Maoist movements (Bhusan 2007). Chandra Bhusan has studied the extent to which India's mineralised tracts are analogous to conflict zones. Company officials, bureaucrats and technical experts, including mining engineers, have not sincerely engaged with the social issues; nor have they sought to revise the legal instruments defining mining law, which are of colonial vintage, anti-poor, and unable to address contemporary realities.[13]

Civil society groups have adopted peaceful paths to use existing control mechanisms such as Public Interest Litigation (PIL) to bring justice to local people. On instance is the PIL matter filed by Mr Haradhan Roy of Colliery Mazdoor Sabha against ECL for failure to maintain an appropriate standard of environmental care. Non-governmental organisations (NGOs) advocate for improved environmental care to the limited extent permissible under the current legal framework. In the absence of legislative requirements for to maintain and ensure specific social obligations – by, for instance, conducting social impact assessments or instituting community engagement systems – resettlement and rehabilitation policies and practices have become the main plank of action undertaken by NGOs to give voice to local people and attempt to secure justice.[14] The complaint made by a small local NGO working with tribals, Chotanagpur Adibasi Seba Samiti (CASS), to the inspection panel of the World Bank regarding the Parej East open-cut mining project's lack of social concern provides an example. Parej East was one of the 25 mining projects funded by the World Bank under its Coal Sector Rehabilitation Project.[15] However, although the panel censured CIL and the bank's management,

---

13   CIL cannot be aware of its poor social performance, but it tends to either bundle the social issues with those relating to the environment or is unable to include the social issues within the current compliance regime that has little or no social content. Consequently, the Coal Sector Rehabilitation Project was split in November 1995 in agreement between the Government of India and CIL to separate the environmental and social components. Apparently it was meant to assist CIL in the implementation of high priority environmental and social mitigation programs, while providing the GoI with the time to take the necessary legal steps for the implementation of reforms.

14   The Samatha judgment (11 July 1997) is one of the few exceptions, in which the Supreme Court accepted the right of local populations to obtain compensation, not only for monetary and rehabilitation relief, but also compensation for a fuller life.

15   The panel report identified several instances in which the World Bank did not comply with its policy on involuntary resettlement, particularly during the project preparation and appraisal phases, but also with its policies concerning environmental assessment; consultation and disclosure of information; and timely provision of land titles to resettled persons. The panel also found that the resettlement action plan for Parej

neither has amended or bolstered civic ethical standards or responded to the panel recommendations (Lahiri-Dutt and Herbert 2004). Measures such as these have been largely unable to advance social issues in useful ways or prosecute mineral resource rights for landowners.

CIL's Monopoly in the Context of Global Change

The Indian coal industry is currently the third greatest producer of coal globally, mining 585 mtpy. The United States is second, producing approximately 1,000 mtpy, and China is by far the largest, mining 3,470 mtpy in 2011 (World Coal Institute (2012) Over the past 5 years or so, China has greatly increased production by ~50 per cent and India has gone up by about 25 per cent, whereas the United States has actually decreased a little. Domestically, coal mining is regarded as commensurate with the national interest, a crucial means for achieving economic growth and industrialization, and for meeting the rising aspirations and comforts of urban residents; nearly two thirds of Indian electricity derives from coal. India's coal mining industry is perceived internationally as highly inefficient in terms of productivity: the mining cost of coal in India is 35 per cent higher than other coal exporting countries such as Australia, Indonesia and South Africa; only 3 tonnes per man-shift are produced, as compared with 12,000 in Australia; and as a consequence the cost of mining is not recovered by coal sales.

State ownership of coal mining in India is important because it exerts significant control over the volatile mineral sector and ameliorates booms and slumps. It was presumed at the time of nationalisation that state ownership would effectively modify the negative roles played by the innumerable private owners, but the role of the state-owned company in the post-liberalisation economic context has not been clarified. Currently, India is experiencing a volatile combination of rising coal demand; rising coal prices in the international market – in response to which the MoC has deregulated coal pricing; and the interest of multi-national companies (MNCs) intending to invest in India. Currently only sub-contracts are awarded to large international companies such as Thiess. Rethinking illegitimacy becomes important at this crucial juncture in India's coal industry.

To deal with the challenges facing the Indian coal mining industry, greater awareness is needed of the changes in the extractive industries sector elsewhere in the world, mainly in established mining countries – those with a significant proportion of their Gross Domestic Product (GDP) deriving from mining – such

---

East provided for a subsistence allowance which has not been paid to eligible families. In response, World Bank Management proposed to the Government of India that it provide additional support to address this shortcoming. The Government of India noted that this allowance was not included in 12 other resettlement action plans prepared under Coal India's corporate Resettlement and Rehabilitation Policy and was reluctant to establish a precedent.

as Australia, Papua New Guinea and Canada. The challenge of mining coal stems from the needs of nearly 500 million or more people thirsting for access to electricity. No matter what the NGOs suggest as alternatives to coal fuel energy sources, coal mining will continue to play a major part in satisfying the demand for electricity for the foreseeable future. Coal consumption in the country is predicted to reach between 800 mtpy and 2,000 mtpy by 2030 (Grover and Chandra 2006; IEA 2006: 600; Reuters 2007). There are obvious global implications concerning climate change and the Kyoto Protocol, enhanced imports, and investments by Indian companies in coal mining in other countries. The implication is a definite increase in informal collieries, which will continue to meet the demands of small, local consumers.

Social License to Operate

In order to evaluate CIL's performance as a company, it is necessary to return to the larger backdrop of changes occurring in the global minerals industry. Increased and concerted global efforts have been underway since 1998, beginning with the formation of Global Mining Initiative (GMI) and the subsequent creation of the Mines, Minerals and Sustainable Development (MMSD) project, supported by nine major mining companies. The International Council for Mining and Metals (ICMM) has been established as an industry peak body; and the recently completed Extractive Industries Review (EIR) findings have led to the Extractive Industries Transparency Initiative (EITI), although the World Bank has not been fully supportive of these initiatives. These processes were in direct response to increasing charges of environmental destruction and the irresponsibility of mining projects to care for the social and cultural changes caused by mining. These global processes have forced some of the international mining companies to accept that legal compliance alone is not enough; they also require sanction in terms of social responsibility for mining operations, referred to as a 'social license', particularly in developing countries where mining projects are burgeoning (MMSD 2002).[16] The main objective is trust building; the MMSD report noted (2002, 5–6): The mining and minerals industry faces some of the most difficult challenges of any industrial sector, and is currently distrusted by many of the people it deals with. It

-----

16 A social license fundamentally changes the manner in which mining companies conduct business. It embodies a set of expectations regarding social obligations on the part of mining interests, including publicly accessible information derived from monitoring mining projects throughout their life cycle. It is an extra-legal, abstract and ethereal concept that is largely outside the core business of mining. It has been defined as 'the recognition and acceptance of a company's contribution to the community in which it operates, moving beyond meeting basic legal requirements toward developing and maintaining the constructive stakeholder relationships necessary for business to be sustainable. Overall, it comes from striving for relationships based on honesty and mutual respect' (DITR 2006).

has been failing to convince some of its constituents and stakeholders that it has the 'social license to operate' in many parts of the world.

The Bougainville rebellion in Papua New Guinea has achieved mythical status in mining lore; poor attention to community development and engagement with the landowners caused the closure of the large copper mine at Panguna (see Filer 1990). Such examples are available closer to India: the Phulbari coal mine project in Bangladesh was shelved in 2006 because of community agitation to respect and maintain alternative livelihoods. Yet the mining industry – unused to examining the complexities in social relations and the territorially embedded nature of communities – has remained insensitive to increasing public resentment against mining projects. For the mining engineers who design project plans, the mining project assumes an importance that easily neglects people as disposable burdens to mining operations.[17] This is exemplified by the views of an ex-Director of the Indian School of Mines, S. P. Bannerjee, who in a 2004 article suggested the erection of boundary walls around the leased mining land after acquiring it. In the absence of regulations requiring appreciation of the livelihoods and rights of members of the society within which the mine operates, CIL falls back upon the legal system and the rationale that 'The LA [Land Acquisition] Act or the CBA [Coal Bearing Areas] Act does not provide any assistance for' the local people affected by mining. Mining projects are characterised by heightened cash-flows, the influx of migrants, rapid urbanization, and the formation of new social alliances. Unused to analysing and addressing these social changes, mine planners complain that 'not all people who live in communities occupying or using land required by the mine are land owners' and resent that these people often 'provide the leadership to those opposed to the land acquisition programs'.

At the level of policy, language, and organisational processes, the global mining industry is changing its approach. Many corporate policies now explicitly address broader social justice objectives such as local and indigenous employment, security and human rights, sustainable livelihoods, culture and heritage, the need for undertaking social impact assessments, ethical procurement, and stakeholder and/or community consultation (see Kemp et al. 2006: 391–2). Many international mining companies now regularly hire social scientists, anthropologists, and gender specialists to advise on good practice and the integrated management of social and environmental issues around their mine sites.

---

17 An example from Bannerjee (2004: 8) would be sufficient to exemplify the tunnel vision of mining institutions: 'As Coal India, already suffering from a load of surplus labour, could not offer so many jobs to the land oustees as demanded, a large number of dissatisfied land oustees were formed and many CIL projects started facing land acquisition problems. many mines have not been able to advance in the way it was planned, many others had to resort to muffled or controlled blasting and many new projects were delayed and the economic impact of these disruptions were enormous.'

Illegal Coal and Illegal Mining

It is important to differentiate between illegal mining of coal and the illegal marketing and distribution of illegally – or legally – mined coal: not all illegal coal is illegally mined. Often legally mined coal is lost from rail wagons or sourced by scavenging from mines and/or sale dumps,[18] and can become illegal. Before prescribing any measure, one needs to remember that there are two separate but related aspects of illegal coal: illegal mines and mining without license; and illegal marketing/distribution of coal. The two are neither entirely homogeneous, nor are they similar in history and organisational structure.

Illegal mining takes three main forms in eastern India: small shallow-dug village mines on private land; mining on re-opened, abandoned, or orphaned government mines; and scavenging on the leasehold land of official operating mines. There are a number of unregistered mines: those that escaped enlistment during nationalisation and became illegitimate. Though uncommon, such cases of oversight by officials include, for example, Pahargora, a mine that has never been brought under government ownership, and which continues to thrive; and the Saltora mine in the Purulia district of West Bengal.

The lands on which mines are dug illegally are usually privately owned in Raniganj, but in Jharkhand these are often village commons (*gair majurwa*[19]) lands. The mines are dug into outcrops exposed at the sides of steep hills or rivers. Coal may be extracted through a series of small open-cut holes that can extend underground. Alternatively, they may be shallow underground operations, entered via drifts or shafts to a depth of about 10 to 15 metres, and which can extend for up to 200 metres horizontally. Small brickworks – the customers – are invariably located nearby. Coal is removed by pick-axe by the coal cutter, after which loaders put the pickings into metal dishes or baskets carried on the head about 25 kg at a time. Some of these mines operate throughout the year; others become unstable during monsoons; generally, the rainy season is the slack time as workers tend to take up employment in the fields. Some of these village mines can be extensive, with four or five thousand people working them each day.

Scavenging from abandoned mines is another important source of coal. The eastern colliery tracts have a 200-year, poorly documented history of underground and incline mining. The entire Raniganj–Jharia region is dotted with small abandoned mines, some of them orphaned by mining companies owned by individuals. As these collieries were brought under state ownership, governmental responsibility for rehabilitating the remains of old mines was conveniently ignored. The shafts of these pits provide ready access to underground remainders of coal.

---

18   Sale dumps are depots where the mining companies store their coal.

19   *Gair majurwa* literally means 'deedless land', that is, land that is not officially recorded and has no legal ownership. Those living or cultivating such land may have de facto ownership, although such customary ownership is not recognised by the law when mining companies claim the land.

More importantly, poor environmental care in rehabilitating the mines encourages scavenging. It is common for CIL subsidiaries to neglect backfilling voids with sand as per regulation (Lahiri-Dutt 1999). Breakage of sealed underground mines is common, and the carbon monoxide present in such abandoned sites often kills those entering to scavenge coal. Because the 'board and pillar' system of coal mining is employed in most underground collieries, the entire amount of coal can never be lifted, thereby leaving significant amounts of coal behind. Mining companies tend to vacate a mine as soon as it becomes uneconomic, thus leaving the opportunity for local villagers to take up extremely dangerous work. The ecological footprint left by open cut mines is more serious; so far there have been only a few cases of backfilling and rehabilitation, re-contouring and re-vegetation of large pits at the completion of mining projects.

Scavenging can occur in both underground and incline mines, but it assumes greater significance in open-cut mines. First of all, there are cases such as the Samdi and Sangramgarh collieries in Raniganj, both among the oldest collieries of this area, where mining operations have existed since the late 18th century. Sangramgarh is an open-cut operation where a two metre-thick coal seam near the surface has been left by the ECL, choosing instead to work on the lower 6 metre-thick layer. This upper, thinner layer of coal has been cut into a maze of honeycomb-like labyrinths, often extending to considerable distance under the surface. Scavenging of small amounts of coal, stealing and pilfering occur regularly from nearly all open-cut mines. Poor security in mines, storage and transportation areas provide opportunities for scavenging. In functioning underground mines this normally occurs in the coal loading area: coal is loaded by head baskets into waiting trucks. Scavenging during transportation is not only small pilfering: it can reach a significant scale. Coal India delivers coal both to the local sale dumps located near the mines and to large central dumps, and pilferage from long-distance trucks and railway wagons can occur en-route. Trucks are hijacked regularly, with the opportunity for significant amounts of coal to be offloaded from rail wagons.

Pauperisation in Mining Areas

The significant amount of social and environmental transformation in colliery tracts stems largely from the monopoly of CIL over coal ownership, mining and marketing. Because the coal industry is nationalised it is assumed that coal is being put to use for the greater national good; besides relying on the power of the institution as a centralised body, government and CIL have also used on this assumption as a means of evading both its obligations to society and the implementation of mechanisms allowing for independent and public scrutiny of the industry's performance. The social transformations effected by coal have come to characterise mining regions: decay of the social fabric; changes in power relations; the erosion of traditional livelihoods; migration from surrounding regions; and rising levels of urbanisation. These impacts vary according to the physical proximity of the mine to those it affects, and are felt differently within the society according to gender, class and caste. For example, in the Raniganj region between 1971 and 2001, both agricultural land and the representation of peasantry in the workforce have steadily declined, even in non-colliery villages. Women – especially those from poorer, lower caste and indigenous (*adivasi*) communities – have been negatively affected to a greater extent as a group than non-*adivasi* women. In Jharkhand, a process of gradual pauperisation of the local residents has taken place whereby the traditional land and water rights have been lost and few of the benefits of mining have accrued to local indigenous communities.

The following example typifies the government's and the coal industry's neglect of their environmental and social obligations. On the highway to Ranchi, amid a procession of cyclewallahs loaded with sacks and bags of coal chunks I meet Nirjal Birhor. I first met Nirjal in his leaf hut a short distance from Hazaribagh town in the early 1980s. At the time (2006), his livelihood continued to be based on hunting and catching animals and making rope from *chop* creepers, despite the dwindling reserves of the surrounding Chotanagpur jungle. In the last 20 years his world has been gradually destroyed by the advancing coal mines. He has been evicted from his home; in the absence of the forest resources that had provided him with subsistence, he has now turned to digging for coal. He describes his living as 'coal collection', but to others Nirjal is a petty thief, stealing and delivering stolen coal.

The links between legal and illegal coal mines go beyond this simple example of pauperisation. The mining companies are the largest owners of land in the coal tracts, the prime employers of people and movers of resources. They choose to either overlook the coexistence of illegal mining with their operations, or see it as a law-and-order problem to be dealt with by the district administration. To complaints of theft from company premises, the bureaucratic reply is usually that the company should secure its property with its own considerable resources. More often than not, the matter ends after the exchange of letters or, at best, a non-committal airing at a committee meeting. As an example of this overall reluctance to accept responsibility for tackling the problem, in letter no. 9627-P, dated 20 September 1999, the Superintendent of Police of Burdwan district wrote to the Chief Managing Director of ECL:

> The response of the ECL authority in mentioning that the responsibility and authority in curbing down antisocial activity rests with the district administration and state police being true but the ECL authority should also not avoid its responsibility of guarding its property lying abandoned in the open especially when ECL is provided with over 1,000 armed Central Industrial Security Force personnel and over 5,500 security personnel. Nearly 1,700 guns/revolvers are available with the security staff also.

In Jharkhand in recent years illegal coal mining has followed the expansion of formal coal mining. This is primarily due to the effect of the *CBA Act*, which overrides existing protections for poor and indigenous peoples that preclude the transfer of tribal land (Bengara 1996). In Raniganj and Jharia, occupational displacement due to the degradation of cultivable land has caused people to turn to this profitable economic activity. Formal and informal co-exist in this part of India; at times the legal collieries have had to adapt their practices to the co-existence of illegal mines. In Khaerbad colliery in Raniganj, leakage of oxygen into the underground coal seams – caused by locally dug shafts that are called 'rat holes' – has caused extensive mine fires. To keep the fires under control Jambad – previously an underground colliery – has now been turned into an open-cut mine that nonetheless requires periodical drenching to extinguish new fires (Lahiri-Dutt and Gangopadhyay 2007).

The links are also evident in the subtle tolerance of illegal coal mining – both on private lands and in scavenging from official mines – as well as small-scale transport and distribution of illegal coal by company officials and district administrators. The metropolitan-based media has been concerned in recent years about the possibility of subsidence of the main railway line passing through the region. Concern evinced by the district administration and CIL is usually driven by the degree of media exposure of major accidents. District collectors view the problem of illegal coal mining as one of law and order, yet avoid taking direct responsibility for preventing theft from company-owned land. They also tend to ignore larger illegal operations on privately owned land as long as the owners accede to local power structures. Police officials vary in their views and actions regarding illegal coal; the district Superintendent of Police often tries to control larger operations, both mining and truck transportation, but tends to ignore the cyclewallahs. Mine managers also appear to be fully aware of the exact locations of large illegal operations.

Understanding these perceptions is necessary in order to suggest functional prescriptive measures. For example, according to a journalist based in Hazaribagh, both large-scale transportation by trucks and small-scale haulage by cycles occur in a centrally controlled manner that resembles the illegal gambling industry (*satta*), operating in large metropolitan areas. However, although mafia omnipresence is noted by everyone in the coalbelt, the mafia operations appear to comprise a different – though related – system of coal production and distribution than that performed by the cyclewallahs. In my interviews with coal cyclewallahs, it was clear that the mafia-controlled system of illegal coal transport operates in trucks. Part of the problem also lies in how coal is marketed by CIL: coal is not freely sold to small and domestic consumers, and throughout the entire coal-producing region in eastern India there are no distribution depots designed to cater to small and domestic consumers. Factories procure coal by means of linkages with the mining companies. Typically, a local coal-based industry owner applies to the central government for a grant of coal to fire its furnaces. Until recently this permission was difficult to obtain and might require multiple bribes at various levels. Once granted the permits can be used repeatedly to obtain tax-free coal from sale dumps. In interviewing the cyclewallahs I repeatedly asked whether their supplies were intended for regular and fixed customers or not. Local brick kilns (*bhattas*) can be major consumers of this coal throughout the dry months. In most cases, the cyclewallahs are itinerant peddlers, selling coal to local *bhattas* and even smaller consumers such as individual homes. Thus, a fundamental reason for delivery of coal by bicycle in the coalfields is the absence of an officially recognised system of regular delivery to small local users. In the eastern parts, up until the 1960s urban households situated around the coalfields cooked with coal. Small coal dumps were established and licensed within a town, the coal being delivered by truck or rail. When liquid fuels (LPG and kerosene) became routinely available, use of coal in middle class households was phased out. Being demand-driven in this case, the coal supply-chain extended as far as Kolkata or beyond.

Regulation, Regularisation: Revisiting the Laws

As the district administrations in Burdwan, Dhanbad and Hazaribagh established committees to control and curb illegal coal mining, debate has primarily centred upon the options of regulation, regularisation and formalisation. Regulation would mean total blockade of all illegal mining – on private lands, operational and abandoned mines. A mine manager suggested that the entire upper 20-feet layer of coal be razed as a preventive measure. As noted earlier, policing has so far been the preferred choice for CIL and its subsidiaries, although responsibility remains a thorny issue.

The possibility of regularisation has been discussed recently in the Raniganj fields of West Bengal, where it has been proposed that the local illegal mines be brought under cooperative management. This is not new; in Indonesia, Soekarno's government recognised the long artisanal tradition of mining and created space for informal mining in 'People's Mines' which are allowed to operate at low production levels. In people-friendly China, such village cooperative mines exist in designated areas. However, responses to West Bengal government's proposal have so far have been unenthusiastic. Although the most viable so far, the proposal is unrealistic; cooperatives are only permitted to include 'illegal mines on privately owned land'. Cooperativisation would also not challenge the exploitative production structures within these mines, thereby perpetuating inequalities and reinforcing existing production relations in the absence of a sound understanding of coal distribution. The important topic of scavenging, pilferage and such other sourcing of coal – often involving the poorest of the poor – would continue to attract policing attention. With coal prices rising and demand soaring, the opportunity cost of illegal mining would remain favourable to the diggers. Blocking the top end of the market by restricting market access or certification would also not help under the dualistic market situation in which small consumers predominate and large-scale sales are centrally controlled by CIL. Above all, isolated measures targeted at stopping illegal mining would be ineffective as long as the state and its representative – CIL – are not perceived as fair and efficient.

Neither regulation nor regularisation would be possible without altering the current legal framework around minerals. The High Level Committee of the Planning Commission devotes a full sub-section to the implications of illegal mining. However, although suggesting many changes, the committee focuses on the revenue losses to the state and suggests that 'for checking illegal mining there can be no substitute for improved standards of governance' (Government of India 2006, 131). Existing mineral ownership and land acquisition laws are antiquated and unresponsive to concepts of social justice; the regulatory system must be revised if it is to address emerging social change.

This need for changing the regulatory system can is explicable in purely economic terms. As noted before, rising prices in the international market and the recent deregulation of coal prices within India mean that in the near future coal prices will be on an upward trend. The demand is arising from the domestic sector as much as from small industries, not only the urbanising classes demanding coal, but in many parts of eastern India where the degradation of forests forces villagers to use coal instead of conventional biomass fuels. Besides rising demand and prices, the other driving forces for increasing coal consumption and increasing informal coal production are the radical transformation of society and degradation of the environment – both of which are forcing people to turn to illegal coal mining for a living.

On the larger, industrial scene, India will turn increasingly toward imports of coal. The nation will not be able to mine enough coal to meet forecasted demand. India currently imports 35 mt coal largely from Australia and Indonesia (ICRA 2006). An example of this new trend of Indian capital moving out of the country to secure coal sources is Tata Power's recent purchase of Bumi Resources, an Indonesian coal company with mining leases in eastern Kalimantan.

It is likely that in future the Indian coal industry will operate in three distinct layers instead of the current formal–informal binary. The three layers will be comprised of a top globalised sector in which multinational companies will operate and Indian companies will secure coal from abroad for their industries and power plants; a middle tier of CIL changing its modus operandi only marginally or refusing to change quickly enough; and a lower tier of unrelenting local private entrepreneurs investing funds at the cost of poor people's labour, yet providing a critical livelihood base for the masses, thereby playing an important social role. As things presently stand, it is difficult to predict the respective shares of these segments, but in terms of numbers and livelihoods involved, this lowest tier will play such an important role that there it will force the rethinking of the Indian legal framework for mineral resource governance.

Limits of Justice

The 2006 High Level Committee on National Mineral Policy (NMP) agrees with the above assessment, albeit indirectly. Although the report unabashedly intends to open up the minerals sector to foreign investors, it also notes in critiquing the NMP (GoI 2006, 20) that:

> the issue of compensation for local tribal populations as a primary charge on the minerals extracted from their land needs to be built into the policy and given primacy along with the issues of deforestation, pollution, and other disturbances caused in the ecology by mining activity.

Quite rightly it points out the various defects in the existing legal framework, particularly the lack of clarity and transparency – on such issues as the basis for grant or denial of concessions; conflicting laws at the federal and state levels – such as that of states imposing additional taxes on top of royalties – that confuse and deter investors; and cumbersome and time-consuming procedures for obtaining a mining lease. In adopting a path of proscribing procedural complexity it explored how a *single window* system could be formed and suggested that coordination committees should be set up at the state and central government levels to hasten decisions on applications. The report refers to a submission by one mining company that had to pass through 77 desks, taking 485 days. However, the report frames the issue of local community injustice in terms of Corporate Social Responsibility (CSR[20]), and contradicts itself in noting that 'a soft state apparatus amounting to a virtual absence of mining policies' will lead to greater illegal and unregulated mining, while recommending the retreat of the state.

Effects of Privatisation

India is currently passing through a volatile time leading to the eventual divestment of the nationalised coal mining industry. How the nation will deal with the hundreds and thousands of people making a living from illegal coal mining is not certain. Unless it is understood as a livelihood, and the rights of people succeed those of mineral resources, India will neither be able to strengthen its democracy nor uplift the enormous number of citizens languishing and threatening our economic prosperity. For this, more robust and socially informed mineral resource laws need to be developed based on an agreed set of broad principles. The foremost of these principles would be respect for the rights and interests of all those involved. With

---

20   Rajak (2007, 2–3) has made a scathing critique of the idea of CSR: 'Corporate Social Responsibility claims a radical break with the legacy of corporate philanthropy – charity replaced by the technocratic rationalism of 'responsible competitiveness' and 'sustainable development', while stakeholder engagement and participation take the place of the mine manager's paternalism…. CSR claims the capacity to change the way in which business itself is done. It asserts the happy coincidence of economic imperative and moral injunction – the convergence of economic value and ethical values. It does so by drawing on a powerful paradigm of social investment through partnership.'

India's poor track record of policing environmental performance, it is dangerous to invite foreign investment or to open up the coal sector to private investment. The key question is whether foreign investors ought to be able to rely on national governments to protect the interests of local communities, or if this should be undertaken by those investment companies through CSR? Is legal reform the best approach, or will the rhetoric of CSR be proven effective in India? Other liberalising developing countries that have opened up their minerals sector have banked on CSR largesse, with significant resources and attention being devoted to developing comprehensive CSR company policies while regarding corporations as vehicles of good governance and sustainable development. For example, in Indonesia, receiving CSR awards has become an attractive deal for mine managers. Describing this as the 'gift of the market' in South Africa, Rajak (2008) observes that most mining companies now have a package of policies covering various areas which fall under the broad spectrum of CSR or socio-economic development (SED). CIL, which at best dispenses philanthropic paternalism, remains far from such measures. Given the history of small-scale entrepreneurship in the Indian coal sector, it will be impossible to expect anything but a mushrooming of small coal mining leases in an open market.

The current laws are focused only on mitigating the negative impacts of mining on the environment and people. Instead, India needs to frame policies and laws that can deliver sustainable benefits for local, regional and global communities. This can be achieved only if the laws emphasise the need for a more participatory and inclusive approach to mining – at the process level – that mining companies can adopt. India is far from establishing such processes. First of all, more needs to be learnt regarding the production organisation of illegal mines, as well as the marketing chains and chains through which legally mined coal may be illegally distributed. It is also necessary to determine the extent of involvement of coal smugglers and mafia, and the level at which they operate. It is crucial to identify the stakeholders of this black economy, and the linkages between the formal and informal coal mining sectors also need to be understood in order to identify the social, political and economic forces driving these unintended collieries.

A revisit of the laws surrounding mineral resource extraction in India would thus involve a fuller understanding of the role of the community in local economies, thereby providing access to resources for local people, and integrating community interests in mine management plans. Access to the land and its natural resources and food security are at the centre of illegal coal mining. If the coal resources of India are vested in the national interest, they must help its citizens build and live in a society where opportunities and benefits are equal for everyone.

The other urgent needs include the protection of common pool resources that help poor communities to survive in rural economies in colliery tracts, and to find ways to vest the power to co-manage minerals with the local communities. If working arrangements can be developed for joint forest management and integrated water resource management systems, a process of rethinking can begin that will ultimately realise the civic co-management of collieries.

Rights of Local Citizens

India needs to accept local and indigenous communities as equal and integral citizens in order to acknowledge their rights over local natural resources. This will enable society to develop according to its needs, as opposed to the dominant mining–urban–industrial economic structure impeding social recuperation. It will also help to identify inclusive, egalitarian forms of decision-making. In India, especially in mining regions, people such as Nirjal are tolerated while engineers plan what investors they think is best for them.

It is important for broad public debate to be conducted between social scientists, planners, international agencies and civil society concerning the issue of justice in coal mining areas. To find socially just and forward-looking resolutions which will enable the growth of Indian society is a challenge which has yet to be solved.

Illegal coal mining provides an important entry point to public debate on the precedence of human rights over mineral resources – one that is of far more significant than might at first appear. Collectively the debate implicitly involves the lives, livelihoods, and futures of a significant number of people inhabiting the mineral-rich tracts of all developing countries. This is not only an enormous population, but it is also among the poorest and most exploited peoples in the world. Mainstream society has avoided accepting the poor and disadvantaged as an integral part, isolating them and flaunting the environmental impacts of illegal mining as a cause for concern. It is necessary to question the limits of justice. Notions of legitimate and illegitimate economic practices are grounded in traditional conceptions of social norms and obligations, in the economic functions of numerous parties within the community. As the formal coal industry continues to isolate and exclude local communities from the formal economy, poor peasants and others denied civil and humanitarian justice need to assert their rights.

References

Appel, P. 2004. 'Small-scale Mining in Mongolia: A Survey carried out in 2004.' *Denmarks Og Gronlands Geologiske Undersogelse Rapport* 2005/4. Denmark.

Bannerjee, S. P. 2004. 'Social Dimensions of Mining Sector.' *IE (I) Journal–MN*, 85(August): 1–10.

Bengara, R. 1996. 'Coal Mining Displacement.' *Economic and Political Weekly* XXXI(II): 647–49.

Buhaug, H. and J. K. Rod. 2006. 'Local Determinants of African Civil Wars: 1970–2001.' *Political Geography* 24: 315–35.

Chakravorty, S. L. 2001. *Artisanal and Small-scale Mining in India*. MMSD India Report, No. 78.

Communities and Small-Scale Mining (CASM). 2005. *The Millennium Development Goals and ASM*. Communities and Small-scale Mining, World Bank. Available online at www.artisanalmining.org accessed 29 January 2010.

Department of Industry, Tourism and Resources (DITR). 2006. *Community Engagement and Development: Leading Practice Sustainable Development Program for the Mining Industry*. Canberra: Australian Government.

Filer, C. 1990. 'The Bougainville Rebellion, the Mining Industry and the Process of Social Disintegration in Papua New Guinea.' *Canberra Anthropologist* 13(1): 1–39.

Government of India (GoI). 2006. *National Mineral Policy: Report of the High Level Committee*. New Delhi: Planning Commission.

———. n.d. *Eligibility to Coal Mining*. http://www.coal.nic.in/eligibility_to_coal_mining.htm accessed 2 July 2007.

Grover, R. B. and S. Chandra. 2006. 'Scenario for Growth of Electricity in India.' *Energy Policy* 34: 2834–847.

Hilson, G. and C. Potter. 2005. 'Structural Adjustment and Subsistence Industry: Artisanal Gold Mining in Ghan.' *Development and Change* 36(1): 103–31.

Investment Information and Credit Rating Agency of India Limited (ICRA). 2006. *Coal Sector Analysis*. New Delhi. http://www.icra.in accessed 27 March 2010.

International Energy Agency (IEA). 2006. *World Energy Outlook 2006*. International Energy Agency, Paris.

International Labour Organization (ILO). 2002. 'The Evolution of Employment, Working Time and the Mining Industry.' *Document TMMI/2002*. Geneva: International Labour Organization.

———. 2006. 'Informal Gold Mining in Mongolia: A Baseline Survey Report Covering Bornuur and Zaamar Soums, Tuv Aimag.' *Informal Economy, Poverty and Employment Mongolia Series* (1). Geneva: International Labour Office, 101–29.

Kemp, D., R. Boele and D. Brereton. 2006. 'Community Relations Management Systems in the Mining Industry: Combining Conventional and Stakeholder-driven Approaches.' *International Journal of Sustainable Development* 9(4): 390–403.

Lahiri-Dutt, K. 1999. 'State and the Market: Crisis in Raniganj Coalbelt.' *Economic and Political Weekly* XXXIV(41): 2952–56.

———. 2001. *Mining and Urbanization in the Raniganj Coalbelt*. Calcutta: The World Press.

———. 2006. '"May God Give Us Chaos, So that We Can Plunder": A Critique of "Resource Curse" and Conflict Theories.' *Development* 49(3): 14–21.

Lahiri-Dutt, K. and D.J. Williams. 2005. 'The Coal Cycle: Small-scale of Illegal Coal Supply in Eastern India.' *Resources, Energy and Development* 2(2): 93–105.

Lahiri-Dutt, K. and P.K. Gangopadhyay. 2007. 'Subsurface Coalfires in the Raniganj Coalbelt: Investigating their Causes and Assessing Human Impacts.' *Resources, Energy and Development* 4(1): 71–87.

Lahiri-Dutt, K. and T. Herbert. 2004. 'Coal Sector Loans and Displacement of Indigenous Populations: Lessons from Jharkhand.' *Economic and Political Weekly* XXXIX(23): 2403–9.

Le Billion, P. 2007. 'Geographies of War: Perspectives on "Resource Wars".' *Geography Compass* 1: 1–29.

Mining, Minerals and Sustainable Development, 2002. *Breaking New Ground.* MMSD Final report, London: IIED & Earthscan.

Mongolian Business Development Agency (MBDA). 2003. *Ninja Gold Miners of Mongolia: Assistance to Policy Formulation for the Informal Gold Mining Sub-sector in Mongolia.* Ulanbataar: Eco-Minex.

NISM News. 1993 and 1994. *The National Institute of Small Mines.* Bulletin No. 11 and 12, October 1993 and January 1994. Calcutta, India.

Rajak, D. 2008 '"Uplift and Empower": The Market, The Gift and Corporate Social Responsibility on South Africa's Platinum Belt.' *Research in Economic Anthropology* 28: 297–324.

Rao, N. 2005. 'Displacement from Land: Case of Santhal Parganas.' *Economic and Political Weekly* Jharkhand Special Issue, 8 October.

Reuters. 2007. *India's Coal Demand May Quadruple by 2031 – Minister.* www.planetark.org/dailynewsstory.cfm/newsid/40808/story.htm accessed 29 April, 2009.

Sahu, N.K. 1992. 'Mine Reclamation and Afforestation: Different Methods.' Lecture notes of Training Course on Management of Small Mines: Techno-administrative Aspects, organised by TISCO, Noamundi, 1–10 February.

Sen, J. 1976. 'The unintended city', *Seminar, 200.* Available from http://www.india-seminar.com/2001/500/500%20jai%20sen.htm accessed on 28 March, 2013.

Vagholikar, N., K. A. Moghe and R. Dutta. 2003. *Undermining India: Impacts of Mining on Ecologically Sensitive Areas.* Pune: Kalpavriksh.

Watt, M. J. 2005. 'Antinomies of Community: Some Thoughts on Geography, Resources and Empire.' *Transactions of the Institute of British Geographers* 29: 195–216.

Watts, M. 2004. 'Resource Curse? Governmentality, Oil and Power in the Niger Delta, Nigeria.' *Geopolitics* 9(1): 50–80.

Zwick, E. 2006. *Blood Diamond.* Bibliographical details: http://www.imdb.com/title/tt0450259/ accessed 27 March 2013.

Chapter 3
Coal in Colonial Assam: Exploration, Trade and Environmental Consequences

Arupjyoti Saikia

The history of coal in Assam is less contested than the history of tea and petroleum, the two other important imperial ventures of the colonial period. It had been suggested as early as 1825, during the climax of the Anglo-Burmese war, that Assam would be an important source of coal supply in British India. Coal was yet to be productively extracted from other parts of British India. The situation changed rapidly, and in three quarters of a century the British Empire maintained a substantial coal mining industry in Assam: coal was an important source of its political and economic strength. The empire proudly claimed:

> Many things have contributed to Britain's progress as a manufacturing nation, but none is more important than her great natural wealth in coal and iron. Without an abundant supply of these minerals she could not have made use of the great inventions which more than a century ago placed her far ahead of other nations in manufacturing industry (SBE 1904, 250).

Statistical data reinforces this view. For instance, in the first decade of the 20th century, the British Empire produced roughly 245 million tonnes of coal. At the same time, the world output of coal was estimated at about 545 million tonnes, of which nearly half was extracted by the United States (US). This meant that the total coal production of the world was 790 million tonnes per year. Britain was the second country after the US in maintaining significant access to coal and coal mining. British India made a moderate contribution to the total output of coal for the Empire, although within British India, Bengal led the total coal output.

By 1882, the total consumption of coal in India amounted to approximately 1.5 million tonnes, of which about one-third was imported from Europe and Australia, and the remaining two-thirds were raised from Indian mines (Iron and Steel Institute 1882, 414–16). Official estimates in 1897 suggest that there were 145 coal mines in operation in British India, of which 128 were located in Bengal. The output from the Bengal mines was reported to be 3,142,497 tonnes; 131,029 tonnes were raised from the Central Provinces collieries; 185,533 tonnes from the

Assam collieries, with contributions from other localities (Great Britain Board of Trade 1898, 49).

Notwithstanding the importance of coal in the political economy of the British Empire, exploration for coal was more than an issue of imperial economic investment. This essay – divided into three sections – deals with aspects of geological exploration, securing rights on minerals, and the political contest over coal as its economic value increases. The first section deals with the early days of coal exploration and geological science. Since the second decade of the 19th century geological science has made rapid advances in mapping potential repositories of coal and other minerals throughout South Asia, including the province of Assam. The commercial viability of coal production drew imperial capital to the region through similar routes to those of tea. Over the decades, the government developed legal rights, both to land and natural resources that enabled it to secure supreme imperial rights over coal. The second section examines how the imperial economy acquired and established its rights over this resource. Unlike the contest over petroleum and the national imagination regarding tea production, coal has meant little to the larger political landscape of the region. Coal extraction and the transfer of rights to the imperial capital caused the loss of livelihood to thousands of peasants and other employees who once lived in these areas. The third section briefly considers how the extraction of coal was made possible by recruiting labour from beyond the province.

End of Anxiety: Finding Coal

At the time that coal was found in Raniganj in Bengal the Empire was uncertain about the future of coal mining in British India (Hunter 1885, 591–3). Possessing training in geology, Lieutenant R. Wilcox visited various parts of eastern Assam. Wilcox, responsible for surveying mineral resources, noticed coal in several places. This finding was sufficient enough to allow David Scott, the first commissioner of the newly occupied region, to suggest that coal would help to introduce steam navigation to Assam – a region otherwise difficult to reach – via the Brahmaputra.

Further investigations resulted in confirmation that coal beds existed in both upper Assam and the Khasi Hills. Wilcox was followed by Charles Bruce, the tea explorer, in 1828. His areas of investigation were in the vicinity of eastern Assam, near the river Suffray, a tributary of the Disung. Bruce reported inferior coal and difficulties in reaching these areas. He extracted an estimated 5,000 pounds of coal from upper Assam. While coal would begin to be extracted within years of Bruce's investigations, its quality remained to be verified. Anxieties persisted in parallel with exploration for appropriate areas for tea plantation and likely reserves of oil. Demand for coal mining in the region began to increase slowly in subsequent decades, driven by the establishment of tea plantations in large numbers and the parallel founding of the commercial venture the Assam Company (Antrobus 1957).

The East India Company officials posted in the region did not forego the opportunity for coal exploration. Many of them, particularly Francis Jenkins, were convinced that coal could be obtained from the southern belts of the Brahmaputra valley. Such conviction was mostly based on rudimentary geological understanding of the topography, field reports from a number of junior officers, and local people's accounts. The persuasive factor was the discovery of shell limestone in the Sylhet and Khasi Hills. Further discoveries of shell limestone in parts of the central Brahmaputra valley was exciting news for the East India Company officials, and formed the basis for the discovery of more coal beds in the region.

Officials were optimistic regarding the capacity to export the coal, as the most probable sites of coal beds were conveniently located and were easily accessible. East India Company officials' experience of navigability in Assam was relatively poor. Until the middle of the century, bulk transportation of goods to the region was low, and there was very little experience of dealing with goods carriage in the region. Nonetheless, easy access would make these and other constraints insignificant. In the meantime, the Bengal government granted a paltry sum for the purpose of opening a coal mine in Assam.[1]

As early as 1835, Jenkins was highly optimistic of being able to secure abundant quantities of coal in the region: 'It now becomes almost certain that we shall find very large supplies of this invaluable mineral on the south bank of the *Brahmaputra*' (Jenkins 1835: 705). He was certain that coal would be available in other areas in the region also. The samples of coal that Jenkins sent to Calcutta for scientific confirmation were determined to be 'slaty and earthy.' Regardless of this, Jenkins believed that they were sufficient to prove the presence of high-quality coal beds, writing that:

> The three specimens of Assamese coal, received with the above note, turn out to be of very respectable quality; they are rather slaty in fracture, and do not coke; but burn with a rich flame, being very bituminous: on this account they would be very suitable for steam engine fires, though unfit for the forge, or for the smelting furnaces (Jenkins 1835, 705).

Several others were instrumental in convincing Jenkins of the presence of coal in the region. Amongst them was Captain Vetch, who sent frequent reports to Jenkins of having found detached specimens of various kinds of coal near several river beds 'from their confluence with the main river, and not far from the foot of the mountains.[2]' Specimens received from touring officials – obtained from both the north and south banks – were useful for Jenkins. Vetch was not alone in making field enquiries. Ensign Brodie, another young geologist working for East India

---

1 The Bengal government allowed expenses for opening of a coal mine in Assam and collecting coal there for its steam ships (IOR/F/4/1308/51976, November 1828–July 1830 [British Library, Asia and Africa Collection]).

2 *Journal of the Asiatic Society of Bengal* 7(2): 948.

Company, was hopeful that coal would found in the region despite his scepticism deriving from the poor quality of the specimens that he found.[3]

Coal discovery in Assam was a matter of some importance for London investors. That coal had been found at separate points along an extended line of landscape resulted in consideration to providing better support to the region. Discussion revolved around the issue of making the region practically useful. Preliminary examination of specimens sent by Jenkins led to mixed reactions. McClelland was hopeful that further investigation would lead to encouraging results for future trade in coal. Jenkins became impressed with this early success. He ensured that several geographical areas be examined thoroughly. The prospect for further investigation improved with the arrival of the scientific mission empowered with the final confirmation for the establishment of Assam tea plantations in the 1840s. Jenkins negotiated with botanist Nathaniel Wallich to detach one of the members of the mission in order to assist in geographical surveying for the purposes of locating coal beds. Despite attempts, the mission did not lead to any practical investigation.

Effort was then made to search for coal in previously unexplored areas. Many agreed that it might be obtained – and that with some difficulty – for local consumption in the province, at a cheaper rate than that at which it could be supplied from Bengal. Jenkins was observed: 'Having thus stated what had been done up to the period at which our last Reports were published, we are the better prepared to show the value of what has since been done in Assam' (Committee Appointed to Investigate Coal 1838: 951).

At the same time, inspiring results came from the exploration of geologist Captain Hannay in far eastern Assam. A committee was appointed to investigate the coal and iron resources of the Bengal Presidency.[4] The committee examined the Assam findings. Hannay had forwarded a 'few hundred maunds to Calcutta for trial'.[5] His samples were found to contain a 'considerable quantity of sulphur'. It was found unfit for use in the government mint for annealing silver. Results also revealed that '40 maunds are only equivalent to 32 maunds of the variety of Burdwan Coal'.[6] Such scepticism did not, however, deter Jenkins to highlight the unfavourable circumstances under which his team had worked, referring to 'the zealous manner in which Captain Hannay, at considerable risk and trouble' had tried to locate coal. The committee had no doubt that

> as far as the Assam coals generally have been tried, their qualities have been found to be so good, that we may regard the small cargo transmitted to Calcutta by Captain Hannay, as chiefly valuable in showing the facility with which the

3   *Journal of the Asiatic Society of Bengal* 7(2): 949–950.
4   Committee Appointed to Investigate Coal 1838: 953.
5   Committee Appointed to Investigate Coal 1838: 953.
6   Committee Appointed to Investigate Coal 1838: 956.

article may be raised and transported (Committee Appointed to Investigate Coal 1838: 954).

The possibility of exploration of natural resources and consequent economic benefits suggested a positive outlook for the region. Several places in eastern Assam came to be regarded as suitable for European habitation. Demands were made to improve means of communication between Bengal and the rest of the region. An official committee noted:

> Of the healthiness of Assam generally, people now begin to form very favourable notions compared with Bengal. Boorhath and Jypoor are said to be situated in one of the finest quarters of the province. In the present state of things, perhaps, the Boorhath and Jypoor Coals are only to be regarded as the elements of local improvement; the intercourse between Upper Assam and other parts of India must assume a better footing, before its Coals could be supplied to Calcutta at a cheaper rate than Bengal Coals (Committee Appointed to Investigate Coal 1838: 955).

The question of the supply of coal from Assam was again reported on by a committee in Calcutta, appointed by government in 1845. The committee expressed its hope that 'Perhaps the most important results from opening coal mines in Assam for the supply of Gangetic steamers, until more convenient sources should become better known than at present, would consist in the assurance of an unlimited and steady supply at all seasons'.[7] The committee was of the view that in the near future coal would emerge as major source of capital investment. The committee's recommended investigation was carried out in 1848, determining that coal could be extracted in the Saffrai Valley (Government of Assam 1877, para 81). Early in the ensuing decade, Thomas Oldham, the first superintendent of the Geological Survey of India (GSI), examined several places in the Khasi Hills during 1851–52 and found evidence of coal distribution.

The next crucial step was taken by the GSI in 1865. Henry Benedict Medlicott, the deputy superintendent of the GSI, carried out another survey of Assam in search of coal fields (Medlicott 1865). His survey did not result in discovering any new coal fields but reaffirmed the economic value of the two existing coal fields at Tirap and Jaipoor. He also outlined the relative strengths of both, submitting that coal from the Tirap fields furnished better coal than that of Jaipoor, though the latter had a relatively advantageous geographical position.

Despite hope, not much progress occurred in terms of integrating commercial production with coal exploration until 1882. The period between 1838 and 1882 was marked by amateurism in exploration, limited attempts to extract coal by opening collieries, and slow increase in local consumption. The tea industry was yet to make a full shift to coal-based boilers. Given access to free timber-based fuels,

---

7 Committee Appointed to Investigate Coal 1838: 959.

tea planters had not determined to use coal. The introduction of a steamer service through the Brahmaputra created limited local need for coal. As coal exploration had not progressed, the blame was placed on hostile communication, scarcity of labour, or the unhealthiness of the places where the coal was located. Renewed surveys were conducted between 1874 and 1876. Meanwhile, speculative interest in Assam coal increased. A number of applications seeking grants to lease territory waited for official confirmation, which came at the initiative of the GSI and the task was given to GSI's future superintendent F.R. Mallet. Mallet made extensive physical surveys of various places which were identified by previous explorers as possible sources of Assam coal. His primary range of investigation was the southern hill ranges of eastern Assam.

Mallet's report indicated two important issues. Mallet concluded that coal found in most places had low ash quality, which was accepted as good for use, and meant that coal mining in Assam posed immense commercial possibility. Mallet advocated financial investment to explore these coal deposits. He estimated the total volume of coal deposits in eastern Assam at 9 million tonnes. His report also provided practical guidance for extracting coal and ways of revenue management involved therein. He laid to rest existing doubts about the economic value of coal in Assam, leading to a favourable market amongst speculators to invest in Assam coal fields.

Coal Mines and Private Investment

Though it was after Mallet's key report that possibilities of private capital investment became a reality, limited private speculation had already begun in the 1850s. For a long time, this venture was confined to a single individual. A grim local market forced the lone private investor to supply a limited quantity to official steamers plying the Brahmaputra. In 1861, the Bengal government allowed another private speculator to supply coal to three depots in return for a commission. This lease went on for three years and was a moderate financial success.

On expiry of the lease, officials at different levels considered the idea of inviting private speculators into the coal business. The model was a successful tea venture – all leases would be ninety-nine years long. While the idea found favour with the Bengal Governor, the Chief Commissioner Henry Hopkinson opposed it. The Commissioner emphasized that the previous disposal of large-scale un-surveyed land as wasteland to the tea-industry had a crippling effect in the government treasury. As an advocate of the *raiyatwari* system in Assam – whereby land revenue collection for taxes avoided the Zamindari middlemen and was imposed directly on individual cultivators – he was against any growth of private monopoly in coal mining. With a view to maximising profit from the future coal industry, he suggested a thorough scientific survey of the coalfields before undertaking any private investment. Medlicott, whose survey in 1864 is mentioned above, was also in favour of a liberal view towards the speculators, despite cautioning against any growth of monopoly trade.

Notwithstanding such apprehension, the Bengal government finally allowed private investment in coal mining in 1867. Cecil Beadon, the Lieutenant-Governor of Bengal argued:

> The object of the government is to promote the production of cheap and good coal, and this can best be done by creating an intermediate preparatory interest in limited tracts of coal field so as to encourage the operation of private capitalists, and at the same time to prevent anything like a monopoly (Quoted in Barpujari 1993, 82).

Within a decade, private speculation came forward without bringing in investment. Early speculation resulted in a mixed fate. The numbers were low and competing interests in the region such as oil helped to dilute potential investment. The government, meanwhile, imposed conditions for allowing leases – 6 annas/acre and Rs 1/100 mounds of outturn (Mallet 1876: 308). Despite the high number of applications, the government doubted whether many speculators would take the leases, creating concern that monopoly trade would result.

By the 1880s it became clear that European speculators had schemed to wrest control of the coal resources and trade in the region. Unlike the Bengali *zamindar*s' entry into the coal trade (Rothermund and Wadhwa 1978), circumstances persisted in which potential wealthy Assamese backers evinced no interest in the coal trade. This failure could be largely attributed to the absence of an Assamese bourgeoisie. A similar fate happened to the tea industry. While a small number of Assamese petit bourgeois showed interest in the tea industry, this never became a success. The most-cited example is that of Maniram Dutta Barua, the prime minister of the pre-colonial Ahom kingdom; Barua was hanged for conspiring from Calcutta in the Sepoy Mutiny. Prior to this, he established two tea plantations, gaining experience as manager of the Assam Company (Guha 1968, 202). This failure is largely ascribed to the imperial politics of the time. The Assamese gentry began to benefit from liberal land policy to reclaim land only towards the last years of the 19th century. However, this did not endow them with enough capital or political leverage to challenge the European companies' control over mineral or other resources, which would have established them as future entrepreneurs.

Demands for coal increased slowly within the province. For fuel the tea gardens were dependent either on coal brought up from the Bengal collieries, or timber procured inexpensively in the region. The Doom Domma Tea factory brought 3,000 mounds of coal from Bengal after ferrying it 1,000 miles. Mallet estimated that the cost of coal, thus arrived, was ten times higher than coal mined in Bengal (Surita 1981, 17).

The newly formed Assam Railway and Trading Company (ARTC) Limited capitalised on the condition of a slow, yet highly speculative local market; functioning in a similar way to the existing tea garden companies, the ARTC maintained dual trade interests – exploration and trade of coal, and the railway. Capital was raised though public auction of its shares in London (Surita 1981). In 1881, the ARTC acquired concessions to explore several coalfields. In the next year the company commissioned the Ledo colliery. The company had a combined interest in building and operating railways in this area, which soon became a major factor in promoting the coal trade. Coal extraction gained momentum slowly. In 1885, the ARTC produced 75,000 tonnes annually: a result far less substantial than that of the Bengal coal fields (Hunter 1882, 620–21).

Despite shareholders in London investing in the ARTC, the success of the coal market remained a major concern. The tea industry remained a major possible customer. Yet, not everybody was enthusiastic about the future of the coal trade in Assam. London's *Railway Times* published a sceptical note on the grim future of coal for the ARTC. It suggested that coal from Assam would not be able to compete with the low-priced high-quality coal from Bengal, and that there was no extensive local market in Assam for coal, despite possible increase in demand from the tea industries (*Railway Times* 1884, 1077–1078). The journal further argued that the total requirement from tea gardens was far below than what could be possibly explored in Assam. Private investment poured in and within a short period of time an estimated £1,000,000 was spent. This also coincided with intensive petroleum exploration and private investment in acquiring leases for it.

The ARTC's mineral concession gave exploration of coal and its trade a formal character, and the prospects for coal mining grew. The company acquired four separate collieries. It began to excavate coal effectively by use of a system that relied upon a shallow inclination, then popular for working the thick coal-seams of South Staffordshire.

Alongside increasing trade in coal, drilling for oil was about to gain ground. The advent of oil exploration facilitated technological collaboration as well as negotiation with the government for acquiring firm lease rights (Saikia 2011, 48–55). Meanwhile, the land laws in the province quickly underwent significant changes. The province had been primarily endowed with a *raiyatwari* system since the 1860s. The new regulation entrusted the ownership of mineral resources to the state, making it easier for the government to negotiate leases to speculators on a long-term basis.

Over the years, the gradual mechanisation and importation to Assam mining of British technology – the introduction of rail carts and haulage ponies – replaced dependence on manual labour (Turner 1896). The coal trade also improved due to governmental support and a developing import market. From an estimated 16,493 tonnes in 1884 and 75,000 tonnes in 1885, the ARTC's coal output grew to a modest 200,000 tonnes in 1899 (Surita 1981, 29). The high quality of the region's coal was feted. The *Black Diamond*, a London-based coal periodical announced that the coal from Assam was an excellent steam coal, and undoubtedly the best found in India, its calorific value being considerably higher than any other known Indian coal. It coked freely, the coke being of excellent quality (Harris 1901, 573). Speculators and engineers alike agreed that Assam coal could be described as highly bituminous; it was very soft, caked freely when burnt, and in practical tests was found to be equal to Welsh coal. It yielded 10,900 cubic feet of gas to the tonne. At first, objection was taken to the Assam coal on account of its smallness grain: fire-bars – needed for placing the coals in a boiler- had to be placed closer, and – on account of the great heat given off – processed in shorter lengths to allow of the use of the coal. Once these adjustments were undertaken, all prejudices disappeared, and it was acknowledged it to be the best coal in India.

Despite multiple challenges, arrangements were made to increase coal output significantly. By the end of the century, the industry claimed that there were ample markets for coal among the tea estates in eastern Assam, growing steamer consumption on the Brahmaputra, and expanding railways. The industry supplied in small volumes to ocean steamers and jute factories in Calcutta. The ARTC, as the primary coal industry, had the additional benefit of controlling the railway system in the province: the railway and coal industry expanded alongside one another. Soon the bulk of the coal was conveyed from the mines by railway to Dibrugarh, and beyond to a network of depots. Steamers and flats – coal barges – were used innovatively to suit the local topography. The steamers, as the major vehicle of transportation, carried coal for their own consumption, and several hundred tonnes for the markets. The flats carried coal for the market – up to 1,100 tonnes each. One steamer towed between one to three loaded flats, and took about six weeks for the round journey.

However, this moderate speculation never developed to the extent of the central Indian coalfields. With the availability of petroleum, tea industries again shifted to petroleum-based fuel. The ARTC's attempt to import to ocean-based ship companies also failed to expand, mainly due to its relatively high cost compared with cheaper Bengal coal. Local markets nonetheless required a continuous supply of coal, and this requirement was sufficient for the ARTC to open seven further collieries in eastern Assam between 1893 and 1940 (Surita 1981, 30).

Although the European trade interest was instrumental in creating the coal trade, the coalfields in the Khasi Hills had been surveyed, but followed another trajectory. Unlike the eastern Assam coalfields, rights over the coalfields in the Khasi Hills remained mostly within the traditional community institutions. The administration negotiated directly with the community chiefs, and petty trade grew around the coalfields. The method and mode of extraction essentially remained dependent on the capacity of the Khasi villagers.

Overcoming Labour Crisis

The scarcity of labour in Assam was the dominant theme of colonial governance. The nineteenth century was a contested time in which the powerful tea industry sought to procure labour. Multiple reasons, that is, the non-availability of local labour or unwillingness of skilled labour from other parts to come here, have been cited as hindrances to local labour recruitment. The coal industry faced similar problems, albeit on a different scale, and of a type predicted before viable trade in coal was realised. The 1838 Coal Investigation Committee was of the opinion that 'the population is certainly scanty, but then it is composed of a class of people, Mikeers and Kacharees who can be taught, and will willingly put their hands to anything that will afford them a moderate remuneration for their labour' (Committee Appointed to Investigate Coal 1838, 950). However, unlike the tea industry, the trouble did not surface for a long period. Meanwhile, the tea industry had put in place means of overcoming this crisis, which found hardly any in support in the press and public discourse. Nevertheless, its approaches helped the future coal industry, which relied on similar mechanisms.

The coal speculators' attention was drawn to a number of local communities which could be engaged in the works of coal extraction and clearance of jungles. Most early attempts were unsuccessful, resulting in public disquiet throughout the coal industry. Thus George Turner, an influential engineer in the ARTC, condemned the Naga population, whose habitat covered landscape earmarked for future coal mining:

> Around the mining area, and back in the hills are found aboriginal tribes, called Nagas, a fairly strong race of people. At first, these tribes threatened to exterminate us if we did not go out of what they claimed as their country, but since then they have become more or less civilized and friendly. These people, as a race, are incorrigibly lazy, having no industries of any kind, and cultivating only sufficient grain for their own consumption. When wanting money to purchase opium, salt, or other necessaries, they will work for a short period in cutting or clearing jungle or similar work, but nothing will tempt them to commence mining.

This note of condemnation towards the local population for their disregard for mine work was no different from the previous attitude of tea industry owners towards the Assamese peasantry. However, the coal industry swiftly adopted methods adopted by the tea industry at a much earlier stage in development: labour was imported from the Bengal province. In terms of their social background, the new workers at the Assam collieries were no different from their tea industry counterparts. Extraction began with the help of Naga workmen. The company considered them as having the requisite expertise in clearing dense and hard jungles. However, was discouraged in its attempts to secure technical or methodical conduct from these workers in extracting coal with minimum damage to the product. The simplest alternative was to procure preliminary support from a small number of British colliers.

Bringing workers from the other provinces did not resolve the crisis. Work at the collieries required training to develop various indispensable skills. One way of resolving the crisis was to train the Indian workforce under European supervision. For many workers the work experience was different from everyday life. They needed to work underground and rarely had any exposure to such work (Turner

1896, 356–64). In making choices in labour recruitment, the coal industry preferred to use similar methods to those of the tea industry. Workers from the collieries of Bengal were also recruited into Assam: this resulted in an uneasy relationship between the Assam coal industry management and skilled workers arriving from Bengal. Accepted tools and methods for work became the new bone of contention: the Bengal-trained miners preferred to work using a crow-bar – known as *sabol* and pointed at both ends – for cutting the coal. The management instead sought the use of a double-ended pick increasingly used in South Staffordshire. As it was difficult to force the Bengal workers to use another tool for extracting coal, the industry argued that the easiest alternative would be to train non-skilled workers. Turner asserted that 'it would be better at once to begin at the beginning, and train up to the work young natives, who had never seen a coal-mine, under selected thick coal miners' (Turner 1896, 360). Years later, the industry boasted how 'this was done, and there are now hundreds of natives working in the mines, careful and good workmen, who were agriculturists a few years ago.' (Grundy 1899, 26) Workmen arrived from Britain in December 1882, as a means of providing technical support.

The industry employed both men and women. The men would be 'cutting and getting the coal, timbering, etc.,' while the women would be 'filling tubs, tramming, etc.' (Turner 1896, 361). Working conditions in the collieries invited criticism and resulted in the gradual improvement of wages and conditions for miners. Since the last decade of the 19th century, the industry began to engage a sizable number of Nepali workers who were recruited from the foothills of Nepal and above (Gurung 2009, 259–75). The industry and its European masters increasingly extolled the ability of the Nepalese as a skilled workforce compared to others from eastern and central India. Slowly, other emerging mineral-based industries also began to recruit Nepali workers. This did not sit well within emerging Nepali nationalist politics, as they thought this would deprive a sizeable section of the Nepali population from being recruited into the British Army, which was crucial both for their financial and social status.[8]

Assam's coal share within Independent India's total coal output was small. In 1957 Assam produced only 1.3 per cent of the total Indian coal output. No significant discoveries were made regarding new coalfields. The industry remained confined to limited areas of eastern Assam. Similarly, the industry's total share of labour force was less than 2 per cent (Government of India 1957), contributing to factors that caused the market to remain localised. An economist who studied the industry as part of modern Assam's larger story of economic development blamed this poverty of the coal fields on inaccessible transport and the poor quality of the coal (Goswami 1963, 159). Yet, he was hopeful that because coal in Assam was endowed with a low ash content that desulphurisation at cheap cost would enhance its economic value significantly.

---

8 P 3725/1923, Political and Secret Annual Files, IOR/L/PS/11/236 1923 (British Library, Asia and Africa Collection).

The total number of workers at the collieries never exceeded a few thousand: an official account in 1899 for the ARTC was 1,311 (Grundy 1899, 26–7). This number was much lower than the workforce employed by the tea industry or those who migrated to western districts of Assam to expand jute cultivation (Behal 1985, PE19).[9] The tea industries maintained a highly rigid labour socialisation regime, which greatly benefited their relations with the local peasant population. The collieries did not differ, but with tiny populations they did not significantly influence the demographic fabric of the province. Their marginal role, for instance, could not evoke attention from the larger Assamese nationalist political programme. However, the industry as a whole had already begun to push tribal populations living in the neighbourhood of coal mines to more distant places (Grundy 1899).

Conclusion

Since the 19th century, use of mineral resources has increased exponentially. Gold lost its relevance as labour-intensive mining techniques disappeared. Newly significant minerals found importance in the social organisation of production. Coal's importance was reinforced not only by private speculators but also by its ability to answer local demand. Significantly, the confluence of coal with steam locomotion and steel production throughout the industrialising world also played its part in boosting the prospects of the coal trade in the colonial era.

However, human cost and the practical challenges of coal mining have always caused setbacks in the industry. Increasing deaths and severe injuries in the mines became a major concern by the end of the nineteenth century. In an enquiry made for the period 1896–98, an estimated 37 labourers were killed in these collieries (Grundy 1899). Concerns were raised that it was necessary to increase workers' safety in the collieries. However, very little was discussed regarding the possible environmental consequences of coal mining. Not only did the urge for development refuse space for such discussion, but ecological change was also yet to be addressed as a serious issue in the academy or in the public domain. By the time such considerations began to emerge, serious and irreversible damage was already commonplace (Government of India 2005, 9; Sarma 2010). A technical study conducted during 2001–04 revealed that coal mining caused significant harm to waterways by increasing the level of sedimentary heavy metals, and that mining localities were increasingly contaminated. Critical detrimental changes have occurred in the coal mining areas of Meghalaya, where coal mining began only towards the end of the third quarter of the 20th century. Studies have shown the degradation of land cover, unfavourable habitat conditions for plants and animals, substantial decrease in the numbers of tree and shrub species, the

---

9   The total number of workers in the tea gardens of the Brahmaputra valley was estimated at 247,760 in 1900.

conversion of agricultural fields into mining areas, the loss of dense forest areas, and a simultaneous increase in open forest areas (Sarma 2005). Attention has been drawn to possible loss of nutrients in the soil and their subsequent impact on agriculture. Studies conducted in Assam coalfields also pointed towards increased localised pollution (Tiwari 2001, 191). Water in the close neighbourhood of the coal mines was found to be highly acidic (Singh 1988, 50). Such acidic waters failed to support aquatic life and caused land damage. Long exposure to coal mining has also led to significant health hazards (Sahoo and Bhattacharjee 2010). Environmental degradation has widely impacted local livelihood practices such as devastating fish stock numbers in the neighbourhood of coal mining areas (Bharali 2007), and destroying the habitat of wild animals, elephants in particular (Choudhury 2004).

Since the late 20th century open-cut coal mining has drawn increasing criticism from various quarters. It has caused damage to forest coverage. In 1985, the open-casting method was adopted in Makum collieries, inviting protest from both the public and the state government (Borah 2007).

After Independence the industry was largely nationalised. In 1973 the ARTC transferred its lease and operation to the newly created nationalised North Eastern Coalfields Limited. Several pre-existing petty trading houses, whose interest was confined to grain trade or money lending, came to play an important role in coal distribution. In 2011, the Assamese press reported increasing interest expressed by members of United Liberation Front of Assam – the political insurgents who had critiqued the Indian government for exploiting Assam economically and having continued colonial policies – in asserting control over the coal trade in the state. This essay cannot speculate the future of this political and cultural desire of the Assamese ultra-nationalists to wrest control of the coal trade from the state. Rather, this constitutes a broader reflection on how the imperial economy created institutions of mineral exploration – in this case coal – too far from local institutions to remain sustainable or profitable: since its inception in Assam, the coal economy has failed to inspire Assam's residents or to galvanise the surrounding agrarian economy.

References

Antrobus, H.A. 1957. *A History of the Assam Company, 1839–1953*. Edinburgh: T. and A. Constable.

Barpujari, H.K. 1993. *The Comprehensive History of Assam*. Vol. 4. Guwahati: Assam Publication Board.

Behal, R.P. 1985. 'Forms of Labour Protest in Assam Valley Tea Plantations, 1900–1930.' *Economic and Political Weekly* 20(4): PE19–PE26.

Bharali, G. 2007. 'Development-induced Displacement: A History of Transition to Impoverishment and Environmental Degradation.' Paper presented at the

Seminar on Ecology, Department of History, Dibrugarh University, 27–8 March.

Borah, A. 2007. 'Open Cast Mining Threatens Assam.' *Down to Earth* August.

Committee Appointed to Investigate Coal. 1838. 'Report Upon the Coal Bed of Assam Submitted to the Government by the Committee Appointed to Investigate Coal and Iron Resources of the Bengal Presidency.' *Journal of the Asiatic Society of Bengal* 7(2): 948–67.

Goswami, P.C. 1963. *Economic Development of Assam*. New Delhi: Asia Publishing House.

Government of Assam. 1877. *Report on the Administration of the Province of Assam*. Shillong: Government of Assam.

Government of India. 1957. *Indian Coal Statistics of 1956*. Delhi: Government Press.

———. 2005. *Database of Summaries of Final Technical Report*. New Delhi: Ministry of Forest and Environment.

Great Britain Board of Trade. 1898. 'The Production of Coal in British India.' *Board of Trade Journal* 25(July): 49–50.

Guha, A. 1968. 'A Big Push Without a Take-off: A Case Study of Assam, 1871–1901.' *Indian Economic and Social History Review* 5(3): 199–221.

Gurung, T. 2009. 'Gorkhas as Colliers: Labour Recruitment and Racial Discourse in the Coal Mines of Assam.' In *Indian Nepalis: Issues and Perspectives*, edited by T. B. Subba. New Delhi: Concept.

Harris, G.E. 1901. *On the Makum Coalfield, Assam*. With descriptions of the collieries at present worked by the Assam Railways and Trading Company, Ltd. *Manchester Geol. Soc. Trans.*, 26: 572–90.

Grundy, J. 1899. *Report on the Inspection of the Coal Mines Belonging to the Assam Railways and Trading Company Ltd*. Calcutta: Government Printing Press.

Hunter, W.W. 1882. *The Indian Empire: Its People, History, and Products*. Ludgate Hill, London: Trubner and Co.

———. 1885. *The Imperial Gazetteer of India*. Vol. 4 (Calcutta: Government of India).

Iron and Steel Institute. 1882. 'India Mineral Resources.' *Journal of the Iron and Steel Institute* 20: 414–16.

Jenkins, F. 1835. 'Further discovery of coal beds in Assam.' In *Journal of the Asiatic Society of Bengal* 4(48): 704–7.

Mallet, F.R. 1876. 'Coal Fields of Naga Hills Bordering Lakhimpur and Sibsagar Districts, Assam.' *Memoir of the Geological Survey of India* 2(2): 269–363.

Medlicott, H.B. 1865. *The Coal of Assam: Results of a Brief Visit to the Coal-fields of That Province in 1865, with Geological Notes on Assam and the Hills to the South of It*. Calcutta: Government Printing Press.

Railway Times. 1884. 'The Assam Railways and Trading Company Limited.' *The Railway Times* 1077–8.

Rothermund, D. and D. C. Wadhwa, eds. 1978. *Zamindars, Mines, and Peasants: Studies in the History of an Indian Coalfield and Its Rural Hinterland.* Delhi: Manohar.

Saikia, A. 2011. 'Imperialism, Geology and Petroleum: History of Oil in Colonial Assam.' *Economic and Political Weekly* 46(12): 48–55.

Sarma, K. 2005. 'Coal Resource Potentials, Utilization Possibilities, Environmental Implications and Eco-friendly Mining of North East India.' Unpublished MSc thesis, ITRS, Dehradun.

———. 2010. 'Coal Resource Potentials, Utilization Possibilities, Environmental Implications and Eco-friendly Mining of North East India.' *Ecotone* 2(1): 9–13.

SBE. 1904. *A Survey of the British Empire, Historical, Geographical and Commercial.* London: Blackie.

Surita, P. 1981. *The Story of the Assam Railway and Trading Company Limited, 1881–1981.* Centenary Volume, Calcutta.

Turner, G. 1896. 'Coal Mining in Assam.' *Trans. Fed. Inst. Mining Engineers* (10)2: 356–64.

Chapter 4

Border Mining: State Politics, Migrant Labour and Land Relations along the India-Bangladesh Borderlands

Debojyoti Das

Introduction

This chapter looks at coal mining in Meghalaya from the historical perspective of colonial state establishment and governance over 'non-state spaces' (Scott 1998). Identified by terms such as 'frontier', 'disturbed', and 'excluded', this region of British India has relied on a form of colonial administration adapted to the sophistication of native land tenure and a political system based on kinship and lineage. The central theme of this chapter links India's colonial legacy with the cultural politics of resource use in contemporary Khasi-Jaintiya society, where ethnic identity is used as a cultural tool valorising indigeneity and the capitalist production of coal by native mine owners.

There is strong historical evidence of population mobility between the Khasi-Jaintiya hills and the Sylhet plains of present-day Bangladesh. However, a contemporary crisis of ethnic differentiation and identity formation is growing in this area; it results from cross-border immigration of primarily non-tribal labour into the coal mining sites of Meghalaya, negating the historical role of mobility and exchange between the highland tribes and the lowland peasantry. In order to resolve such a crisis it is necessary to revise colonial mineral and extractive policy in these excluded areas, as they are the primary drivers of contestations and ambiguities between the mining lease owners, syndicates, local elites (village headmen), the Khasi subjects and the colonial state. More importantly, it is worthwhile reflecting on the means by which colonial capital accumulation was realised through the grant of mining leases to syndicated and local entrepreneurs, businessman and state officials.

The existing territorial boundaries of Assam, Meghalaya and Tripura with Bangladesh were non-existent during the pre-colonial and colonial period. Sylhet was once part of colonial Assam (Van Schendel 2005). The post-colonial construction of borderlands and boundaries demarcating these states is the result of the British Boundary Commission and survey parties, producing an irregular and complex border. Nonetheless, in the post-colonial territorial space dividing the two nations ofIndia and Bangladesh – formerly East Pakistan – the peoples of the

Khasi, Jayantiya and Garo Hills comprising the state of Meghalaya maintain their pre-partition and pre-colonial trade and cultural exchanges with the Sylhet plains. However, one of the radical changes over time have been the development of coal as the major item of export this has had significant effects on cross-border labour mobility to access mining areas. Post 1971, the export of coal to Bangladesh has attracted mining labourers to Khasi, Jaintiya and Garo Hills due to labour shortage in the region and the opening up of innumerable mines. During the colonial period mining leases were granted to colonial as well as local entrepreneurs, and to mining syndicates with registered offices in Dhaka and Kolkata that acted on behalf of private and international interests. There was intense competition between mining syndicates to acquire mining leases in the Khasi, Jaintia and Garo Hills.

Colonial revenue records and grants for mining leases clearly document the extent to which private capital was involved in extracting coal to feed colonial tea gardens in Sylhet. The colonial office was equally interested in reserving the Khasi, Jaintiya and Garo Hills coal deposits for the expansion of Assam–Bengal Railway across Chittagong to Sunamganj town, in Sylhet, and hence to Dhaka city. There were elaborate plans to extend rail lines across the Arakan frontier to Rangoon and mainland south-east Asia. The sea route was also explored. It was through these routes that coal from the remotest part of the Garo Hills was transported to the Sylhet plains, where colonial capital was heavily invested in tea gardens as a means of breaking the dominance of Chinese tea production. Thus during the late 19th and early 20th century we see two sides of the Meghalaya plateau – one towards the north (Brahmaputra Valley) and the other towards the south (Sylhet foothills and plains) – becoming centres of colonial capitalist production through plantations and mining leases.

In the post-colonial period the construction of political boundaries restricted the free flow of resources towards the newly emerged nation state of India and East Pakistan. These new borders fostered a brief lull in hostilities; however, this was later forgotten on the creation of Bangladesh in 1971. Prior to the birth of Bangladesh, age-old cultural exchanges between the highland and lowland people were re-established. Oranges, wax, and rice were once again exchanged as part of a vibrant trade in coal and limestone, the most important item of export from Meghalaya. However, with mineral export and the exchange of commodities occurring mostly through the use of currency, barter and cowry shells began to decline sharply as a medium of exchange. Cross-border labour mobility continued despite patrolling and restriction by border troops. The history of population mobility is embedded in the cultural and commodity exchanges that pre-date the arrival of British. Colonial capital explored coal for the expansion of railways and for the sustenance of its tea industry in Sylhet as well as the Upper Brahmaputra valley. However, in the post-independence period the drive for coal and limestone export is fuelled by the rapid growth of townships and infrastructure in neighbouring Bangladesh. The opening up of cement plants in Bangladesh has attracted foreign capital from multinational firms and corporate enterprises. The coal and limestone reserves of Meghalaya have once again become a sink for capital accumulation.

However, given the land tenure arrangements and the rights attached to land use, controversies surrounding ownership and control have caused conflicts between the state government and local communities. Such phenomena I would argue are not the result of recent socioeconomic and geopolitical circumstances, but rather derive from the area's long history of colonial capital production.

Meghalaya is a Sixth Schedule borderland state of the Indian union presents a unique case study of mining in which small-scale informal coal mining forms the mainstay of the vast proportion of all economic activity. Because of the protected land transfer regime (*Meghalaya Transfer of Land Regulation Act 1972*) and Meghalaya's location as a frontier territory of the Indian subcontinent, mineral exploration is considerably under-developed. Until very recently the image of the frontier in the early days of coal exploration has defined its limited and interrupted exploration. However, in the past two decades mineral exploration has gained momentum, with competition among local entrepreneurs to extract increasing quantities of coal. Today coal mining is a lucrative business, controlled by an oligopoly, generating substantial profits (De 2007).

With the proliferation of mining activity throughout the mining belt in Meghalaya, the ecological costs have shocked local communities surrounding mining sites. Civil society groups have fostered a growing awareness of the social and ecological impacts of coal mining on their environment. Activists of the Meghalaya Mountaineering Association, a local NGO filed for Public Interest Litigation in 2006, appealing for a ban on all informal mining practices, following which the Supreme Court of India asked for a status report from the state government on all mining activities conducted in the state (*The Shillong Times* 2006, *Telegraph* 2006).

These ecological concerns are coupled with other kinds of pressure group activities at the local level that are linked to questions of ethnic identity and indignity.[1] Local pressure groups – mostly student unions, the largest being Khasi Students' Union (KSU) and the Federation of Khasi and Jaintiya People (FKJGP) – claim to be the custodians of Khasi culture and tribal identity. They have successfully established links between ethnicity and indignity of the Khasis' with questions of livelihood as the central thread in their resistance to state regulation

---

1　The Khasi, Jaintiya and Garos are Scheduled Tribes and are enlisted in the Sixth Schedule of the Indian Constitution. They are recognized as the indigenous sons of the soil. However with India's independence they have frequently come in conflict with immigrants from Bangladesh, Nepal and internal migrants who have come to settle in this land for their livelihood from other parts of eastern India. The indignity movement has gained moments in the last decades with the growth of green NGO's and pressure groups that have successfully campaigned against uranium mining in Meghalaya. The indigenous Pressure Groups and green NGOs have evolved a narrative that questions non-tribal presence in the state as a threat to their culture and identity. They rarely question the human insecurity caused by pollution from unscientific coal mining. Their struggle is often subsumed within the narrative of ethnic identity and the nationalist discourse of illegal immigrant pressure on their land.

of mining activity. The immigrant, non-tribal presence among mine workers is perceived in the local domain as a major problem that triggers ethnic tension in coal mines.

Issues of ethnicity and labour in the Khasi-Jaintiya and Garo Hills relate to the perceptions and interests of local stakeholders who actively participate in coal mining. The coal mines are operated by local elites who maintain oligopolistic control over production. In the past, ecological concerns regarding rat-hole[2] mining have encouraged the larger political economy of coal mining in Meghalaya. What needs to be considered in this regard is the history of patron–client relations that were legitimised by the colonial and the post-colonial state. The colonial government recognised and legitimised local power hierarchies by formalising the posts of *Siems* and *Dolois*[3] in the Khasi and Garo Hills. In the post-colonial period the establishment of Sixth Schedule status of Meghalaya has marginalised the role of traditional chiefs. However, their elevated social status within the community persists. As my ethnographic field research has revealed, the village headman (locally known as *gaubura*) still enjoys primacy among his kinsmen, although his political role in village decision-making and development programmes has been replaced by the District Council and its elected bodies. These powerful actors at the village level (District Council members) control mining sites and frequently their blood relations hold high positions in the state bureaucracy and district planning boards. Thus these networks of patronage and power are vital in sustaining informal and illicit mining activity. The local power relations within the community continue to operate through kinship bonds and clan affiliations.

As (Khan 1999) has observed, in the highlands patronage plays a very important role in exercising state control over local resources. State patronage is established through these historically established structures comprising kulaks that receive royalties, rent, and tributes from the state as acknowledgement of their social function. In determining the production permits and ownership of mines, the village chiefs and democratically elected District Council members act in connivance with the state bureaucracy to promote their class interest. The state earns revenue from coal export: its role is restricted to revenue sharing. However, in the production of coal regarded by the state as illicit, government officials and the political class refrain from policymaking on mining, relying on the pretext of contravening communities' customary rights over land and forest, as legitimated by constitutional amendment to protect native land rights (e.g., *Land Transfer Act*). In recent years such protections have been breached by the state government,

---

2   Rat hole or pit hole mining is a crude mechanism of surface mining adopted by mine owners in Meghalaya as it involves least capital intensive mining. The drawback of such mining is seen during the monsoon season when these mines cannot be worked properly.

3   *Siem and Doloi* are the traditional village heads in Khasi and Garo society. They acted as local chiefs during the colonial times.

which has granted corporate leases to explore for minerals in order to feed large cement plants across the border.

In Meghalaya human insecurity has become a turning point in defining development and levels of impoverishment. An intensive fieldwork-based study on human insecurity and civil society in Meghalaya by McDuie-Ra (2007b) posits that human insecurity is the outcome of privatization of land holding and its concentration in the hands of a few rather than the constructed notion of insecurity caused by cross-border immigrant settlement in coal mining areas. These immigrants – mostly Bangladeshis, Nepalis, and people migrating from the Hindi-speaking belt of central India – are migrant wage earners. In *Durable Disorder* (2003), Baruah has termed such immigrants 'denizens'. Denizens are people who do not enjoy political or entitlement rights and yet in the Northeast they are framed by indigenous community patrons as a potential threat to local culture and tribal identity.

Migrants are primarily denizens who do not have property rights and are often illegal immigrants. These floating populations who desperately seek jobs for survival have come into conflict with local communities, who regard large-scale immigration as a threat to their cultural identity. The identity politics in the past has led to community outrage and ethnic cleansing in the coal mining areas and in labour camps.[4]

Emerging class difference within the community is an effective instrument of social change. Thus coal mining in the state presents an insight into how communities are organised and how they represent their interests in a democratic and decentralised polity. Civil society patrons and pressure groups in the state today claim to playa significant role in delivering social services, organising resistance, and framing public opinion. However, in the case of Meghalaya they also represent elite class interest.

Although discourse on the ecological destruction caused by mining has become overwhelmingly driven by facts, I argue that the complexity of the problem is inherently structural, embedded in the power relations of local communities represented through claims to land; and, further, that land relations and the patronage of local community elites have developed from the post-colonial state. The central and the state government appear to have no control over mining operations, which are almost exclusively under private oligopolistic control. Nonetheless I suggest that the state is embedded in the day-to-day struggles over mining ownership and trade in minerals, controlled by a powerful coal lobby constituent of sitting Legislative Assembly members. This lobby prevents the government from taking steps to legalise unplanned mining activity. Meghalaya is perhaps the only state yet to draft a state mineral policy. However, the former understanding of why the state fails to act upon the needs and abuses of coal mining is explained by an

---

4 See *The Statement* report dated 13 April 2007. In this news report the FKJGP branded all mine workers as criminals and pledged to flush out all immigrants working in coal mines in Meghalaya.

underpinning of patronage: of reciprocal alliance between the state bureaucracy and the local mine owners who form part of the coal lobby.

The networks of patronage in Khasi-Jaintia society have a colonial history of administrative arrangements in the highland that is not unique to Meghalaya, but can be seen all over South-East Asia (Li 1999). Coal mining in Meghalaya has to be perceived in the context of the broad social history of colonial and post-colonial state intervention in tribal frontier areas, rather than merely concentrating on the discourse of failed developmentalism and ecological destruction that has prevailed in civil society critique of coal mining activity in the state.[5] Such discourse claims to critique state failure and highlight its inability to control hazardous mining, but fails to scrutinise the social and political complexities that underlie the political economy of coal mining in Meghalaya.

Studies of highland communities by anthropologists in Southeast Asia do not necessarily consider coal mining, but are implicitly engaged with state, community and extractive industries in the highlands. Some of the most significant contributions have been made by Hafner (1990), Tsing (1993), Li (1996, 1999, 2000 and 2007) and Khan (1999), which look at the discursive politics of the local people embedded in production structures. As Li (1996) argues, in the highlands, which are often seen as marginal, under-productive and backward spaces, some of the most lucrative livelihoods are not based on agriculture, but on extractive industries, trade and wage employment.

However, my analysis will locate these narratives within the broader picture of capitalist mass production, oligopolistic control over mines, exploitation of wage earners, land consolidation triggering landlessness, and the growing convergence of interests between politicians and mine owners that presents a clear picture of privatised land-holding and oligopolistic control. The more pressing problems of peoples' livelihoods, degraded environments surrounding mines, wage rates of miners and their living conditions are subsumed within concepts of 'culturalization' and 'environmental' as a means of addressing the local problem. Local civil society and pressure groups draw popular support from challenging the state's desire to regulate mining activity by claiming indigenous peoples' customary right over land and forest resources. These claims do not necessarily articulate the resistance of civil society, but often represent people's stakeholder interests, the interests of mine owners, or the interests of cultural patrons, political and indigenous leaders. Such claims depoliticize basic questions of human insecurity and do not question problems of identity politics and indignity.

Similarly, by implicating state control over day-to-day life of the indigenous communities, proponents of green development models have declared time and again that indigenous tribes are the appropriate custodians of the resources of their territories. The example of 'sacred groves' – collectively owned forest resources of the community – is often cited to validate local prudence in conservation and

---

5   See 'Preservation of Meghalaya Caves Withdrawn', dated 2 August 2006: www.indialaw.com/guest/DisplayNews Accessed 20 August 2008.

management of resources (Gadgil and Guha 2004). It is commonly also suggested that such communities possess values counter to those held by the state: tribal silvicultural values arguably pre-date colonial practices and benefit from long habituation of the same environment; they are relatively homogeneous in class and caste; they are not unduly driven by motives of exploitation and greed; they have limited consumption requirements; and they focus upon long-term management of natural resources. However, the transitions in these values are brought about by liberal democracy, market interventions, monetary inflow, privatisation of land ownership, commercialisation of farming, and changing attitudes towards community life with the growth of services and trade activities in the highlands.

Mining activity provides locals with gainful employment which immigrant groups would not experience the benefit of in mainland India. Such rhetoric directed against workers is commonly used by pressure groups as legitimate reasons to expel all immigrants from the state. The standard complaint is that mine workers disrupt the local economy by claiming all low-paid and hazardous jobs,as these are not performed by local tribes. By branding the immigrant labour force as illegal, the local elites maintain a discretionary means by which to control labour, protecting immigrant mining jobs on a quid pro quo basis of providing security in the form of shelter from state prosecution in return for labour. Interviewed mine owners have claimed that these workers are delinquent and dangerous to associate with. This has also appeared in a series of magazine reports that featured coal mining in Meghalaya and the condition of laborer's working in these mines (*Frontline* 2006, 2007). Such image construction of mine labour advantage mine owners' who maintain power through bargaining terms of labour contracts for exploiting coal, and benefit from the enormous payoffs it creates.

Instruments of social change – education, the free flow of goods, and the circulation of cash in a previously barter economy have modernised traditional communities and directly impacted their consumption patterns. A study of the Bemba shifting cultivators in the northern province of Zambia reflects that with the inflow of monetary exchange and market products, Bemba consumption patterns changed radically from only purchasing daily necessities towards the capacity to save and spend disposable incomes on manufactured products from industries for clothing, utensils, soap, transistors, and so on (Moore and Vaughan 1994). Similarly, in the case of Meghalaya, the capital produced from the export of coal to nearby Bangladesh and mainland India allows conspicuous consumption, and has led to the growth of real estate businesses in Shillong and renting structures of capitalism among the indigenous community elites who own mines and maintain control over production processes. The production of coal continues to be conducted prevailingly through human labour, as mines are not mechanised. Immigrant labour power is put to use under minimal wage rates with no social security. My fieldwork revealed that the wage earners work under extreme risk and are provided with no social security benefits. The dwellings in immigrant settlements are suffocating: each hut houses a dozen people. In the cities and district headquarters, mine owners live in modern houses with luxurious facilities and glitzy cars are parked in the apartments and bungalows.

The following discussion considers these issues from the perspective of different stakeholder interests in mines and how the cultural factors of indignity and tribal identity function within the predominant ethno-political discourse, which blurs the distinctions between the state's indirect control over mines and the various interests it serves in the production of capital.

Mining in Meghalaya

Mining dates back more than one-and-a-half centuries in the Khasi Hills. However, commercial exploration only began during the post-independent period; prior to this, colonial powers were skeptical as to the economic viability of coal mining in frontier regions of the Assam.[6] No written document records the use of coal by the indigenous Khasis before the arrival of the British. Nonetheless, oral accounts by village elders reveal that gold and other minerals, essential in making ornaments, were mined by them. In traditional Khasi society, ownership of land was very important. The imposition of colonial governance over the Khasi hills and the political subjugation of *Siem* (the king) under the patronage of the Raj transformed the customary and cultural practices of the Khasis.[7] The British introduced the house tax (*hoe* tax) in every household practicing shifting cultivation. The *Siems* (kings) were required to collect tax and pay an annual tribute to the colonial government.[8] The British also introduced the lease system on land by which people outside the community could assume de facto control over tribal land. The development of the timber trade and the felling of trees to meet colonial demand started during this period. These changes had a serious impact on building upon the power and prestige within the Khasi

---

6   See 'Report on the Coal Mining in Jaintiya Hills', 1850, British Library, India Office Record, IOR/V/23/93.

7   The native title was officially recognised in 1867. It comprised 25 petty states in the Khasi hills, 15 of the first-class category presided over by *Siems* who, though deriving from one family, are chosen by popular election; one confederacy under elected officials styled *Wahahahars;* five under *Sirdars*; and four under *Lyngdohs*, which are entirely elective. See 'Political Proceedings', Khasi and Jaintiya Hills, March 1867, No. 14. Assam State Archives, Guwahati.

8   The collection of *hoe* tax was considered a necessary step for improvement in the condition of the natives' livelihood. Writing on the Khasi hills, B.C. Allen stated that light and judicial taxation would contribute to the preservation of tranquility and good order in the Jaintiya Hills. While comparing this with the LurkaCoolis of the Shingbhum district, Allen observed that moderate taxation had a beneficial effect upon the savagery of the LurkaCoolis. Taxation made the natives less turbulent and aggressive, and more thrifty, diligent and submissive to the authorities: see Mackenzie 1884, 240–41.

community, particularly among the ruling clans.[9] Revenue and royalty payments also promoted land as an economic good in a predominantly lineage-based, non-monetary,barter exchange community. Slowly these incentives caused land to be valued in an economy, at the centre of which accreted the developing state machine of clan politics, property accumulation, and control by powerful clan elders.

The emergence of Meghalaya out of Assam as a separate state in the post-independence period (1971 onwards) necessitated the development of mineral resources within the state to sustain its domestic economy. As Meghalaya became a Sixth Schedule state, land came to be owned by the local people; thus, the local inhabitants could develop their resources. Coal production boomed as infrastructure was enhanced and border trade with Bangladesh established in the late 1970s and early 1980s. In less than three decades of unplanned mining, land degradation, aquatic pollution and the disruption of the local ecosystem became entrenched and unavoidable circumstances for the region. Public debate in policy and administrative circles now focused on the legal status of coal mines in Meghalaya that needed to be regularised (*The Pioneer* 2006). The absence of a state mineral policy necessitated that the Supreme Court order the state government to submit a status report on all mining activities in the state. This caused pandemonium in the state legislative assembly, fuelled by public outrage. Local and general public opinion was that such assessments would lead to the closure of mines, which would adversely affect the local economy and the lives of locals.

Confusion emerged from the contradictory laws governing mineral resources in Sixth Schedule areas, where historically political and administrative powers were diffused by tribal chiefs and the village headman. The coal boom that began in the late 1970s changed the cultural landscape of the Khasi Hills, as local entrepreneurs who were community stewards or people's representatives pursued private self-interest, privatised communally owned land, and engaged in reckless and destructive mining activity. This resulted in a major mining crisis that destroyed the commons and its heritage. The extant mining regulations are ambiguous and contradictory: the Sixth Schedule status of Meghalaya makes local communities custodians of their own resource, while the *Mines and Mineral Development and Regulation Act 1957* declared all private mining activity in Meghalaya to be illegal.[10]

---

9   As early as 1853, colonial officials posted in these frontier territories observed that the simple character of the 'Khasias' (Khasis) had, to some extent, become corrupted by civilisation and increased wealth; civil wars which continually distracted the country in older times had been put down; trade had been expanded; and an increased demand for hill territory products had set in: see Mackenzie 1884, 239.

10   According to the *Mines and Mineral Development and Regulation Act 1957*, amended in 2002: 'No person shall undertake any reconnaissance, prospecting or mining operation in any area, except under and in accordance with the terms and condition of the reconnaissance permit or of a prospecting license or, as the case may be, a mining lease, granted under the act and rule made there under.' Substituted by MM (RD) Amendment Act, 1999, vide Go I Ext. Part ii, Section 1, No. 51, dated 20.12.99 (No. 38). In the state

The state has historically been reluctant to develop mineral resources in frontier territories because of its sporadic availability and limited economic value. However, it claims to protect the interest of small miners in scheduled areas. This has also been clearly spelt out in the *National Mineral Policy, 2003*, which states:

> Small and isolated deposits of minerals are scattered all over the country. These often lend themselves to economic exploitation through small scale mining. With modest demand on capital expenditure and short lead-time, they also provide employment opportunities for the local population. Efforts will be made to promote small scale mining of small deposits in a scientific and efficient manner while safeguarding vital environmental and ecological imperatives. In grant of mineral concessions for small deposits in Scheduled Areas, preference shall be given to the Scheduled Tribes.[11]

The policy on mining and land tenancy regulation is beset with a number of loopholes. As Mukhim (2005, 3) writes:

> Meghalaya has a Land Transfer Act (LTA) which does not allow non-tribals to purchase land in the state. But the Act has a clause, which states that land can be alienated in favour of a non-tribal if it is in the interest of the tribal. Several companies have circumvented the Act by claiming that they would provide jobs to unemployed tribals.

The transfer of scheduled land to the French multinational corporation Lafarge in order to explore limestone to feed its cement plant across the international border was undertaken by the state government as a step towards opening trade and investment in the region. However, all this occurred in consultation with the district council and local stakeholders, partners in Lafarge India limited. In contention are the mining rights given to Lafarge Umiam Mining Private limited in the Shela–Ningtrai village in East Khasi Hills. The gross violation of Environment Impact Assessment (EIA) regulations and the access to forest land for mining activity were objected to by the regional office of the Ministry of Environment and Forests after its findings in 2006 that brought mining to a halt in July 2007. However, Lafarge Umium limited had earlier obtained environmental clearance from the district commissioner of East Khasi Hills in 1996 and the district forest officer in 2001. Lafarge filed a petition in the Supreme Court in July 2007. In a judgment

---

more that 95 percent of the coal mines are illegal if one goes by the act of parliament passed by the legislature. All private mines operate without a reconnaissance permit from the state government.

11   As seen in the *National Mineral Policy 2003*, Section 7. 'Strategies on mineral development – small mines.' Ministry of Mines. Government of India. www.gsi.gov.in/minpol.htm.

passed on 22 December 2007, the Supreme Court has allowed Lafarge to resume mining in the state (*The Hindu Business Line* 2007).

The regulatory contradictions and bureaucratic connivance with multinational interests in mining districts of Meghalaya expose the micro politics of resource use in which local tribal elites becomes co-partners in the appropriation of mineral resources at the expense of local communities' rights to exploit their own mineral wealth. The illegality of coal mines in Meghalaya creates new ideologies of bureaucratic corruption and misappropriation of resources at the expense of artisan small scale miners' rights to use these resources to maintain sustainable livelihoods. As Lahiri-Dutt observes (2007), Meghalaya, where coal has been classified as major mineral, keeps nearly thirty thousand people who are dependent on mining in extra-legal or non-legal circumstances because coal can only be legally excavated by large organisations such as state-run corporations and multinationals.

The contradictory legal framework governing mineral resource prospecting, use and exploration has created a practical situation in which illegal mining is governed by tribal elites who have learnt the art of co-option and ways of safeguarding their interests – often in the name of customary right – while declaiming the complex and ambiguous legislation as the main shortcoming in protecting tribal peoples' rights to resource use. The Khasi community is no longer egalitarian, but rather is stratified with interest-ridden community stewards – money lenders, coal mine owners, state bureaucrats, rent-seeking landlords, political intermediaries, party workers, and contractors. Since the formation of Meghalaya state in 1972, tribal elites have consolidated their power through an expansion of public sector reserve seats in Parliament, and the capture of development funding allocated to the region from the central government. New democratic institutions such as the District Council, a substitute to the Panchayat Raj system in the schedule areas (Sixth and fifth Schedule states), where the traditional institution heads – the *Siems* and *Dolois* – who preside over the *Darbars* (village councils), have played a critical role in shaping the cultural politics of land use in the tribal state.[12] The question of access – discussed later – is important because it is critical in maintaining equality in the distribution of resources (Ribot and Peluso 2003). It is also important in understanding communally owned land, community forest and mineral tracts, subsequently privatized, so that certain groups claim permanent rights over what were once commons, disenfranchising others in the process (Bell 1998). In Jaintiya hills, the transformation of community land to private land has occurred in subtle ways, as commons only exist in the collective memory of the people. Land is tightly controlled and tied up in private hands. This has been made possible

---

12   The *Darbars* are the traditional institutions of decision-making in Meghalaya and lie at the bottom of the hierarchical ladder, at village level. However, no political actor tries to violate their judgment. *Darbars* are heterogeneous. In some areas of the Khasi Hills they can have as many as five different levels: clan level, village level, clusters level, 30 village levels, and executive level.

by changes in the local value system; by people changing their attitudes towards land; and by the growing economic importance of coal in the life and culture of the indigenous community. Social prestige has given way to cash income and the value system has changed as a result. The people who own mines are also representatives of the community and act as brokers in translational coal trade with multinational companies. For example, In the Lafarge Umium limestone Mining Project, 29 percent of the stakes are owned by two tribal chiefs (*Down to Earth* 2007).The illegality of mining creates spaces in which local actors can define strategies that eliminate access for other members within the community; customary laws are modified to suit the interests of vested groups who represent powerful patrons in the tribal community (Baruah 2003; McDuie-Ra 2007b).

The political economy of coal mines is closely associated with access control and access maintenance (Ribot and Peluso 2003, 156). In Jaintiya Hills, access and maintenance of coal lie in the hands of the community, but are held by the community institutions – the village councils or *Darbars*. Here, maintenance and control are complimentary, since the community members who maintain communal land also retain the greatest access to land (Dev and Baruah 2003). Legality and legislation cannot fully define who can use the commons. In Meghalaya, it is ecological decline that has become the primary discursive term that regulates perceptions and debate regarding coal mining.

Small-scale artisan mines located in frontier territories have historically been neglected. However, today these mines have become the centre of contestation between the state and the community. Defining the legal status of these mining claims and counter claims over user rights has been complicated by ill-defined legislation and indiscriminate exploitation of resources; the resultant pressure to regulate coal mining by state legislation and autonomous communities' resistance are underpinned by overlapping boundaries of the legality and illegality of mining in these areas.

It is my hypothesis that the extra-legal space created by the national mineral policy in Meghalaya breeds new ideologies of cash inflow and capital formation that contributes to the mushrooming of pit hole mines. The ambiguity in legislation, conflicting institutional regulation and flexible land tenure rights runs the risk of serving the interests of a few while risking the livelihoods and ecology of the Khasi village communities. The emergence of civil society institutions in the form of pressure groups after the formation of Meghalaya as a separate state in 1971 has had a major impact in local decision-making. They often maintain a close association with state bureaucrats and political elites and emergences of civil society, institutions in the form of pressure groups after the formation of Meghalaya as a separate state in 1971 has had a major impact in local decision-making. They often maintained a close association with state bureaucrats and political elites and are influential in public policy-making. My contention is most simply enumerated in three parts. First, given the extra-legal nature of Meghalaya's coal mining activity, benefits accrue to powerful merchants and coal mine owners rather than to forest villagers and mine worker immigrant labourers who bear the ecological and social cost. Second, this skewed distribution of benefit results from manipulated customary practices by the village chiefs and the elected bureaucracy. Third, resistance to state control over local resources in light of their degradation is a misrepresentation of subaltern voices: such resistance is invariably constituted by the co-opted voices of the poor reflecting the protection of elite interests in extra-legal mines.[13]

Coal mining in Meghalaya has evolved as a vibrant economic activity: it appropriates migrant labourers, who in the discourse of cultural politics are redefined through the narratives of identity crisis and indignity. The relationship between the local people and state/multinationals is complex. The hegemony of the state is not well-defined. People's resistance is co-opted by vested interests which ostensibly represent the community as guardians of cultural identity. The misrepresentation lies in the local politics of resource-use and the dilemma of who owns those resources – the state or the community? and not on how the resources are used whom they benefit.

Coal Mining and Miners: In the Pit Holes of the Plateau

Besides the practice of swidden cultivation (*jhum*), coal mining is the primary economic activity of Meghalaya. The history of coal mining in Meghalaya dates back one-and-a-half centuries to when coal was first mined in the Cherra area, locally known as Sorha. Today these mines are abandoned and mining activity has moved on to other parts of the state. Colonial records suggest that coal mining was begun in the Khasi Hills in 1840 by Captain Lister, an agent of the East India Company.[14] Since then, coal mining has been conducted using traditional ways – primitive, hazardous and crude. In rat-hole mining, land is first cleared by cutting and removing the ground vegetation. Pits ranging from 5 to 100 meters are dug to reach the coal seam. Thereafter, tunnels are cut into the seam sidewise to extract coal, which is first brought into the pit by using a conical basket or a wheelbarrow, and is then taken out and dumped. The Constitution of India guarantees indigenous peoples' rights over control to land and forests, including natural resources found within that land. Coal reserves are thus intended by the Constitution to be under the direct control of the indigenous communities who extract and export it to Bangladesh and mainland India.

In the past three decades exponential growth in the export of coal has led to rampant and unorganised extraction by private mine owners seeking further coal resources, resulting in severe ecological constraints in the plateau region that houses more than 600 plant species. The toxic litter from abandoned coal mines pollutes the fresh water bodies. Much of the natural science research into

---

13   See also J.C. Ribot. 1998. 'Theorising Access: Forest Profits Along Senegal's Charcoal Commodity Chain.' *Development and Change* 29(2).

14   See E.P. Phillemon. 1995. *Cherrapunjee the Area of Rain*, 23.

the region examines the implications of contamination caused by gas and acid spills over land and water, and its effect on human settlement surrounding the coal mines (Rai 2002; Singh 2005; Singh 2003). Corrective solutions to land pollution and issues concerning the proper environment impact assessment of coal mines have also been considered (Tiwari 1993; Shankar, Boral, Pandey and Tripathi 1993; Swer and Singh 1994). However, there is no significant work on the mining communities engaged in the informal small scale mines of Meghalaya. It is important to understanding their socio-economic status and livelihood strategies, given the complex land tenure and private ownership of coal mines. Often the mining ridges owned by communities as commonly perceived are occupied by local elites who lease out their land to contractors who arrange for extraction of the coal and who pay a fixed share to the mine owner. In such contractual agreements the mine workers are disadvantaged. The contractor shares the profit with the owner and the labourers are not paid their wages. According to the survey conducted during my fieldwork in Laitrumbai, Jaintiya Hills, for extraction of 5 tonnes of coal, which may take a month, a labourer is paid around INR 2000–4000. Small cottage-type coal mining is prevalent in different areas extending from Jaintiya Hills, East and West Khasi Hills in the east to East Garo Hills in the west.

The following section narrates what contemporary research has seldom sought to consider: the life stories of coal miners and their relationship to the mines.

The Inside Story: Mines and Workers

The story of coal miners who work in informal mining is harsh; the work often violates basic human rights. A close ethnographic fieldwork based on field interviews and participant observation during my visit to Laitrumbai reveals how mining communities comprising labourers' migrated from Bihar, Nepal, Assam and Bangladesh, worked under terror and threats of expulsion. While the local mine owners ensure their safety and lure them for work at much cheaper rates than the prevailing market wage rate, the local population provoked by public rhetoric agitate against the non-tribal presence in their territory. The common perception of demographic transformation and colonisation of local territory by immigrants leads to frequent violent flare-ups. The local elites capitalise on the customary rights and mobilise this rationale to prevent state intervention, thus securing their private interest and the status quo of indigenous people's rights to manage local resources. In turn, there are no governmental regulations, safety measures; basic minimum protections from hazards, minimum wage rates or protection from the emerging landed gentry' class.

Mining communities, comprised of mostly migrant labourers, slowly become denizens.[15] They enjoy no political or entitlements rights and are vulnerable to expulsion when the local community sees them as a threat to their cultural identity. The mine owners exploit them, paying wages far below the prevailing market rate and requiring them work in debilitating conditions. Our study in Laitrumbai area revealed that most of the people, especially the Bangladeshi migrants, are willing to work at low wage rates because they are greater than what they would receive for similar kinds of work in their own country. The minor ethnic groups of Rabhas and Hajongs living at the fringe of Meghalaya find no work in their own territory and are attracted to mining during the lean agricultural season.

What needs to be addressed is the long standing debate over who should own resources in tribal domains where colonial encounters were restricted to revenue generation and political administration (Bryant 1996; Guha 1991). This question was raised in the parliamentary debates of 1988 when a local member of parliament representing Meghalaya contested state claims over forest wealth and natural resources, claiming that customary rights cannot be appropriated by state authority. Although the debate ended in favour of the indigenous community, their customary rights to exploit coal were restricted to domestic use alone. However, in spite of such regulations, illegal and unorganised coal mining has increased over the years because of the huge profit earned through its trade. Revenue records of the coal trade with Bangladesh represent only a minor proportion of that trade. Recognition of the illegality of coal mining by the state government motivates the rhetoric of legalizing what is presently an informal economic activity. The illegal market has attracted mafia syndicates that extort money from traders. The people who are sent demand notes are no longer the non-tribal *mahajans*, but the tribes people themselves. This has increased instances of violence, corruption and unrest in the coal mining areas. The local tribal elites and coal mine owners adopt multiple identities of mine owners, community stewards and local politicians. The free trade of coal in the post-liberalisation period has further entrenched neo-liberal politics in the coal mine, where production and exploration are determined by demand and supply variables. All this is achieved by sacrificing the hard labour of mine workers, presented as antisocial criminals who work in the coal mines to escape from state surveillance.

---

15 'Denizens', Sanjib Baruah writes, is not a contemporary legal category. The term goes back to the power of denization that British monarchs once granted to aliens to replicate some of the privileges of natural-born subjects. Denizens, for instance, could buy land, but not inherit it. While restrictions of the rights of the non-tribal population have a very different history and rationale to 'denizens', the particular limitations to the tribal population, e.g., right of property relationship, access to public employment and elected office, are also applicable to denizens.

Ethno Nationalism and Cultural Politics

The colonial categorisation of tribe and non-tribe established divisions between highland and lowland regions and was internalised in the post-colonial context: tribal egalitarianism as opposed to class and caste-ridden peasantry in the plains. The ethnographic representation of tribes has undergone remarkable change both in structure and political practice. With the diffusion of modern education and the involvement of non-state actors such as the church, traits of modernity has infused frontier societies. As a consequence, the post-colonial history of the Northeast is even less served by the misrepresentation of the past as a golden age than it is of plains society (Guha and Gadgil 2004). The protectors of forest and sacred groves are now themselves the perpetrators of environmental violence. While modernity has imbued the tribal peoples with cosmopolitan cultural habits, class-consciousness has also grown. The egalitarian aspects of tribal culture are fast fading. Typical examples of this are the privatization of land and the emergence of landed gentry; and the intervention of commercial and settled plough farming in the Garo Hills as a panacea for developing the swidden (*jhum*) economy. This has led to de-facto control of women's land by male landlords, thereby substantially altering the gender balance in civic and social authority in what was traditionally a predominantly matrilineal social structure (Agarwal 1994). These transformations are significant, but they are easily obscured by the populist ethno-politics and ethno-national discourse prevalent in contemporary khasi-Garo society (McDuie-Ra 2007c).

Emerging inequalities in the tribal communities concealed the micro politics attaching to resource alienation and elite profit-making. Although with the establishment of the District Council the traditional tribal chiefs in Khasi society have been politically disenfranchised from administering their territories, they claim social sanction through democratic and decentralised electoral practice. Nonetheless, any development initiative can only be initiated with the consent of the village headman and village elders. In many cases the social authority wielded by traditional village headmen has been translated into practices of landlordism and control over mining and forest resources. The transformation of the social order no longer hinges on matters of equity, as is commonly portrayed in the popular media. The lack of well-documented studies on changing social relations in Khasi society preserves the notion of a community that is undifferentiated and egalitarian. The pre-colonial political system and social relations were defined by reciprocity and redistribution between the village chiefs (*Siems, Doloi*) and their clan members. With the extension of colonial administration over these territories in the late 19th century, the prevailing hierarchies based on kinship have been redrawn through relations based on monetised exchanges of goods and services and the replacement of the traditional political structure with one based on colonial practices of patronage and power.

Much early anthropological and sociological studies have shaped our understanding of tribal societies through static, decontextualized historical representations. Leach's research into Kachin social structure is a path-breaking study that contested earlier ethnographic representations of indigenous tribes of highland Burma. Leach proposed that primitive societies are constantly transforming and that a conceptual model of equilibrium exists only in theory (Leach 1964). The sharp demarcation of tribes and non-tribes in the history of the Northeast has been a construction of colonial rule, which shaped discourse at the margins and the centre of political power. Anthropological knowledge was structured to serve the interests of empire. Post-colonial anthropological research has carried on the colonial tradition of identifying tribal identities along ethnic lines and advocating protectionism (Elwin 1964). During this process a significant transformation that has been taking place within the community has been undermined. As Baruah (2005) writes with reference to the Khasi community, what is formally clan-controlled land has been appropriated by powerful individuals. This transformation from egalitarian social formation of primitive accumulation to more broad-based capital accumulation and class consciousness in the tribal economy has been sustained by a political culture based on patronage and power in contemporary Khasi society.

Colonial Intervention and Customary Laws

In the context of Meghalaya, Baruah (2004) argues that it is the allocation of modern political function and land distribution powers to the *Darbar*s that has allowed elites to appropriate village land. While others such as Nathan (2000) trace it to the colonial administration, which elevated the status of *Siems* among the Khasis to that of landlords by holding them accountable to collect tax. This was not unique to the Khasi Hills: in 1876 tribal chiefs of the Chittagong Hill Tracts were made responsible for collecting revenue from their subjects (IOR 1876). This new practice introduced the monetised economy to Khasi society. Whether land relations in the Khasi society were predominantly feudal remains contested; colonial ethnographic works do not reflect much on the economic relations of the monarchies that existed within Khasi society. What can be determined is that relations were shaped by the power inequalities within families. In the post-colonial period, the attachment of economic interest to land became evident with the exploration of minerals, timber and charcoal trade, which initiated private *Darbar* ownership of land. However, this was achieved through manipulation of the customary laws by the rural as well as the emerging urban elites who represented the educated English-speaking white-collar job seekers in the public services. The vesting of rights to grant land to individuals on the Village Council by the Sixth Schedule allowed District Councils and the *Darbar*s to determine the beneficiaries. The process of privatising community-owned land has culminated in exclusive collaborative arrangements whereby powerful rural elites work tactfully and discretionarily to direct the proceeds of wealth.

There are several ways that community-owned lands can be privatised in the state of Meghalaya. Usually, if the occupier and/or cultivator of the common land can show that they have improved the land over a period of three years, they are entitled to claim ownership at the discretion of the *Darbar*. In future, they receive every right to personal claims and transfer of land by sale or de facto deeds (*benami* transactions). Much of the communal land in coal belts of the Khasi and Jayantia Hills have been privatised through such acts of customary transaction and claim to land-use.

Direct outcomes of this change in land relations are rural indebtedness, landlessness within the Khasi community, and the prevalence of absentee landlordism. As McDuie-Ra (2007) shows, in Laitrumbai, the busiest coal mining region of Meghalaya, land is leased out to contractors as the owners have resettled in the urban centres with the flow of cash from coal money, while a major section of the rural population who could not develop their land are denied rights to permanent settlement by the *Darbar*. With the rise in private land ownership, farmers are losing their claim over common land, leading to landlessness. However, there are no official statistical figures to substantiate such outcomes. This is so because of the resistance of the *Darbar*s to any kind of revenue survey that challenges their hold on land. Nongkynri (2002) observes that in the absence of proper state or district-level data, generalisations are difficult to make. However, some small surveys indicate the persistence of the growing crisis in rural Meghalaya and areas where coal and limestone mining are conducted. The study by the Kharot Snong Social Organization (KSO) on landlessness and poverty conducted in 1999 in the Kharot Snong area of the East Khasi Hills is revealing. According to the survey – conducted across 2000 households in 40 villages – a quarter of the villagers used common land. A further study by the KSO in 2003 in one village from each of the seven districts of Meghalaya reveals that landlessness was increasing; the price of land was increasing to lease and buy, and these factors were increasing rates of deforestation.

The changes in land ownership from communal land to private land lordism has resulted in rent seeking, share cropping, land mortgage, landlessness, and other forms of private control (Karna 1990).This has also given rise to a prevalence of middlemen among the rural elites who engage in buying and selling land and in gaining loans at exorbitantrates. As Baruah (2003) writes, the moneylender is no longer the cunning non-tribal urbanite taking advantage of the simple tribal farmer; today, the *mahajan*s (traders/moneylenders) are as tribal as the village folk and as cunning as the non-tribal moneylenders of the past. Questions of marginality, disempowerment, ecological destruction of the Northeast, and the concentration of power in private hands remain deeply contested. However, what remains is to examine how capitalism and the roots to inequitable social order are created and sustained under the rubric of neo-liberal economic order. It is the task of ethnographic knowledge to reconstruct the socio-cultural history of indigenous society, its evolution and structural transformation at the point at which it engages with modernity.

The cultural politics of Meghalaya foster resentment as rural elites connive to secure indigenous peoples' rights over local resources. As has been highlighted by Duncan McDuie-Ra, local cultural groups are contesting the ownership rights to control over land and its resources. Given a chance for private ownership over land, community stewards acknowledge that mining will no longer be a problem (McDuie-Ra 2007b). This has been the case in the coal mining areas of Khasi and Jaintia Hills, where arrangements for de facto and de jure individual ownership of land have led to the creation of private property and the subsequent proliferation of mining activity. The end result has been the creation of private property – through land consolidation and individual ownership – where the earlier structure of community owned land has been lost.

The lack of regulation over privately-owned mining has transformed what was once community owned land into mine-damaged landscape necessitating institutional intervention. The state sees these mines as illegal. The definition of legality in relation to local practices is relevant here. However, the differentiation between legal and illegal is complicated by the legitimation of such practices by the state government, which collects revenue on exported coal.

Mining Legalities in the Pit Hole Mines

The issue of legality in mining occupies a unique position in Meghalaya, where no large-scale state mines exist. As Scott (1998) writes with reference to shifting cultivators as illegal migrants, the coal miners of the state are also illegal migrants as their territories do not fall under cadastral revenue settlements. To understand this issue we need to consider the colonial encounter with native societies and how colonial authorities appropriated territory through the discourse of protectionism. In the Sixth Schedule states, mineral exploration is primarily concentrated in the hands of the local people with customary rights to land-use. This departure from mainstream large-scale coal extraction in the Northeastern frontier should be seen from a historical perspective of colonial strategies in managing the Assam frontier tracts. In the Sixth Schedule states the processes of mineral extraction are primarily concentrated in the hands of the local people with customary rights who govern land use and land relations. This unique state of affairs in Meghalaya is the result of special administrative arrangements initiated by the British for military strategic reasons rather than for reasons relating to revenue settlement in the plains. These strategies of governing the frontier were continued by the post-colonial state, which today confronts seemingly irreparable contradictions in authority structures, definitions of rights to use subterranean and surface resources,and the civic role of communities who until recently have remained marginal to the processes of post-colonial state-formation.

Until recently the strategies of governance used in the region permitted only limited interference by community members and did little to improve the living conditions of the local inhabitants. Past improvements were strongly linked to revenue generation that could sustain colonial administration. When coal mining first began in Meghalaya in the 1840s the colonial government was reluctant to improve the local mines due to a lack of incentive. A British official by the name of Darley – the acting superintendent of the Cherra mines – carried on excavations until 1951. An 1856 report stated that coal mining in Lakadong, Jaintiya hills, could only be made profitable if ancillary activity such as ceramic production were developed. The result was that coal mining ceased completely. The same plight was faced by Assam's tea gardens when Robert Bruce developed Chinese tea plantations in the Brahmaputra valley (Sharma 2006). The colonial Frontier Regulations of the 1880s – which demarcated the highland territories from the lowland – established the seeds of identity politics that embroil the contemporary ethno-landscape of the Northeast, and Meghalaya in particular.[16] The Inner Line Regulation of 1873 was a political necessity for the colonial frontier administration that helped to create secured enclaves in the lowlands where colonial capital could safely transform virgin tropical forest into profitable tea gardens, wastelands, reserves and mining tracts. This historical background is significant to the status and interrelations of legality and illegality for coal mines in the post-colonial environment.

While inflexible and stringent mining regulations determine the illegality of coal mining in Meghalaya, they rarely intervene in the cultural landscape of the tribes people that restricts state intervention in their daily practices. To a significant degree, the emergence of informal coal mining is the outcome of illicit trade. However, this does not explain the reason for the proliferation of small-scale informal mining in the Meghalaya plateau. The small scale rat-hole mining that has developed relies on the protection of the Sixth Schedule that does not allow land to be alienated from the tribes. The colonial government found it difficult to grant mining leases to private individuals as if often encountered contestation for surface land rights claimed by individual owners. Thus the status of ownership and the rights for use have remained ambiguous and have never been properly defined. The situation continues, placing the small mines of Meghalaya and the rest of the Northeastern highlands in an undefined, extra-legal category.

Coal Calculations and Changing Land Relations

In contemporary Khasi society the rights to ownership of land are tightly bound up with the customs of the local people, individual clans and landowners. The existence of community owned land in the coal-mining belt is at best marginal and virtually non-existent. McDuie-Ra (2007) has concluded that a significant transformation has occurred in the land tenure system of Meghalaya that can only

---

16   Racialist assumptions regarding the capacity of hill peoples to obey colonial regulations led to the *Frontier Track Regulation II, 1880*, specifying that the operation of suitable laws might be barred in all the hill districts: see E.A. Gait, *History of Assam*.

be observed in the field. The traditional nomenclature of land as either *Ri-Kinti* or *Ri-Raid* no longer exists in practice. The general trend has been a shift from *Ri-Raid* land ownership (community owned land) to *Ri-Kinti* land ownership (private land). How private ownership to land has evolved to supersede community land is described by a complex historical process in which colonial administrative intervention acted as a catalyst (Nathan 2000). Some argue that the present situation is a continuation of the process initiated by the British, who elevated the *Syiems* to a type of landlord and introduced land taxes, while others have attributed current structures to the power exercised by the lineages and the power dynamics within clans, which has resulted in some clans becoming more powerful landowners than others.

Human insecurity and growing marginality is not embedded in identity crises or threats of demographic ruin from immigrant denizens, but rather is located in growing land alienation, resulting in economic disparity in mining territories (McDuie-Ra 2006, 2007 a, b). The emergence of new elites within a presumably egalitarian tribal social system is not unique to Meghalaya. A wealth of ethnographic studies among the indigenous Indians in Latin America, particularly in Peru, Ecuador and Bolivia depicts the subversion of community response in resource rich areas where oil and gas reserves in forest territories have led to venture capitalism. In their study of Ecuadorian indigenous Indian communities, Rival (2002) and Suwayer (2004) examine the local politics of the communities and how they understand translational interests and struggles to obtain stakes in their resource-rich land. As Rival argues in the context of her study on the Huaorani Indians, people's resistance may often involve an element of subversion. She looks at the carbon trade phenomenon where large multinationals buy rights to protect forest in lieu of payment to the indigenous communities. The Huaorani have signed such contracts with multiple corporate entities while fully aware of the likelihood of their future violation. The claims of marginalized tribes thus become deeply contested. In Meghalaya, a similar context is reproduced; the Khasi elites are conscious of the ecological constraints that coil mining will bring to the local landscape, yet it is the drive towards easy capital that negates sustainable choices of coal extraction and places the community in danger.

The notion of frontier ignorance is often inaccurate; it is negated by the negotiated micro politics of community stewards, who are knowledgeable regarding global politics and corporate interest, and who compete to gain control over indigenous resources. Scott (1985, 1990) falsifies the classic notion of resistance to state power and coercion: resistance can often dissolve into subversion oft he voices of local people inhabiting extra-legal spaces. In extra-legal spaces and frontier regions the state maintains a diffused, weak inter-locus of administrative control. However, geo-political interest represents the binary of state power and strong border control – a strong military presence with limited access to control over local resources and desire for improvement. However, this simplistic binary can be contested. The extra-legal nature of tribal territory at the frontiers breeds new ideologies of power that are embedded in the daily practices of subjects. As Das and Pool

(2004) argue, the margins are not mere territorial representations of the extent of state power, but they are also – and perhaps more importantly – sites of practice at which laws and other state practices are colonised by other forms of regulation that emanate from the pressing need of the population to secure political and economic survival. The margins are sites of contestation between customary and state laws. The illegibility of state presence in frontier tracts does not constrain state power, but is experienced and undone through the illegality of its own practices. The illegal trade of coal as it crosses the international border represents how legibility through levying of tax is undone in presence of state institutions. However, it does not constrain the exercise of state power.

Conclusion

Given the naivety involved in land tenureship in Meghalaya's coal mining belt, the state, community leaders, village elders, democratic institutions (both traditional and modern), state bureaucrats and pressure groups play important roles in shaping the discourse on rights and access to resource use. While tension exists between these institutions due to overlapping rights of control, coal mining in Meghalaya presents a unique case study wherein local political and bureaucratic practices have attributed new meaning to extra-legal coal mining. Resistance to legislation derives from political and community actors who have been successful in appropriating resources. The coal lobby has initiated political changes in the past. However, as a recent example of the setbacks such lobbying confronts,after the Supreme Court gave its directive to the state government in July 2006,the committee that was constituted to frame a minerals extraction policy for the region has yet to see the light of day. Instead, the Ministry of Geology and Mines is now headed by a local legislator who is himself a coal trader.

Although the extra-legal nature of coal mining has raised public debates time and again, such mining practices have survived the pressures exerted by the central government seeking nationalisation under the patronage of interested local politicians, bureaucrats and community elders. Recent legislation in the post-liberalisation period has drawn increased complexity into the debate. Recent amendments to state industrial policy have allowed big capital to venture into mineral exploration on land leased in tribal areas. The issuing of exploration leases to one such multinational firm, Lafarge-Umium Mining Corporation, in 2006 has exposed the close relation of interests between the tribal privileged class and big capital. While the extra-legal nature of coal mining is fought over and protected by the coal lobby, political actors have allowed foreign investment to flout local customary laws in the name of trade promotion.

The Lafarge Limestone Mining Project, which was recently swamped by controversy, is a case in hand. Lafarge Surma, a subsidiary of Lum Mawshun Minerals, has secured a 35-year, 100-hectare lease agreement with the villages of Shella–Nongtrai formining, and the liberty to use another 26.6 hectares for mining-related activities. However, land lease obtained manufactured environmental clearance and a misguided Rapid Environment Impact Assessment by a private consultancy firm. Even the *Land Transfer Act* was relaxed in 2001 by the state government to allow Lafarge to acquire land in the Sixth Schedule state. The company has acquired tribal land otherwise not accessible to outsiders and has mortgaged to international banks for a loan of USD 225 million to build cement plants across the border in Bangladesh. Today, Lafarge Surma owns 74 percent of this company and the remaining share is owned by two Khasi tribesmen.

In conclusion, I note that population mobility is not new to the Northeast. It has a genealogy, a social history; links of which can be traced in colonial extractive projects that introduced foreign labour, monetized the local economy and produced identities that remain contested and ethnically polarised in contemporary times. In this regard coal mining in Meghalaya has become a lucrative economic activity, and the recent drive to open up the region to its neighbours will have limited success if the social change within the indigenous Khasi community and the lure of oligopolistic control over coal trade are not taken into account. The populist ethno-politics that claims immigration as a threat and challenge to the identity of the indigenous tribes is contested in this chapter. Immigrants' weak social and political locus in Sixth Schedule areas is effectively used by tribal elites to exploit their labour for the production of capital in the coal-production belt. In retrospect, we need to engage critically with changing land relations in Khasi-Jaintiya mining villages and revise the notion of tribal community as egalitarian and undifferentiated social spaces.

References

Agarwal, Beena. 1994. *A Field of One's Own: Gender and Land Rights in South Asia.* Cambridge: Cambridge University Press.

Bareh, Hamlet and R. Kyanti, eds. 2001. *The Economy of Meghalaya: Tribes in Transition.* New Delhi: Osmond Publication.

Baruah, A.K. 2004. 'Ethnic Conflict and Traditional Self-governing Institution: A Study of Laithumokhra *Dorbar.*' Working Paper No. 39. London: London School of Economics Crisis State Programme.

Baruah, Sanjeeb. 2003. *Durable Disorder: Understanding the Politics of North East India.* New Delhi: Oxford University Press.

Bell, D. 1998. 'The Social Relation of Property and Efficient'. In *Property in Economic Context*, edited by Robert C. Hunt and Anthony Gilman, 29–45. Lenham: University Press of America.

Bryant, Raymond L. 1996. *Political Ecology of Forest in Burma.* 1824–1994. New Delhi: Oxford University Press.

Das, Veena and Deborah Pool. 2004. *Anthropology in the Margins of the State.* Santa Fe and Oxford: School of American Research Press.

De, Utpal Kumar. 2007. 'Dynamics of Coal and Limestone Extraction in Meghalaya: Comparative Analysis.' Munich Personal RePEc Archive Paper No. 5678, 9 November 2007.

Dev, Rajesh, and Baruah A.K. 2003. 'Local Liberal Democracy, Traditional Institutions and Politics of Representation: Analysing the Nangkynrih Shynong Dorbar.' Unpublished paper, Johannesburg Crisis State Workshop.

Down to Earth. 2007. 'Meghalaya Limestone Quarries Closed.' Delhi, 25 December.

Elwin, Verrier. 1964. *The Nagas in the Nineteenth Century*. Bombay: Oxford University Press.

Frontline. 2006. 'Coal Calculation.' 23(13): 1–14.

Frontline. 2007. 'Invisible Immigrants.' 24(11): 2–15.

Gadgil, Madhav and Guha, Ramchandra. 2004. *Ecology and Equity: The Use and Abuse of Nature in Contemporary India*, New Delhi: Oxford University Press.

Guha, Ramchandra. 1991. *The Unequal Woods: Ecological Change and Peasant Resistance in Himalayas*. New Delhi: Oxford University Press.

Hafner, Robert W. 1990. *The Political Economy of Mountain Java: An Interpretive History*. Berkeley and Los Angeles: University of California Press.

IOR (India Office Records). 1876. 'Territorial and Fiscal Jurisdiction of the Hill Chiefs of Chittagong: Correspondence of the Revenue Administrator of the Chittagong Hill Tracts', Calcutta, Minute by the Lieutenant Governor General of Bengal, dated 5 July 1876, MF 957, British Library, India Office Records.

Karna, M.N. 1990. 'The Agrarian Scene.' *Seminar* 20(366): 30–38.

Khan, Joel S. 1999. 'Centralising the Indonesian Uplands.' In *Transforming the Indonesian Upland*, edited by Tania Murray Li, 79–104. Amsterdam: Harwood Publication.

Lahiri-Dutt, Kuntala. 2004. 'My God Gives us Chaos, so that we can Plunder: A Critique of Resource Curse and Conflict.' *Development* 49(3): 14–21.

Leach E.R. 1964. *Political System of Highland Burma: A Study of Kachin Social Structure*. London and Atlantic Highlands: The Athlone Press.

Li, Tania Murray. 1996. 'Images of Community: Discourses and Strategies on Property Relations.' *Development and Change* 27: 501–27.

Li, Tania Murray. (ed.). 1999. *Transforming the Indonesian Uplands*, Amsterdam: Harwood Publishers.

Li, Tania Murray. 2000. 'Articulating Indigenous Identity in Indonesia: Resource Politics and the Tribel Slot.' *Comparative Studies in Society and History* 42(1): 149–79.

Li, Tania Murray. 2007. *The Will to Improve: Governmentality, Development and the Practice of Politics*. Durham and London: Duke University Press, London.

Mackenzie. A. 1884. *History of the Relation of Government with the Hill Tribes of the North Eastern Frontier of Bengal*. Delhi, Mittal Publication.

McDuie-Ra, DA. 2006. 'Civil Society Organization and Human Security: Transcending Constricted Space in Meghalaya.' *Contemporary South Asia* 9 (2): 35–53.

McDuie-Ra, DA. 2007a. 'Anti-Development or Identity Crisis: Misreading Civil Society in Meghalaya.' *Asian Ethnicity* 8(1): 43–59.

McDuie-Ra. DA. 2007b. 'Civil Society and Human Security in Meghalaya: Identity, Power and Inequalities.' Unpublished PhD thesis, Department of International Studies and Politics, University of New South Wales.

McDuie-Ra. DA. 2007c. 'The constraints of Civil Society beyond the State: Gender Based Insecurity in Meghalaya, India.' *Voluntas* 18(1): 359–284.

Mitchell, Timothy. 2002. *Rule of Expert: Egypt, Techno Politics, Modernity.* Barkley, University of California Press.

Moore, H. and M. Vaughan. 1994. *Cutting down Trees: Gender, Nutrition, and Agricultural Changes in Rural Zambia, 1890–1990.* London: James Curry.

Mukhim. 2005. 'Industry versus Environment.' *The Telegraph*, Shillong, 1 February 2005.

Nathan, D. 2000. 'Timber in Meghalaya. *Economic and Political Weekly* 35(4): 182–87.

National Mineral Policy. 2003. Ministry of Mines, Government of India. http:// mines.nic.in/nmp.html Accessed 26 December 2007.

Nongkynri, A.K. 2002. *Khasi Society of Meghalaya: A Sociological Understanding.* New Delhi: Indus.

Ribot, J.C and N.L. Peluso. 2003. 'The Theory of Access.' *Rural Sociology* 68(2): 153–81.

Rival, Laura M. 2002. *Trekking through History: The Huaorani of Amazonian Ecuador.* New York: Columbia University Press.

Scott, James C. 1985. *Weapons of the Weak: Everyday Forms of Peasants Resistance.* New Haven, CT: Yale University Press.

Scott, J.C. 1990. *Domination and the Art of Resistance: Hidden Transcripts.* Yale, NY: New Heaven.

Scott, J.C. 1998. *Seeing Like a State: How Certain Schemes to Improve the Human Condition have Failed.* New Haven, CT and London: Yale University Press.

Schendel-Van, Willem. 2005. *The Bengal Borderlands: Beyond State and Nation in South Asia*, London: Anthem Press.

Shankar, U., Boral, L., Pandey, H.L and Tripathi, R.H. 1993.'Degradation of Land due to Coal Mining and its Recovery.' *Current Science* 65(9): 680–87.

Sharma, Kiranmoi. 2003. 'Impact of Coal Mining on Vegetation: A Case Study of Jaintiya Hills of Meghalaya, India.' Unpublished Masters Dissertation submitted to Indian Institute of Remote Sensing, Dehradun.

Sharma, Jayeta. 2006. 'British Science, Chinese Skill and Assam Tea: Making Empires Gardens.' *Indian Economic and Social History Review* 43(3): 429–55.

Singh, O.P. 2005. *Mining Environment: Problems and Remedies.* New Delhi: Regency Publication.

Suwayer, Suzana. 2004. *Crude Chronicles: Indigenous Politics, Multinational Oil, and Neoliberalism in Ecuador (American Encounters/Global Interactions).* Durham, N.C.: Duke University Press.

Swer, S. and O.P. Singh. 1994. 'Status of Water Quality in Coal Mining Areas of Meghalaya.' *Proceedings of the National Seminar on Environmental Engineering with Special Emphasis on Mining*, edited by Inder N. Singh and Manik K. Ghosh, 88–101. Shillong: Shillong College.

The Shillong Times. 2006. 'Save Cave Campaign: Government Apathy Flayed.' Shillong, 28 July.

The Hindu Business Line. 2007. 'Lafarge to Resume Meghalaya Mining this Week.' Chennai, 26 November.

The Pioneer. 2006. 'Meghalaya to Clean Up Coal Mining Act.' Delhi, 10 August.

The Telegraph. 2006. 'Northeast Echoes: Threat to Environment.' Kolkata. 1st of August.

Tiwari, B.K. 1993. 'Studies on Environmental Impact of Coal Mining in Jaintia Hills District.' Sponsored by Meghalaya State Pollution Control Board, Shillong.

Tsing, Anna Lowenhaupt. 1993. *In the Realm of the Diamond Queen*. Princeton: Princeton University Press.

Chapter 5

Slaughter Mining and the 'Yielding Collier': The Politics of Safety in the Jharia Coalfields 1895–1950

Dhiraj Kumar Nite

This chapter is concerned with the problem of mining accidents and the question of safety politics. While reporting on the circumstances responsible for dangerous mining, Chief Inspector of Mines R. R. Simpson summed up the problem of the development of safety wisdom among miners (Simpson 1928, 19):

> It is regrettable that the discipline maintained in Indian mines is far below the standard attained in other countries. During the year at a certain coalmine, a miner went down the mine after absence from work for a fortnight. Instead of going to the *sirdar* for instructions, he drilled a hole and fired a shot in the corner of a pillar of coal, and was the cause of another miner being injured. ... Most of the labourers in Indian mines are illiterate and ignorant men to whom ideas of discipline are something new and difficult to comprehend. They are, however, amenable to discipline, and it is one of the most important duties of a manager to train the labourers in 'safety first' principles both by personal example and through subordinates. At many mines, a great deal has already been done, and at this mine after a change of management, improvements were effected. ... In the course of an enquiry into the circumstances, it was disclosed that during the night shift on which the accident took place there was no underground official on duty, and this had come about because the night shift overman had been acting for the day shift overman who was ill and the night shift *sirdar* had been taken ill and had failed to report his ability to go on duty.

Simpson's anguish draws attention to problems faced by the Indian collier, and presents the need to re-fashion particular mining behaviour patterns. Enquiry into the meaning of colliers' work beliefs, their experience, and their work-behaviour patterns is pertinent in understanding the precarious existence of a mining community. Through a study of this concern and safety politics, connected with an appropriation system, this chapter aims at unraveling the mining regime laid down by social forces in the Jharia fields.

The work-hazard is a manifestation of the discordant feature of an appropriation system adopted by the industry, and it reveals how the mining world, while

offering many livelihood opportunities, saps life energy from these people.[1]
The public counter-action and the necessity of production process in terms
of restorative measures, investment in new tools and enhancement of mining
knowledge to ensure safe mining, provide the conditioning environment for
these people. The prosaic discourse of accidents describes them as 'unavoidable'
geophysical occurrences. Recasting against it and that of the explanation focused
on technical and mechanical factors (Kirby 1933), some scholars concentrate on
the responsibility of management[2] and labourer for 'unsound' mining methods.
The management and the inspectorate argued that 'illiterate,' 'ignorant' and
'agriculturist' mineworkers frequently indulge in 'reckless' and 'unscientific'
mining. Opposed to such a behaviourist argument, some suggest that unfriendly
and despotic terms and conditions of employment in the mines led to faulty mining
practices (Nair 1998; Lege 1992, Onselen 1976). Concentrating on autonomous
functions of outlook and politics, Mukhopadhyay (2001) and Das (1988) point
out that having developed an 'economistic' and 'fatalistic' outlook, the 'ill-fed',
'ignorant' and 'pliable' mineworkers sought only 'ephemeral relief' through
desertion and claimed petty economic compensations. I would suggest that
mineworkers sought to avoid the demands exerted by poor labour processes, the
appalling nonchalance for accident control, and the means by which the corporate
politics of safety subdued workers' militancy aimed at addressing perilous mining
conditions.

A Hazardous Labour Process

At the turn of the twentieth century, industrialists and colliers re-fashioned a
new social life in the coalfields of Jharia once was it connected with wider coal-
consuming areas. Most of the colliers immigrated from rural backgrounds. A small
section of them and other managerial and technical staff gained mining experience
in Raniganj and the Giridih fields, or in England. Proprietors of capital derived
from a wide range of professions, including the older Indian and British mining
areas, trades, agriculture, banking, construction and labour, and the moneyed
professions.[3] Coal companies did not include mining experts on their boards of

---

1   For a discussion on this, see Nite (2010), chapter 'Classifying Work-hazards'. The
statistics prepared by the Inspector of Mines indicates that, on average, one and half to
two mineworkers per thousand employed lost their life, while five to eight mineworkers
suffered serious bodily impairment. The instances of minor injury and occupational disease
did not receive much statistical attention in this period.

2   Basu (1988); Ghosh (1988); Gupta (1988); Padhi (2003); Pathak (1982); Sahu
(1988); Singh (1988); Sinha (1982); Valdiya (1988).

3   A few of them were fresh, humble adventurers – especially Indian proprietors.
See Indian Mines Federation (1963: 17–19, 247). The Bengal Coal Company Limited,
formed in 1843 in the Raniganj field by a few British and Indian merchants, was the first

directors until well into the 1930s. Inevitably, mining was typically an industry in which a small number of professional personnel controlled a vast mass of human labour power. The managing agencies assisted the coalmines proprietor to secure access to markets of finance, technical personnel and goods (Burrows 1937, 9–32; Kling 1994; Nite 2010a).

In the colliery, coalcutters (*malcuttas*), loaders and trammers were employed to cut coal from coal seams, collecting coal pieces in baskets and tubs and carrying them. As colliery depths increased and mechanisation was adopted, various others were required to operate machines, prop up roofs, construct drainage and ventilation systems, conduct geophysical assessments and surveys, as well as supervise work, undertake payment and dispatch, labour recruitment and other related tasks.

The industry relied on a handicraft or artisanal form of production, i.e., employment of predominantly human labour power and archaic ways of mining.[4] As mining practices advanced, a new generation of better-informed young colliers appeared armed with safety-related wisdom. In the late 1910s Kesho Rawani, employed at the Bhowra collieries, explained to his eldest son, Shyamnarain Rawani:

> at the time of my first underground employment my father used to admonish me to avoid a number of things: work at the working face, tearing off coal from pillar, sitting on tram line, riding on tub, and the likes; rather, for taking care of roof condition and timber's sound.[5]

Joint Stock Company in the coal industry. The East Indian Coal Company (1893), Bengal Nagpur Coal Company (1890) and Khas Jharia Coal Company (1893) were launched by European merchant houses as Joint Stock Companies for exploiting the Jharia coalfields. In contrast, the Cutchi businesspersons were former construction and labour contractors associated with the work of the Grand Chord Line of the East Indian Railway and the Marwari businesspersons were former coal merchants.

4   The handcraft type of labour process is primarily characteristic of the piecework economy usually found in a preindustrial modern society, reliant on labour power rather than mechanical power. In this situation, Marx (1867/1977) suggested, labour remained subject to formal subordination to capital. Modern-industry labour processes helped produce relative-surplus value and real subordination of labour to capital. Burawoy (1985) and Kerr (1997) have pointed out instances of coexistence of the two labour processes in the same worksite, operated in different assembly lines. Kerr maintains that intensification of labour with the help of enhanced foremanship and labour discipline could result in real subordination of labour.

5   Interview with Shyamnarain Rawani, 30 March 2008, at the Bhowra Six Number Incline bastee. Kesho Rawani, born sometime in 1907–1909, joined his parent as a helper in 1917–18, and he handed over in 1978 his permanent government job (under Bharat Coking Coal Limited) to his elder son, Shyamnarain Rawani.

However, numbers of new recruits were not as privileged as the Rawanis. Many continued to mine coal in an unsafe manner, raking off coal by sheer force; robbing pillars; undercutting tunnel sides; insufficiently propping shaft roofs; tampering with safety lamps and using them in gas-filled mines; overusing explosives; working beneath loosened roofs; and riding trams to cover distance.

To address some of these issues, the industry initiated two measures: the modernisation of production through mechanization; and introduction of a new class of safety staff to conform to the stipulations of the *Indian Mines Acts* (IMA), 1901, 1923 and 1937–1939.[6] However, these remedial measures remained plagued, as Simpson's earlier-quoted anguish indicates, by insufficient presence and inefficient performance of the supervisory safety personnel. The direct producers – coal cutters, loaders, trammers, line artisan and timber artisan, known as *mistries* – continued to have full authority inside the colliery. The lack of constant, efficient expert safety supervision was an expression of both the colliery's dependence on the mining *sirdar* and his employees responsible for accident control, and reliance on the collier's practical skills of production and protection, which were regarded as more cost effective than any investment in training.

The industry relied on a productivist paradigm and each collier's practical mining skills, thereby resulting in the persistence of outmoded technology in wooden propping, and slow adoption of mechanical ventilation, water-pumping, safety lamps, and shaft and tram machines (see Foley 1920, 193–9). The board and pillar method of mining provides an example: this consisted of coal deposits that were mined by cutting a network of rooms or galleries into the coal seam. Pillars of coal were left behind in a chequerboard fashion to support the roof. The pillars could constitute up to 40 per cent of the total coal in the seam, depending on the geophysical environment. These were subject to excavation in a second working, involving thinning-down and removal of the pillar. During World War I, this method underwent marked transformation in response to increased demand, with attempts to remove the greatest possible quantity of coal in the first working, leading to unsafe pillars. Notably, pillar winning in Jharia and Raniganj accounted for about one quarter of the total output in 1918–19: investigators such as Adams (1919) and Rees (1919) underlined these two changes as responsible for collapses, subsidence and fire. These began to cripple the Jharia and Raniganj fields. Wilson (1908) had categorised this attitude as 'slaughter mining,' 'unscientific mining' or 'unsound mining.'

Slaughter mining hinged on handicraft forms of labour process and relations. In the workplace, especially underground, colliers worked in independent gangs of six to 12 persons wholly responsible for their section of the mine. While executing

---

6 Legislative requirements included supervision by a certified mining manager, an overman, mining sirdar, onsetter, banksman, engineman and surveyor. Later amendments to the IMA made training and trained supervision mandatory. Attempts by the Inspectorate of Mines to discipline and remedy slaughter mining has involved recourse to precaution, training and regulation, and prosecution: see Nite (2010).

a mining plan outlined by the manager and overman, they were responsible for testing for gas, supporting the roof and sides, and assessing the right size of pillars. Colliers in the leadership of a gang headman relied on their collective practical skill. The mining *sirdar*[7] relied on practical experience as a collier or a certificate of training under the 1924 scheme, and on his abilities as a recruiter. With time, increased regular employment of supervisory-safety staff helped extend foremanship to the gallery and the working face; consequently, the artisanal type of labour process gradually gave way to the manufacturing type, which meant an increasing loss of autonomy by colliers. It occurred gradually, and initially only in a handful of the largest collieries. Generally, however, controls of human labour-power over tools of mining and safety remained dominant in the gallery and at the coal face.[8]

One major aspect of slaughter mining was the coercive nature of labour relations. It was responsible for driving workers to place a premium on raising coal over safety concerns. For instance, colliers were coerced to collect stones from dangerous cordoned areas at Katras colliery (Nowagarh Coal Company Ltd.):

> ...miners were getting stones from a goaf for building material underground. A heavy fall of roof took place in the goaf and one man was killed. In accordance with the rules, the goaf was fenced off, but the place where this man received his injury was well inside the fence and it appeared to be a practice of this colliery to send workmen beyond the fence. ...accidents due to persons going into goafs are not infrequent and to send miners to such places is to risk their lives. (Adams 1919, 4–12)

More than anything else, the wage relation between capital and collier was responsible for the vapid attitude to safety and health. The piece-rate wages applied to coal-cutters, trammers and wagon loaders at bare-subsistence rates, and over-long workdays. The workday was further extended due to short or delayed supply of tubs and propping materials (Nite 2005). Such coercion of labour encouraged unregulated and dangerous mining activities to locate easily accessible coal, often endangering lives of colliers. However, management persisted in distancing itself

---

7 Between the 1920s and 1940s, the IMA gradually divested the manual roles of the gang headman, creating a supervisory position (Mining *Sirdar*) typically drawn from a literate, privileged caste and no longer directly involved in coal cutting and loading or attached to a specific working gang. In contrast, gang members belonged to the depressed and 'tribal' classes. The new mining *sirdar* displaced many traditional gang-headmen, who did not acquire the certification on account of their illiteracy.

8 Reliance on colliers' practical skills remained a noticeable feature, mitigated only by the formality of subordination. As late as in 1939, Kirby (CIM, 1940: 25–31) prosecuted cases against owners, managers, overmen and mining *sirdars* in numbers of cases of uncertified management, lapses in effecting safety regulations; and the grotesque negligence of supervisors.

from what it regarded as 'forbidden' labour practices such as slaughter mining, avoiding vicarious liability through use of discourse that depicted 'illiterate' and 'ignorant' miners as responsible for their own ill fates.

Inherent in existing employer-employee relations, these problems reduced colliers' working behaviour to an inchoate culture unable to improve working conditions. However, this could not bear the test of history. Colliers' idiosyncratic ways of executing tasks and approaching safety imperatives were a response to productivist imperatives, and they represented means of negotiating safety within a manifestly unsafe culture. As the mineworker Karpo Rajwar stated:

> The *burbak* [reckless/quixotic] man used to indulge in a risky manner of extraction and gathering of coal; why would a *chalak* [rational/prudent] man... risk his life and limb? Would he do [this]? We did not [practice] such ways.[9]

The *burbak* and *chalak* approaches to mining encapsulated a substantial breadth of tangential approaches: misadventure, ignorance, negligence or foolishness. These were expressions of trial and error undertaken through engagement by mineworkers adapting to the demands made by colliery employers. The notion that a number of mining accidents could not be prevented – 60 per cent in the opinion of Simpson (1928) – seems to treat the lives and safety of workers with fatalistic insouciance.

Few archival sources reveal how colliers survived hazardous conditions in slaughter mining. Neither the simplistic discourse of employers and the inspectorate about mining performance of the 'ignorant' and 'reckless' agriculturist-collier, nor the theory of work relations fully explain the collier's approach to safety and safety politics.

Mining work demanded an elaborate, measured, reflective association between mining individuals and the constraints of a colliery. In defiance to the overwhelming power of production wielded by the employer, the collier's *burbak* approach to mining appears to have been marked by a reference to trial and error principle. For instance, at Babu R. B. Chatterji's Chota Dhemo colliery, the failure on the part of the mining *sirdar* and the timber *mistry* in maintaining adequate propping resulted in an unexpected roof collapse. The *sirdar* had given no definitive instruction about either supporting the roof or removing the overhanging roof coal. The timber *mistry* neither removed the overhanging roof coal nor put up any props. Meanwhile, the loader continued to collect coal near the unsecured work face, and was subsequently crushed. 'The inspection of the mining *sirdar* should have shown him the necessity of having the roof coal taken down and he ought to have given instructions accordingly,' lamented Wilson (1908, 3–12).

---

9   Interview with Karpo Rajwar, 31 March 2008, at his residence in Bodroo Bastee (Chandankiyari Block, Dhanbad District). His mining career spanned between the latter 1946 and 1990.

The boundary between *chalak* colliers as a new class of employees and *burbak* colliers was becoming recongnisable despite the fact that their work followed trial and error principle. The formation of a class of *chalak* collier – i.e., one who is equipped with experience and skills in safe mining practice – was a matter of boasting in the mining community at the tea-stall and the *Kalali* (local pubs). S/he was found to be avoiding riding back and forth by toeing on the rope of shaft-lift, overcrowding the shaft-lift, stepping out of a moving lift-cage, riding a haulage-tub, sitting beneath any overhanging roof, applying over-dose of explosives and tampering with the safety lamp. A very specific instance of this progress and its limitation was discernible in Kanga & Company Ltd. Nawagarh, Jharia, describing a situation in which mineworkers were sent to extract coal from pillars in the Khodo Vallery colliery in 1944–45:

> Sadhuram Samsoye, the supervisor-in-charge, directed a few gangs of miners to strip the pillars.... The *sirdar* did not raise any objection and the undercutting of the side continued even when a mass of stone and coal became overhanging.... No support was erected at or near the place of the accident. Eventually, the place became so dangerous that the coal-cutters were unwilling to work there, but Samsoye persuaded them to do so, saying that there would be no danger. The overhanging side had been undercut to such a dangerous extent that it collapsed without warning. (Kirby 1945: 20)

Unsurprisingly, mining witnessed a decrease in the fatality rate related to shaft and haulage accidents, respectively from the 1920s and the latter 1940s. Hereon, collapses, the explosion of coal-dust and gas, and water inundation claimed a disproportionately high number of fatalities.

Chalak colliers comprised a twofold condition: autodidactic and didactic learning; and new safety awareness as supervisory staff. The practical skills of colliers and of managerial staff inevitably developed through the augmentation of experience and the desire for self-protection. Kesho Rawani's generation was bequeathed with a safety-friendly attitude to mining; by the time Shyamnarain Rawani and Karpo Rajwar ventured into a colliery they were aware of the differential effects of the application of *burbak* and *chalak* approaches. The modernisation of the mining workforce and the disciplining and streamlining of mining practices with accident control measures facilitated the development of a didactic mining knowledge. Three various approaches – precaution; training and regulation; and prosecution – undertaken by the Inspectorate of Mines enforced the regime of disciplining and remedying unsafe work practices: 'Prosecutions, for instance, meted out to increasing numbers of truculent violators impelled the rest to, reflectively, make use of their experiential learning to resist shoddy methods and the temptation for easy coal', reported Balchandra Ravidas on the question of improvement of mining practices in the industry.[10] The certified safety-supervisory personnel marked the addition of safety knowledge to the mining community. This was evinced, for instance, in a firedamp explosion that occurred at the Sitalpur Colliery (Bengal Coal Co. Ltd.; Andrew Yule & Co.): a series of prior examinations of the coalface conducted with safety lamps, and the arrangement for ventilation did not prevent the explosion; nonetheless, the attentiveness towards accident control ensured no loss of life or other casualties, as workers were called to the surface upon discovery of a block of combustible gas (Adams 1919, 6–12). In explaining where he learnt the *chalak* mining technique, Karpo Rajwar observed: 'The presence of a responsible mining *sirdar* helped to educate the rest of his gang members and neighbors; for this reason he was also at times held answerable for fatalities that occurred under his supervision.' The *sirdar* provided mining knowledge to colliers that helped to sustain their mining careers.

The safety-supervisory personnel shared responsibility for propagation and advancement of wisdom for sound mining – especially after the IMA, 1901, which made colliery managers responsible for safety. The National Association of Colliery Managers (NACM: formed in 1906–8 and representing mostly managers of European origin) and the Indian Mine Managers' Association (IMMA: formed in 1923 and representing mainly managers of Indian origin) undertook regular deliberations on safety issues, research and development works and produced and distributed films demonstrating safety methods. As members of the Mining, Geological and Metallurgical Institute of India (MGMII) and the Geological, Mining and Metallurgical Society of India (representing Indians), many of them participated in regular discussions, study circles, lectures and research organized by the associations. A few of the resulting papers are worth mentioning, as were they give an insight into the types of new research findings and the issues they addressed: 1907 papers on 'Fighting a Colliery Fire' (W. H. Pickering and R.R. Simpson), 'Solid Rope Capping' (T. H. Ward); 'Chelsea Electric Power Plant' (T. Adamson); 'Premature Explosion of Powder' (J. Grundy); 'Mine Dams' (W. T. Griffiths); in 1917, 'On Hydraulic Stowing in Mines in Bihar' (J. Hendry); 'The Burning of the Coal Seam at the Outcrop' (L. L. Fermor); and 'Housing of Labour and Sanitation at Mines in India' (J. H. Evans); in 1924, 'Notes on the Coal Dust Danger in Indian Mines' (D. Penman); in 1927, 'Further note on the Ventilation of British Mines' (F. L. G. Simpson); and 'Coal Resources of the Jharia Coalfield' (N. Barrowclough); in 1929, 'Discussion on the future of the Jharia coalfields' (R. R. Simpson); in 1935, 'Report of the Second Subsidence Committee (appointed by

---

10   Interview with Balchandra Ravidas, 20 March 2008, at his residence in one of the *bastees* near Industry colliery bastee (Jharia). His father hailed from Nawada district (Central Bihar), went to the coalfields in the 1930s, and settled down in the Lodna colliery. Balchandra worked as loader since 1962, and got promotion to mining *sirdar* in 1984 to work for next 20 years.

the MGMII).'[11] These findings were disseminated through publications of annual proceedings, and also by organised lantern lectures and research tours among safety staff in the field. These forums, lectures and study groups constituted a 'pedagogical mission' and established the earliest official public sphere in the social life of mining communities.

We earlier identified how mining safety was defined by three institutional contexts – parent and neighbourhood-based knowledge; legislative-derived skilling; and managerial instruction. In his paper 'How the British Got the Railways Built in India,' Ian Kerr (1997) notes the significance of the creation of a new generation of 'informed' labourers and the routinisation of accumulated, transmitted, codified knowledge of managerial and construction techniques. He further emphasises that personal qualities and extra-economic factors help explain the strengths and limitations of performance on the part of contractors, engineers and workers. In this regard it is the issue of the adoption and implementation of mining safety policy and contestation over the routinisation of mining knowledge that we need to address. However, the function of mineworkers' safety approaches was not simply a mediation of innate cultural traits; rather, it was informed by industrialism, by the demands of the employer, and the employee's adaptation to those demands. The following section deals with the form, substance and direction of workers' adaptations.

Ideas of Safety in Coalmines

A noticeable number of early mineworkers tended to avoid accident-prone shaft mines, particularly after the occurrence of an accident. Wilson (1908, 1–2) observed:

> It apparently takes very little to frighten the native worker away from the mine, and it will become increasingly important for mine managers to study the prejudices and customs of those under their charge.

Colonial accounts of the working masses typically underplayed labourers' experiences, whether of the type that benefited mining knowledge, or the experience of trauma. Reference to the problem of inadequate labour supply prevailed in official parlance, to the exclusion of the psychological or social wellbeing of employees. Insights into the long-term effects of a collier's experience of an accident that inflicted irreparable psychological damage can be gained indirectly from the following statement (Wilson 1908, 3–12):

---

11 Coalfield fires were a problem as early as in 1861 in the Raniganj coalfield, while subsidence posed a cataclysmic threat to public life in the Jharia and Raniganj coalfields from 1916 onwards.

As soon as it was known, the place was visited by the officials, but the bodies had already been removed from under the coal by the other workmen. A lamp however was found beneath the new falls. ...The evidence in this case is most difficult to obtain. Of the three surviving men, one was not close at hand at the moment and immediately he knew of the accident he went out of the mine with the two women who assisted them. ...Another man disappeared altogether, and the third, who either could not or would not give any clear explanation of the occurrence, died two days later form excessive drinking.

Many mineworkers evaded the strenuous mining environment of deep-shaft collieries, preferring congenial quarries with relatively small inclines, or surface work. Their relocation to incline-collieries (*Shirmuhan*) and quarries (*Pokharia*) surged during World War I, until collapse of the coal price in 1923, which caused labour-retention difficulties for deep-shaft mine operators (Simpson 1922, 3). Experience of the treacherous workplace environment of the mines finds a bitter, melancholic expression in folklore:

We sad coalcutters,
Our hand, hard and callused,
Our insides dark with dust,
Oh! This (is what) I think.
Once in the lift-cage,
I shivered,
What if the rope snaps?
Oh! This I think;
The cage goes down,
My father, my mother – so far away
Shall I ever see them again?
Oh! This I think
If a chunk of coal falls,
My head will be smashed,
God knows what is due
Oh! This I think
Ghuga Mahto tells you this story
The warm Damodar flows on,
Oh! The heat, the heat,
Tortures me on and on.[12]

Colliers of the time comprised immigrants as well as locals. The migrant labourers, known as *Paschimas mazdoors*, preferred sweating in big mines because they received a higher wage rate and accommodation. The better remuneration was often regarded as recompense for extremely dangerous work. Higher pay ensured

---

12   A Folksong by Ghuga Mahto, cited in Ranjan Ghosh (1992, 372).

that disposable workers were available to the industry without having to accede to arrangements for accident control. The larger mines – Bhowra, Jealgora, Jamadoba, Kustore, Lodna and Katras (Chaitudih and Bhadroochak) – increasingly adopted remedial measures in order to establish settled and experienced mining forces. Such improvements included sanitation, water supply, accommodation, medical facilities and technical training.

Developments in safety and work conditions comprises three subtexts: growth of understanding of the dangers and safety issues of mining through the formation of a class of safety staff; the political and legal struggle to prioritise safety; and deification of the colliery environment with the image of goddess Kali. It is arguable that the new class of safety-conscious staff owed as much to the necessity of production as to socio-political effort. Labourers and their representatives took up the cause during the 1920s at three levels: agitation and representation; legislation; and education and propaganda. Three strands of safety assurance were considered: legislation prescribing safe mining methods; the improvement of working and associated living conditions; and the political and legal restructuring of production relations. Following the burial of 74 labourers in the Parbelia colliery disaster in 1922, N. M. Joshi – a founding member of the All India Trade Union Congress (AITUC) – insisted on the necessity of legislation for 'compulsory certification of mining *sirdar* and overmen with a view to improve their performance as the front line safety-supervisory staff' (Department of Industry and Labour 1924a). As early as 3 November 1920, Simpson pointed out:

> …management of collieries was very much handicapped by the poor calibre of the persons who were appointed to make the daily statutory inspections of the workings. *Sirdars* should be required to have certificates, and certificates would be granted after an examination in which the *sirdars'* practical knowledge of timbering, goafing, ventilation, etc. would be tested and, in the case of gassy mines, the *sirdar* would have to prove that he understood the principles of safety lamps and know how to test for gas. (CIMAR 1921)

Discussing Simpson's Draft Bill (Amendment of the Rules for Coalmines under the India Mines Act 1901 in order to Provide for the Examination and Certification of Underground *Sirdars*) in March 1921, B. Y. Hajibhoy raised the question of tightening up regulations with a view to minimise the chances of accidents, and amendment of the rules in order to provide for the examination and certification of underground *sirdars* (Department of Industry and Labour 1924b). Faced with a renewed insistence by Joshi on the Simpson resolution, the colonial government acceded that 'a prompt measure taken in 1921 would have helped minimise the danger of accidents such as what occurred in Parbelia' [*sic*].[13]

---

13   The 1919 *Montague–Chelmsford reform to the Indian Council (Representation) Act* included regulations requiring nomination of labour representatives to the legislative assembly. Joshi's enquiry contributed to the amendment in 1923–24 to the *Indian Mines Act*

The Politics of Safety

The staggering increase in the late 1920s and 1930s in colliery roof collapses, coal dust and gas explosions, fire and inundation shook the mining community. It brought about thorough examination of the problem and public demand for comprehensive statutory control of the industry in order to avert reckless methods and shortsighted production agendas in Indian mining. In response to public pressure, the Government of India (GOI) constituted the Burrows Coal Mining Committee (1936–37), which made a comprehensive investigation into the reasons for perilous mining practices and the necessity of significant control of mining methods. Consequently, a series of regulations and amendments to the IMA were stipulated in 1937–9. Mandatory requirements under the *Mines Regulations*, 1937 and 1938 included: certification and training of shot firers and mining *sirdars*; employment of certified surveyors for every colliery; the requirement for joint surveys of geographically contiguous collieries; maintenance of a minimum barrier of 12 feet from water bodies; and use of non-flammable safety lamps and electric lights in all gassy mines. The guidelines for the first workings and the design of galleries and pillars were stipulated. If it appeared dangerous to the safety of colliers and miners, work could be stopped. The *Indian Mines Conservation (Stowing) Act 1939* specified the mechanical support essential for stowing and which enabled depillaring. A cess on coal dispatch would finance stowing. It also mandated fencing and safety of abandoned mines (Burrows 1937; GOI 1963). Operators and managers were rancorously opposed to many of these safety requirements. As an example, representatives of the Indian Mines Managers' Association regarded provisions of control over mining methods as an unwarranted and undesirable interference in the functioning of the industry (Evidence to Burrows Committee 1937 and Mahindra Commission 1946).

Contrary to the views of management, colliers witnessed an inadequacy of inspection procedures, evidenced by the growing number of accidents and casualties. Responsibility for such failures was invariably directed at workers rather than at supervisory or management staff, 'wherein false evidences, many times put forward by management to contest cases under the Workmen's Compensation Act, became the basis of judgment' (Whitely 1931). Workers' representative groups sought increased numbers of mine inspectors and for the method of inspections to be implemented according to stringent regulation. Chapala Bhattacharya, representing the All India Mine-Workers Federation, reported to the Mahindra Committee in 1945–6 an acknowledgement of the general improvement in safety since 1937 under the more stringent regime of mining regulations, while

---

1901 in favour of mandatory certification of overmen, surveyors, and mining *sirdars*, and regular evening mining classes for that purpose. At the same time, members of the British House of Commons pressed for the necessary steps to minimise loss of life among men and women underground: see *Industry and Labour* (I&L), Geology & Minerals Branch (G&M). 1925. F/No.-M-366 (6). I.&L, (G&M) 1925. M-366 (7).

nonetheless observing the need for more to be done in terms of safety and working conditions (Mahindra 1946).

In insisting on legislation specifying safe mining techniques, independent presses helped to assert the pressure required to ensure that employees received technical and safety training; that stowing – the practice of back-filling the excavated area with waste material – became standard practice; and that statutory controls on mining processes were enacted. By the mid-1930s they began to argue firmly in favour of universal stowing and a check on pillar cutting to avoid collapses and fires. To reduce casualties, it was argued circa 1938–9 that the bamboo headwear which prevailed should be replaced with steel helmets, and that boots be made generally available to workmen. Similarly, S. K. Bose advised the Burrows Committee that a programme of mass training for rescue operations would help to bring down the scale of casualties (Burrows Committee 1937, vol. II: 25).

The call for universal education and training for all adult members of the mining community was another issue for which workers sought change, but which failed to receive concrete action. The mineworker publicised the egregious reliance of managers and industrialists on current loopholes in the existing laws of negligence. Faced with a lukewarm response on the part of mining operators and state powers to address sought-for changes in regard to illiteracy and safety training, mineworkers' unions by 1926 oversaw development of a scheme of lantern lectures on matters of trade unionism, sanitation, housewifery, health and safety. How much this helped in saving lives is yet to be determined, although some positive effect of these programs is evident in the creation of considerable numbers of *chalak* colliers.

Their politics of safety emphasised links between safety and the necessity of improving working and living conditions, including labour relations. They believed in labour politics and the principle whereby, what SK Bose (secretary of the Indian Colliery Labour Union) called, 'the interest of labour is the interest of industry' (Bose et al. 1931, Vol. I: 184–91; Vol. II, Part I: 145–7). They argued for improved living and working conditions. This included improved wage rates, housing and water supply, reduced working hours, paid leave, social insurance – sickness, old age and maternity benefits – as well as arrangements for proper mine ventilation, adequate supply tubs and sufficient safety materials – timber; permitted explosives; designated separate explosive rooms; electric lights; safety lamps; boots and helmets.

A direct connection existed between inhibiting labour relationships in the workplace and unworkman-like performance: A forced or advance-based bonded-labour relationship, popularly known as *Bandhua Majdoori*, between direct producers and employers characterised the work relation. This led to many inequities. A considerable number of proprietors exploited zamindari rights and service tenancy relations to force service tenants to perform mining work. These arrangements persisted in collieries operated by the Bengal Coal Company (Murulidih, Katras, Sanctoria), East Indian Coal Company (Jealgora, Bararee and

Buggidih), Martin and Company (Sitalpur), Equitable Coal Company (Harilahdih, Dishergarh, Chowrasi), and the Eastern Coal Company Ltd (Bhowra, Amlabad, Kankani and Pootki). Under these arrangements, deductions were made from the employee's wages for paying off purported advances made to the employee, often in the form of regular submission of *salami* – six to eight paisa per tub or rupee of earning – to the *sirdar*.

The effect of coercive labour also impinged on the performance of safety supervisors – mining *sirdars*, overmen and surveyors – who were subject, as colliery surveyor S. K. Bose and B. Mitter observed, to 'despotic,' 'abusive' and 'racial' work relations that obstructed the prospect of competent service on their part. The solution they proposed was employment of separate safety staff that should not be saddled with production tasks, and deployment of separate, trained shot-firers. The vulnerability of workers to the power of colliery operators had a bearing on their performance: They are inadequately paid, which has become a constant source of conflict between them and other workers over the issue of eliciting labour in an uncongenial manner. Truculent personnel were dismissed on flimsy grounds for inability to conform to management requirements. Indian workers were subject to loss of work upon change of management in favour of European labourers. They were frequently abused and humiliated; they suffered racial slurs, and endured a demoralising environment of job insecurity. As Bose observed, '[j]ustice is non-existent and fairness is guided by self-interest...the palatial manager looks down upon down-trodden, clerk, and overman [alike]' (Bose et al. 1931, Vol. II, Part I: 193–4). Consequently, there was little commitment to work.

Likewise, colliery managers, especially those of Indian origin, protested autocratic and racist work relations:

> The colliery proprietor makes undue interference in the working of mines for cost effective mining and compromises with the necessity of scientific method. A commitment to principle is meant loss of favor and replacement. They (Indian managers) are inadequately rewarded and placed in little reassuring social condition. For efficient and rational management, the manager deserves full facilities, reasonable salaries and amenities for enabling him to discharge of all those statutory responsibilities and obligation [sic]. (Burrows 1937, Vol. II: 20)

With time, the number of Indian mine proprietors increased through the expansion of their ownership and shareholdings in European-managed stock companies. Consequently, initial grumbling by native managers gave way to their defense of private ownership, and the discounting of their earlier endorsement of the nationalisation of mines.[14] In contrast, European managers argued against

---

14   'Evidence from the Indian Mine Mangers' Association (Oral evidence from B.K. Bose and S.K. Ghosh)' to the Burrows Committee, vol. II), and 'Evidence from Indian Mine Managers' Association (Oral evidence from SN Mullick and AN Mitter)' to the Mahindra Commission 1946, vol. II.

what they regarded as less-skilled technical direction provided by indigenous managers. To combat such a danger satisfactorily, they suggested the need 'for a strong association of technical men.'[15] At the same time, they opposed the official proposal of regulating mining methods and plans of first working, as it meant 'an excessive interference in the working of mines and management' (ibid). Rather, they proposed that 'proper remedies to the problem of wasteful mining methods lie in increasing the powers of the managers so as to successfully withstand demands of owners' (ibid).

The third strand of colliers' safety views was connected directly with production relations. Prevention is better than cure became their strategic principle. Ideas that equated industrial democracy with labourers' rights to political participation informed their early approach to industrial problems – labour philosophy. For protective legislation including safety regulation, substantive labour representation in the legislative process is needed, Bose argued (RCL (Vol. II, Part I: 147); Burrows Report (1937 Vol. II: 25). Bose and Mitter argued for the participation of labourers at the level of the Inspectorate: 'Some of them [inspectors] should be chosen by labour unions. Accuracy in figures of statistics…can only be obtained by regular inspection and investigation conducted in collaboration with the labour unions' (Ibid). By the 1930s, and faced with the failure of the inspectorate, Bose suggested the direct role of labourers to resolve industrial relations problems; this could be constituted by rights ceded to labour representative to conduct independent inspections and make recommendations to the Inspectorate:

> The rules, Regulations, By-laws and Temporary Regulations, if followed in proper spirit, are quite adequate. More useful rules can be framed but they are of no use, unless they are actually adhered to in reality than in paper. We hold that unless some Trade Union officials are allowed to inspect the mines and report directly to the Mines Department about the observance of the laws, the Mines Department with their best efforts cannot humanly detect all violations. They can at best investigate and come to conclusion after some catastrophe has happened. Nobody in owners' employee can have the audacity to report about any violation. It is apparent that bona fide Trade Union officials should be allowed to cooperate and inspect the mines for finding the condition of the mines as far as safety is concerned more or less on the line of the British Mines Act. [sic]. (Burrows Report (Vol. II: 25)

Bose's call for direct involvement of labour in inspection and investigation remained unrealised for the next two decades in the face of stiff resistance from mine owners and stolidity shown by the Mines Department. By the 1940s, radical

---

15 'The National Association of Colliery Managers' (Indian branch) (Oral evidence of R. Roberts Arnold and W.M. Burch)' to the Mahindra Commission (1946 Vol. II: 276). See also 'Evidence from the National Association of Colliery Managers (Oral evidence from J. Brook and R.J. Pothecary)' to the Burrows Committee Vol. II.

reformers were convinced of the need for removal of resistance from industrialists, contractors and managing agencies to the proper functioning of remedial and protective legislative measures by enacting state ownership – nationalization – and accompanying socialisation of the industry. Initially, they found fault with the contract system in the industry:

> The contract system is evil responsible for all problems, such as extracting maximum works at the minimum cost, and skewed work relations are its governing principle. This system should be abolished; the company should be responsible for employment of and payment to miner, managerial and other safety staff. (Evidence to RCL (1931 Vol. I: 181–91)[16]

However, S. K. Bose observed that 'given the necessity of rationalisation of the industry for scientific mining, for protection of coal and labour as two "national assets," the mines will have to be nationalised as a way to keep alive this industry' (Bose, 1930, 1936). However, S. K. Bose vacillated over the issue. In contrast, P. C. Bose and C. Bhattacharya firmly insisted on the necessity of 'socio-political takeover of the means of production or industry as a means to effect recommendation of the fact finding committee' (Memorandum to the Mahindra Commission, 1946). Unlike the concept of state control of the industry as benefiting the health of coal and labour as national assets – a notion shared by Nag, Krishnan and Mookerjee ('The Dissent Note of Nag and Krishnan as part of the Burrow's Committee', 1937, Vol. I; Mookerjee in Hindustan Review 1945) – the idea of socio-political takeover – nationalisation and socialization – advocated by P. C. Bose and Bhattacharya – subsumed the necessities of social control within labour processes, social exchange and property relations. That such a comprehensive program was on the public agenda is evidenced by the Vice President of the AITUC Dange's observation of press references to Indian mines as 'Death Pits' and 'Nationalisation as Way Out' (Dange 1945). Furthermore, with their varied views on the ideal political and legal structure required to support safe mining, they adopted different political practices. S. K. Bose and Mitter represented the National Congress-led nationalist movement, while P. C. Bose and C. Bhattacharya were vanguards of socialist and communist politics.

The managerial staff opposed proposals for socio-political takeover of the industry. They recognised the necessity of nationalisation of the royalties from mines in the interests of conservation, statutory control over mining methods such as mandatory stowing, and training support in modernising mining labour. Unlike their Indian counterparts, members of the NACM R. Roberts Arnold and W. M. Burch acceded by 1946 to the point that if some owner fails to comply with the legislation of safety and conservation, the state-power should step in

---

16   The Royal Commission on Labour (RCL) agreed to recommend replacement of the contract system by a *sarkari* system of labour employment. Later, the Burrows committee recommended for a need of discouraging the managing agency system.

those particular cases (Mahindra Committee 1946 Vol. II: 279). However, they maintained that '[t]he technical knowledge of managers is sufficient to satisfy the imperatives of safety and conservation' (Ibid). They held firmly to the proposal of no interference of proprietor, and only moral and technical guidance of the Mines Department. These would, they argued, strengthen the hands of managers, and it would suffice. They did not welcome any idea of democratisation of the industrial system, particularly those forms espoused in terms of labour ideology. Their endorsement of the proposal of state control, though conditional in nature, was marked by significant difference from the content encapsulated by Bose, Bhattacharya and Dange's resolution regarding the socio-political takeover of death pits.

The safety resolution articulated by the mining community was decidedly innovative and comprehensive in terms of its social and political features when compared to the technological and managerial resolution shared by members of the mines department and other scientific communities. A worthwhile examination would consider its strengths and limitations by reference to ideas in – and transformative practice of – mining and labour politics. In the industrial period of the present study, such advanced requirements proved of limited effect in practice; in the struggle for improvement, most technical safety elements were not to be found in the core of their agenda of agitation, although they increasingly resorted to contestation for compensation claims against loss of life and limb. Rather, they showed active interest in improving conditions of work and life and representation. As a result, work stoppages became frequent occurrences. Beginning with strikes in 1920, direct actions conducted by the labourers over working conditions continued to grow, with characteristic setbacks and regrouping (see Nite 2010). In their collective action, emphases on issues of sustenance wages, social insurance and labour rights was an expression of their overriding concerns for food and dignity, as Bakshi Dâ – a labour activist since the early 1960s – reported. Responding to a question regarding why the safety movement appeared timid, he replied:

> the keen concerns of those colliery struggles were security of *Pet* and *Izzat* [dignified social standing] regarded tantamount to graduation of the life to the status of *Aadmi* [politico-social humanness / embodied being] and improvement over the afflicted status of *Bandhua Mazdoor* [bonded/forced labour] akin to cattle in the contemporary fields.... Somewhere, in such paradigmatic colliery movements, the issue of security from workplace risk failed to receive due attention and political energy. ... It drew attention after independence.[17]

---

17 Bakshi Dâ, a communist labour-union activist, known since 1966, member of the CITU, interviewed on 20 March 2008 at his residence, Jharia Town (Dhanbad District).

Bakshi Dâ maintained that colliers were progressing from agrarian to civilised societies, seeing progress in programs for child education rather than child labour, and denying women the choice to undertake hazardous work (Nite 2009).

Mukhopadhyay (2001) suggests that the ill fed and ignorant mining classes fought only for petty economic demands. However, he does not identify the particularistic ways of handling questions of workplace safety. The pronounced tendency in the colliers' safety politics was equally to address issues of accident control in terms of social insurance. Mineworkers were cognizant of the need to advocate for safety improvements to the extent that they recognised the necessity of workplace control of the labour process. However, the social insurance form of safety resolution necessitated postponement of everyday agitation over daily accident control measures.

The deification of mines using the image of Kali is an unavoidable subtext of colliers' adaptation to their environment. With time, a two-fold tendency developed in the mining world: besides the pursuit of actions for compensation for afflicted households in the aftermath of mining accidents, workers deified the colliery as the womb of Kali. Such religious sentiment persisted into modern times. As 'the believers in the intervention of god or spirit in the everyday life' (Census Report of Bihar and Orissa B&O 1921, Vol. VII. Part I: 129; Census Report of B&O (1931, VII. Part. I: 246–258), the majority of colliers worshipped *khadan* as a means of ameliorating adjustment to social and psychological strain. As one observer of the time, Karpo Rajwar noted: 'The regular offerings to *khadan*, conceived as the womb and mouth of Kali, helped further the cause of preservation and protection of life of person in and about the mines' (Interview 2008). Karpo Rajwar observed:

> We used to organise offerings to *Khadan* in the aftermath of any accident and the resultant fatality borne out in and about *khadan*. Any fatal accident called on suspension of work for a while, and collective offerings…. All the members of the colliery including the muhammedan and company/management used to make financial contributions for conducting such offerings. We regard *khadan* as the womb and mouth of 'Kali *Maee*,' and working in it is a visit to her womb. The accident is the expression of displeasure or anger of the *Ma/Maee*, and therefore she calls for *Balï* (blood sacrifice). A *balï* assuages the *Maee*, and helps secure, in a renewed way, blessing for preservation and protection. …That of entering her womb, and winning over coal by cutting and blasting it would have caused disquiet, disturbance, and hence displeasure. (Interview 2008)

Colliers bowed before the image or cult site of Kali, usually positioned over the mouth of colliery, and invoked her blessing each day, besides occasional offerings and annual ceremonial *puja*. Shyamnarain Rawani states:

> The rite of offerings/*balï* to Kali *Maee* especially in the aftermath of a serious accident has been practiced since time immemorial… The *Maee* becomes displeased and angry over the wrongdoings in *khadan*, such as reckless coal

cutting and mining ... robbing here and there in a haphazard way ... blasting ... and a few other 'unethical' behaviors; notwithstanding the offerings, accident does not fail to re-visit sooner or later; but people continue to trust her and renew the offerings. (Interview 2008)

The application on the part of colliers of a kind of religo-ritual contract with the mystical mistress of the cycle of life, livelihood and death shares cultural temporality with other societies (see Absi 2006; Taussig 1980; Kosambi 1962/2005; Prakash 1990). Under the sought-after blessing of Khadan-Kali, liable to regular renewal as with any contract, workers undertook the risks of formidable work (Deshpande 1946, 114). However, a denial of consent to hazardous mining, it could be said, was inherent in the Khadan-Kali cult.

An instrumental approach to the modern safety principle – as distinct from the 'Think Safety' approach – devoid of an epistemological break necessary for commitment to the principle of science, characterised mining safety politics. It was, indeed, in tune with the epistemological tradition of *purusarth-karma*: an incoherent combination of beliefs and attitudes constituted by faith in fate, coincidence, and in the necessity of exploiting favourable omens. The narrative of Hori, a north Indian peasant, in Premchand's *Godan* (1936) and the autobiographical narrative of Dr Rajender Prasad (1946) have equally taken cognizance of this aspect found in north Indian life.

Upshot

The coexistence of *Chalak* colliers with practical expertise alongside unformed *Burbak* colliers was a marked development in the coalfield. Informed, initiated colliers applied new safety behaviours and politics to the workplace. The *Khadan-Kali* tradition of colliers was all but impermeable to the modern safety perspective. Contrary to the conventional ideals, safety was anything but a secular and apolitical matter: it was subject to socio-political and religious dimensions in mining life. The industrial juggernaut remained an agent of appropriation, which ushered workers into perilous mining environments.

The safe, better-paid mining personnel – popularly called the *babus* – remained somewhat blithe by the safety agenda of mineworkers. A small number of mining lectures organised by the Mining Education and Advisory Board began to enlighten, by the 1920s, only supervisory personnel, who were considered suitable to attend these lectures in order to gain the benefit of technical education and promotion; the vast majority of workers was unable to satisfy the criteria of literacy and was thus excluded. The *babus* indulged, on the other hand, in discriminating and excluding the children of low-class mineworkers in colliery schools (Deshpande 1946, 98; Nite 2009). The industrial structure of mining society reinforced belief in the hierarchical precedence of the supervisory and managerial workers and industrialism in mining communities. To oppose the proposal of vesting additional power in the Mines Inspector, J. T. Caldwell, Secretary for the Association of Colliery Managers in India, stated:

> If mining operations are to be carried on satisfactorily, certain risks must be undertaken, and we consider that the person best qualified to judge as to whether the risk is a reasonable one or not, is certainly the manager of the mine... the proposed amendments [to the IMA 1901] are not justified, taking into consideration the number of accidents per person employed underground. They will compare favorably with any country in the world in which coal-mining operations are carried out. (Department of Industries and Labour 1923)

This belief was incongruous with the safety-first principle and many of the demands that colliers' associations made in opposition to that belief. This excluded the direct producer from the system of accident control – what Padhi (2003: 119–29) calls the 'traditional system of accident control.' Contrary to the view of labour proponents, the estranged safety behaviour of concept-oriented mining men was an expression of more than vicious work relations; it was traceable to the status quo reliance and the belief in industrialism. The law constrained them from any talks about the preindustrial belief that workers should bear workplace risks.

The efforts at democratisation of modern safety resolution advanced slowly. The form and the way of democratisation were products of the trade unionism that conceptualised safety resolution within forms of social insurance and sank into constitutionalist ways on matters of workplace safety. Furthermore, its everyday insistence on accident control measures – the 'Think Safety' paradigm – was subject to the influence exerted by political and ideological apparatus of the production process. Emphasis, after Burawoy (1985), settles on the ideological ground on which interests were organised and historical conditions were determined by capital labour relations: safety and employment negotiated with the profiteering.

References

Absi, Pascale. 2006. 'Lifting the Layers of the Mountain's Petticoats: Mining and Gender in Potosi's Pachamama', in Jaclyn J. Gier and Laurie Mercier (eds), *Mining Women: Gender in the Development of a Global Industry, 1670 to 2005,* Trans. Michele A. May, pp. 58–70. Palgrave: Macmillan.

Adams, C.F. 1919. *Annual Report of the Chief Inspector of Mines (CIMAR) for the Year 1918.* Calcutta: GOI.

Bakshi Dâ. 20 March 2008. Interviews: Jharia Town.

Barrowclough, N. 1949. *CIMAR 1948.* Calcutta: GOI.

Basu, Reena. 1988. 'J.K. Nagar Fire: An Environmental Problem Created by Coal Mining in Raniganj Coalfield Area', in S.C. Joshi and G. Bhattacharya (eds), *Mining and Environment in India,* pp. 172–186. Nainital: Himalayan Research Group.

Bose, P. C. et al. 1931. 'Evidences from P. C. Bose, B. Mitter, S. K. Bose (Clerical Staff), Shani Cheria (Women Miner), Chotan Kora (Miner) and Gobinda Gorai (Pumpman), Members of the Indian Colliery Employees Association, Jharia.' *Royal Commission Report* (RCL 1931 Vol. II, Part. I and II).

Bose, SK, 1930. 'Representation of Interest Concerned Before Courts of Inquiry Appointed Under Section of 21 of the Indian Mines Act, 1923', 1930, Revenue Dept (Industries), Patna: Bihar State Archive.

———. 1936. 'Rejection of the Proposal of Mr. SK Bose, General Secretary, Indian Labour Union, Jharia, to Provide for the Representation of Labour on Courts of Inquiry Appointed Under Section 21 of the Indian Mines Act, 1923', 1936, F/No-M-1055 (120), Delhi: National Archive of India.

Braverman, Harry. 1974. *Labour and Monopoly Capital: The Degradation of Work in the Twentieth Century.* New York: Monthly Review Press.

Burawoy, Michel. 1985. *The Politics of Production.* London: Verso Books.

Burrows, L. B. 1937. *Report of the Coal Mines Committee,* Vols I and II. Delhi: GOI.

Coleman, McAlister. 1943. *Men and Coal.* New York: Farrar & Rinehart INC.

Dange, S.A. 1945. 'Death Pits in Our Land: How 200000 Indian Miners Live and Work', Pamphlet shared at Miners International Federation Conference, Paris, August.

Das, Amalendu. 1988. 'Dust Hazards in Coal Mines: An Overview', in S.C. Joshi and G. Bhattacharya (eds), *Mining and Environment in India*, pp. 195–199. Nainital: Himalayan Research Group.

Department of Commerce and Industry. 1905a. 'Pickering, W.H: Measures Suggested to Prevent the Practice of Pillar Robbing in Collieries', August. Serial No. 1. Part B. F/No. 136/05. National Archive of India.

———. 1905b. 'Grundy: Procedure to be Observed by the Chief Inspector of Mines in India in Persecuting Persons Under the Indian Mines Act, 1901', August. Serial No. 1. Part C. F/No. 142/05. National Archive of India.

Department of Industry and Labour. 1923. 'Revision of the Indian Mines Act, 1901 (Act VIII of 1901)', May. F/No. M-665. Pros. 1–67. Delhi: Mines Regulations Branch. National Archive of India.

———. 1924a. 'Question in the Legislative Assembly by N.M. Joshi, a (nominated from Bombay) MLA, Relating to the Accident Caused by an Explosion of Coal Dust at the Parbelia Colliery where 74 Persons Got Out Alive, Bihar and Orissa', 22 February. F/No. M-407 (16). Delhi: Geology and Minerals Branch. National Archive of India.

———. 1924b. 'Question By Hajibhoy in the L.A. on the Tightening up the Regulations Governing Coal-mining Operations with a View to Minimizing the Chances of Accidents in Mines and Amendment of the Rules for Coalmines under the India Mines Act 1901 in order to Provide for the Examination and Certification of Underground *Sirdars*', M-498 (11). Delhi: Geology and Minerals Branch. National Archive of India.

————. 1929. 'Fatal Accident on 24th March 1929 at the East Indian Coal Company, Limited's Bararee Coal Mine', April. F/No. M-966/88. Delhi: Geology and Minerals Branch. National Archive of India.

Deshpande, S.R. 1946. *Report of Labour Enquiry Committee.* Delhi: GOI.

'Evidence of the Indian Mine Mangers' Association, P.O. Kusunda, Manbhum (Oral evidence from B.K. Bose and S.K. Ghosh)' and 'Evidence from National Association of Colliery Managers (Oral evidence from J. Brook and R.J. Pothecary)' to the Burrows Committee 1937, vol. II.

'Evidence from Indian Mine Managers' Association (Oral evidence from S.N. Mullick and A.N. Mitter)', 'Evidence from The National Association of Colliery Managers (Indian branch) (Oral evidence of R. Roberts Arnold and W.M. Burch)' to Mahindra Report (1946).

Foley, B. 1920. *Report of Foley Coalfield Committee.* Calcutta: GOI.

Ghosh, Anjan. 1988. 'Environmental Impact of Coal Mining in Eastern India', in Joshi and Bhattacharya (eds), pp. 43–51.

Ghosh, Ranjan. 1992. 'A Study of the Labour Movement in Jharia Coalfield, 1900–1977', p. 372. Doctoral Thesis. University of Calcutta.

Government of India (GOI). 1963. *A Guide to Central Labour Legislation in Mines.* Delhi:

Labour Bureau, GOI.

Gupta, S.P. 1988. 'Nitrous Fumes: Constituent Gases of Blasting Fumes', in S.C. Joshi and G. Bhattacharya (eds), *Mining and Environment in India*, pp. 165–171. Nainital: Himalayan Research Group.

Indian Mines Federation. 1963. *Fifty Years of Indian Coal Industry and The Story Of The Indian Mining Federation: A Souvenir Volume.* Calcutta: Indian Mining Federation.

Kirby, W. 1933. *The Causes of Spontaneous Combustion Underground in the Mines in the Jharia Coalfield, Together with a Consideration of Some Preventive Measure.* Calcutta: GOI.

————. 1940. *CIMAR for the year* 1939. Calcutta: GOI.

————. 1945. *CIMAR for the year 1944.* Calcutta: GOI.

Kerr, Ian J. 1997. *Building The Railways of The Raj 1850–1900.* Delhi: OUP.

Kling, Blair B. 1994. 'The Origin of the Managing Agency System in India', in Rajat Ray (ed.), *Entrepreneurship and Industry in India 1800–47,* pp. 83–92. Delhi: OUP.

Kosambi, D. D. 1962/2005. *Myth and Reality: Studies in the Formation of Indian Culture.* Bombay: Popular Prakashan.

Leger, Jean-Patrick. 1992. '"Talking Rocks" – An Investigation of the *Pit Sense* of Rockfall Accidents amongst Underground Gold Miners'. Ph.D. thesis submitted to the University of Johannesburg. Johannesburg.

Mahindra, K. C. 1946. *Report of Indian Coalfields Commission.* Delhi: GOI.

Marx, K. 1867/1977. *Capital,* Vol. I. Moscow: Progress Publisher.

'Memorandum to the Indian Coalfields Committee on Behalf of the All India Mine Workers' Federation (Affiliating All Mines Mine Workers' Unions under AITUC)', Mahindra Commission (1946).

Mookerjee, HC, 'Accidents in Coal Mines', *Hindustan Review*, December, 1945.

Mukhopadhyay, Asish. 2001. 'Risk, Labour and Capital: Concern for Safety in Colonial and Post-Colonial Coal Mining', *The Journal of Labour Economics*, 44(1): 63–74.

Nair, Janaki. 1998. *Miners and Millhands: Work, Culture and Politics in Princely Mysore*. New Delhi: Sage Publication.

Nite, Dhiraj K. 2009. 'Family, Work and the Early: the Jharia Coalfield 1890s–1940s', in Marcel van der Linden and P. Mohapatra (eds), *Labour Matters: Towards Global Histories*. Delhi: Tulika Press.

———. 2005. 'Work and Culture in the Mines: Jharia Coalfields 1890s–1940s', M.Phil. Dissertation, School of Social Sciences, Jawaharlal Nehru University (JNU), New Delhi.

———. 2010. 'Work and Culture on the Mines: The Jharia Coalfields 1890s–1960s', Doctoral Thesis, Jawaharlal Nehru University, New Delhi.

———. 2010a. 'Precarious Life, Terrible Work: the Jharia Coalfield 1895–1970', a research report submitted at the VV Giri National Labour Institute Noida, Gurgaon.

Onselen, Charles Van. 1976. *Chibaro: African Mine Labour in Southern Rhodesia 1900–33*. London: Pluto Press.

Padhi, Surendra N. 2003. 'Mines Safety in India: Control of Accidents and Disasters in the 21st Century', in A.K. Ghose and L.K. Bose (eds), *Mining in the 21st Century – Quo Vadis?* Vol. II., pp. 119–29. New Delhi: Oxford.

Pathak, Manmohan. 1982. 'The Death of Workers', in Nirmal Sengupta (ed.), *Fourth World Dynamics: Jharkhand*, pp. 65–73. Delhi: Aurthers Guild Pub.

Penman, D. 1936. *CIMAR for the Year of 1935.* Calcutta: GOI.

Prakash, Gyan. 1990. *Bonded Histories: Genealogies of Labour Servitude in Colonial India*. Cambridge: Cambridge University Press.

Prasad, Dr. Rajender. 1957. *Autobiography*. Bombay: Asia Publishing House.

Premchand. 1936/1995. *Godan*. Jaipur: Rajasthan People's Publishing House.

Rajwar, Karpo. March 31, 2008. Interview: Bodroo Bastee.

Ravidas, Balchandra. March 20, 2008. Interview: Industry colliery bastee.

Rawani, Shyamnarain. 2004 and 2008. Interview: Bhowra bastee.

Rees, T.R. 1919. *Report on the Method of Mining in Indian Mines*. Delhi: GOI.

Renu, Phanisewarnath. 1953. *Maila Anchal*. Delhi: Rajkamal Prakashan.

Sahu, K.C. 1988. 'Environmental Impact Assessment of Mineral Exploitation', in S.C. Joshi and G. Bhattacharya (eds), *Mining and Environment in India*, pp. 3–14. Nainital: Himalayan Research Group.

Sanjeev. 1986. *Savdhan Niche Aag Hain*. Delhi: Radhakrishn.

Simpson, R.R. 1922. *CIMAR for the Year 1921*. Calcutta: GOI.

———. 1928. *CIMAR for the Year 1927*. Calcutta: GOI.

Singh, B. 1988. 'Environmental Pollution, Fly Rock and Vibration During Blasting', in S.C. Joshi and G. Bhattacharya (eds), *Mining and Environment in India*, pp. 147–155. Nainital: Himalayan Research Group.

Sinha, Arun. 1982. 'Struggle against Bureaucratic Capitalism', in Nirmal Sengupta (ed.), *Fourth World Dynamics: Jharkhand*, pp. 111–36. Delhi: Aurthers Guild Pub.

Taussig, Michael T. 1980. *The Devil and Commodity Fetishism in South America.* Chapel Hill: The University of North Carolina Press.

Valdiya, K.S. 1988. 'Environment Impact of Mining Activities', in S.C. Joshi and G. Bhattacharya (eds), *Mining and Environment in India*, pp. 29–42. Nainital: Himalayan Research Group.

Whitely, J. H. 1931. *Report of the Royal Commission on Labour*. Delhi: GOI.

Wilson, W. 1908. *CIMAR for the Year 1907*. Calcutta: GOI.

Chapter 6

Stranded Between the State and the Market: 'Uneconomic' Mine Closure in the Raniganj Coal Belt[1]

Kuntala Lahiri-Dutt

Introduction

Coal India Limited (CIL) is a profit-making Public Sector Unit (PSU), a 'Maharatna', literally, a 'fine gem', for the Indian state. Its high status gives rise to the question: 'how can a corporation renowned for corruption and inefficiency be a profitable enterprise?' To answer this question, one needs to explore not just where the costs are concentrated but also go below the claim to investigate the extent of substance these claims have. To that aim, this chapter explores in detail how the costs of protecting the environment are overlooked by the industry, and the ways in which CIL identifies which mines under its management are inefficient and uneconomic in order to remove them from its profit-seeking agenda. It is an important task to explore the simple economics in order to consider how this state-owned enterprise, and hence the state, values the environment.

To investigate how 'costs' are constructed in Indian mining, this chapter explores the decision by CIL and one of its subsidiaries, Eastern Coalfields Limited (ECL), to close down 64 collieries that were identified as 'uneconomic' in late 1990s. These collieries are located mainly in the Raniganj coal belt of West Bengal, an area that has been mined since the early days of the coal industry and which houses a significant urban population that depends heavily on coal and its ancillary industries for economic activities. The Raniganj area roughly covers 1,260 square kilometres where ECL is not only the largest landowner, but also the largest single employer. Collieries that were regarded as uneconomic were invariably those that depend upon underground mining technology, usually labour-intensive but often manually operated in much the same manner with pick-axe and hand shovels as in the early days of coal mining in India. The decision to close certain mines was apparently based upon a report provided by an external commercial firm, the Industrial Credit and Investment Corporation of India (ICICI); the report suggested that the financial losses of ECL are concentrated

1 An earlier version of this paper was published in *Economic and Political Weekly*. I thank the Editor of EPW for his permission to publish the revised version.

in underground collieries in the form of sand stowing to back fill the voids. If this decision of closure was to be implemented, it would have cost as many as 72,000 positions, and would have sent shock waves through the economy of eastern India. One recognizes that like everything else, all mining operations have a definite life cycle: starting from exploration, to development, processing to post-mining and closure. Guidelines of better practices suggest that throughout these stages baseline and ongoing conditions are to be studied and monitored (DRET 2011). Contemporary guidelines also suggest that although the amount of detail on specific issues may vary, preparations for mine closure should be undertaken progressively throughout an operation's life cycle In some countries, a preliminary closure plan is required by the regulatory authorities as part of the approval process in order to reduce the impact of closure on social and economic lives of the area in which the mine operates. In Raniganj, where some of the mines have existed for a very long time, the impact of a sudden closure of such a large number of mines would be devastating.

In discussing the social and environmental issues surrounding the coal mining operations of ECL and the region it operates in, and this chapter argues that poor environmental practices undertaken by the coal mining company are responsible for a host of downstream social and environmental problems. More specifically, it argues that if the proper cost required for environmental care is taken into account in open cast (also called open cut, internationally) mines, they would not be considered as economically viable as they are currently held to be. It further suggests that the agenda of cost avoidance conceals active efforts of ECL towards the disposal of working collieries to private operators, a process that has been ongoing since the liberalisation of the coal sector. Not bothering to propagate its own research and development (R&D), CIL has been obsessed with the importation of technology, often imported from overseas such as in the case of the open cast Kottadih mine. As evidence from the symbolic disastrous results of longwall collapse in Kottadih, the use of labour-saving technology has failed to yield anticipated results. To start with, it is useful to note that the consequences of a technology-intensive approach have been beneficial neither for the company as a whole nor for the older coal-producing regions such as the Raniganj coal fields (for the full story of Kottadih, see Lahiri-Dutt 1999).

Until the 1960s, resource management decision-making addressed questions such as whether the project was technically feasible; whether it was financially viable; and if it was legally permissible. Since then a paradigm shift has occurred: instead of pure cost–benefit analysis in economic or technical terms, resource planners all over the world now include impacts – both environmental and social – in their analyses. Stand-alone economic cost–benefit analyses are no longer considered adequate. Nonetheless, Indian public sector companies continue to use such reports as an excuse to lay off substantial numbers of employees. Moreover, although some of these reports are written by economists trained to put monetary costs on the environmental degradation and social disruptions caused by mining, they tend not to use these skills, preferring to stay within the limited view of financial costs as those related to operation. One can presume that much is hidden within these analyses. Understanding such reports and the mindsets that produce them requires an consideration of how they are intended to be used and the possible alternatives that might have been taken by interest groups involved. However, in order to place these considerations in their context it is important first to look briefly at what was occurring in the coal mining sector globally, and how the state-owned coal mining industry of India responded to these changes.

Competition, declining mineral grades, higher treatment costs, privatisation and restructuring are issues that have been exerting pressure on mining companies to reduce their costs and increase their productivity. Employment is falling in many mining areas as a direct result of increased productivity, radical restructuring, and privatisation. These changes affect mine workers, who are forced to find alternative employment. Therefore, finding the right balance between the desire of mining companies to cut costs and those of workers to safeguard their jobs has become a major issue. Since 1988, coal mining industries in most of the world's coal-producing countries have undergone considerable change. The driving forces and the results have been varied, but the central objective has been to make the industry increasingly competitive, self-sustaining and viable as technological, environmental and energy policy changes impose considerable burdens on the production and use of coal. The objectives of restructuring according to the International Labour Organization (ILO) are to ensure energy security and meeting the demand for coal rather than attaining a pre-determined, inflexible production target, while achieving increased productivity, lower costs, higher quality and increased worker safety, and lessening the impact of coal mining on the environment. The ILO recognises that rationalisation will necessitate the closure of mines, freeing up of coal imports, a review of subsidies, divestment of non-coal mining operations and, in some cases, movement towards privatisation. However, it has outlined two complementary lines of action: ensuring that an adequate safety net is in place; and promoting non-coal related activities in the concerned regions. The CIL is now caught between the welfare approach of the state and liberalising effects of the market; it has begun to avail itself of the opportunities offered by these recent changes and has begun to use them in ways other than those prescribed by the ILO. Controlling all major management decisions including coal pricing, transport and allocation, CIL has for many years relied on a decision to pursue open cast mining as a means of lowering the cost of production and increasing output. It has initiated technology changes that have led to an increase in production from about 90 million tonnes in 1977 to the present rate of over 250 million tonnes per year.

At the time, ECL was one of the 11 subsidiaries of CIL, and due to its losses was becoming a burden on the parent company. It might be noted in passing that due to several reasons, the old age of the collieries, poor labour management and safety records being amongst them, ECL has never made significant profits ever since its inception. In 1997, ECL was referred to the Board of Industrial and Financial Reconstruction (BIFR), but through financial restructuring it was able to avert closure. Although CIL was unable to solve the chronic ailments of ECL, it had started offering the newer collieries to private contractors. The collieries in the Raniganj region therefore present an incongruous picture; older labour-intensive underground mines, burdened with centuries old legacies of various kinds, are owned and operated by the state whereas the newer open cast and technology-intensive mines have been given over to private operators. The report can be seen as a bid to formalise the process of handing over the gems, but in order to legitimse the process a complex and convoluted path of divestment was required. Obviously, the intentions were not immediately apparent – and which necessitates closer consideration of the region, the coal mining industry, how restructuring is being implemented, and at what cost.

In July 1997, CIL hired ICICI to create a 'Development of Restructuring-Cum-Revival Plan for Eastern Coalfields Limited.' This plan (the ICICI report) was presented to CIL in August 1998. At a meeting in Calcutta in the last week of October the plan was suggested to ECL and sent to the Coal and Labour Ministries for final approval. On the basis of this report, ECL planned to close down 64 mines mostly located in the Raniganj region of West Bengal and retrench nearly 72,000 employees working in these mines. ECL's cumulative loss from the 64 mines producing 7 million tons of coal was Rs 4,200 million in 1997–98.

Mining in the Raniganj Coal Belt

In late 1990s, ECL operated 121 coal mines which were located in the Burdwan (or the Raniganj coal belt), Bankura, Birbhum and Purulia districts of West Bengal, and the Dhanbad and Godda districts of Bihar. Of these, as many as 80 mines were inherited at the time of nationalisation of the coal mining industry. During 1997–98 these mines collectively produced about 27 million metric tons (Mt) of high-quality thermal coal, at around 75 per cent capacity. Raniganj coal is unique in that it is of 'long-flame high volatile matter' with quick heat-release capability. This quality has specific applications in the glass and ceramics industry, for example.

Nearly half of this production comes from underground collieries – a much higher figure than the 29 per cent of the national share of underground coal production – though in recent years, ECL has concentrated on the establishment of open cast mines. Underground mining is usually conducted by the board and pillar method, whereby a mesh of tunnels or galleries is first driven into the coal seam, leaving coal pillars for roof support. In this development phase, 20–30 per cent of the coal reserve is extracted. In the final or 'depillaring' phase, coal contained in the standing pillars that hold the roof up are also extracted. If the surface area is free from human settlement, it is allowed to subside. If the mine is under a town or a village, the voids, according to the rule, must be back-filled with sand mixed with water. This process is referred to as 'sand stowing' and is a critical input in maintaining the post-mining environmental integrity of the area. Not all of the coal that lies underground can be extracted in this labour-intensive process. Therefore, in recent years ECL has initiated new open cast projects (or OCPs). These mines generally have a much a higher degree of mechanization, and are usually located where coal occurs reasonably close to the surface. Heavy machines are used to remove soil and rock material, collectively called overburden, depositing it in a nearby area. If sand is stowed properly, underground mines ultimately have a lesser impact on the environment than OCPs. As they are operated in India, open cast mines leave permanent scars on the land, disrupt the natural processes of rivers and soil formation, cause air and noise pollution, displace indigenous and peasant communities, and destroy valuable farmland in adjacent areas. Nowhere in the world is open cast mining recommended for densely populated areas for these reasons. Although OCPs in India yield better returns than underground mines, their life span is shorter. In ECL, the share of production from OCPs has increased substantially from 1.23 Mt in 1975–76 to 14.78 million Mt in 1997–98.

In terms of capacity utilisation, however, underground mines are much better than open cast mines. ECL has some of the oldest underground mines currently in operation in the country. The deepest coal mine in Asia is at Chinakuri, 3,000 feet under the Damodar River. The main technical problem in underground mines is that in the process of extraction they extend deeper, thereby increasing the distance which the coal, material and labourers need to traverse as the mine gets older. Additional transport adversely impacts on the productivity and capacity of the mine. As a direct consequence, resources required to sustain the production to the rated capacity simultaneously increases. This, in turn, affects the economic viability of the mine and this has turned into a significant issue since most of the underground mines of the Raniganj coal belt are quite old.

One must note here that with its heritage of early mining, the ECL has been a labour-intensive mining company since its commencement. Until now, it has remained the single largest employer among public undertakings in the Raniganj coal belt. Since underground mines employ 89 per cent of the total manpower of ECL, the per capita output (tonnes per man-year) of its mines is low at 138 tonnes. The productivity (output per man-shift) in some of these mines varies from a low of 0.30 tonnes to a maximum of 0.5 tonnes. The losses of ECL, therefore, *appear to be* concentrated in underground operations, when an economic cost–benefit approach is adopted without incorporating environmental and other consequences of mining.

ECL has been a loss-making concern since the time of nationalisation, a case of the public sector failing to sustain itself economically. There has been no thorough enquiry on capital spending at ECL since nationalisation. Along with the coal mines owned formerly by private companies, ECL inherited a range of problems – environmental, social and economic – some of which are specific to the region. The long history of mining in Raniganj has left its imprint on the economy, the physical environment, as well as on the attitudes, values, beliefs and perceptions of those involved in mining, and the consequent socio-political circumstances of the region. The region has problems of land dereliction; underground void-land subsidence; mine fires; dying agriculture and decaying commons; and various conflicts over resource use (Lahiri-Dutt 2001). These problems are difficult to ignore while creating a future plan for restructuring and revival. Unfortunately, as has been noted, it appears that the ICICI report considered only the economics of mining to the exclusion of its broad effects. Environmental costs were not included in its purview, nor were not the social costs of mechanised open cast mining, ensuring social and economic distress and the displacement of local communities.

Environmental Concerns in the Raniganj Coal belt

Any report that uses the oxymoron 'sustainable development in mines' as its basic tenet is likely to limit itself to superficial economic considerations. The entire range of precautions and procedures followed to make mining the least harmful to humans and the environment is described as 'best practice mining.' In ECL's various plans for the environment there is no mention of these practices and no direction is given in clear terms by the administration to those involved in mining on how to carry out the business of protecting the environment. The best so far that has been done by ECL is carrying out periodic programs of tree plantings – sometimes on old overburden dumps, or on leasehold land lying vacant and near the residential quarters of its executives. Many of these saplings obviously die since they cannot survive on the rocky overburden material; those which survive are of little use to local communities either as food, fuel or fodder. The valuable topsoil – which best mining practice makes a point to store and return to local communities after mining operations are completed – is lost wherever open cast mines are operated.

In the Raniganj coal belt, the level of urbanisation is exceptionally high – as much as 67 per cent – considerably greater than that of Burdwan district – 35 per cent – or state levels – 27 per cent. The long history of mining has attracted a large number of people into this area. Besides Kolkata metropolis, it is the most urbanised region of West Bengal. Such a high density of population poses a serious threat of subsidence in the case of underground mines and limits the scope of open pit mines. The 19 existing open cast mines are not only environmentally hazardous; the average residual life of each such mine is as low as 2.1 years. Together, these mines employ about 12,200 people. The reason for such short life spans of open cast mines in the Raniganj coal belt is that these mines are based on outcrops and patches of the top-most seams in the area. The ICICI report notes: 'In order to fully exploit the economic benefits within the leasehold area, exploitation of these patches may be important. However, such operations can be undertaken more economically on a contractual basis.' ECL appears to follow this approach.

The costs in operating a mine are incurred in several main areas: wages, power, timber, petroleum-oil lubricants (POL), explosives, and costs for stowing. In the Raniganj coal belt there is a heavy burden of wage costs, as is evident from Table 6.1.

Table 6.1   Comparative Cost Structure (as % of sale value): Underground and Open Cast Mines under ECL

Heads	Underground	OCP
Wages	102.51	17.17
Power	17.22	3.51
Stores	7.93	14.44
Transport	2.64	1.80
Administrative exp.	7.07	4.81
Miscellaneous exp.	9.17	5.64
Interest exp.	5.22	7.82
Depreciation	10.20	16.27
Total	161.96	71.27

Source: ECL and author's own.

ECL's Measures to Improve Productivity

ECL had adopted the following two measures to expedite production turnaround: an exercise in which the manpower requirement of each mine has been assessed and 12,580 persons have been identified as 'surplus workers'; and a Voluntary Retirement Scheme (VRS) in which an employee who has completed 10 years of service or is 40 years of age is retrenched. However, VRS has not been very successful. This is mainly because alternative sources of livelihood in the region have largely ceased to exist. Traditional agriculture and forest-based activities have decayed to an alarming extent, and the sizeable number of migrant employees from nearby states has no attachment to local agriculture. Indigenous communities who have for generations laboured in the collieries have also lost connection with the local ecosystem which has degraded considerably. The large manufacturing sector of the region – consisting of steel plants, locomotive works and such other units – is over-saturated and offers few employment opportunities. Several medium-sized units, such as the paper mill in Raniganj or the Sen-Raleigh factory, have closed down in the last two decades, further restricting the job market in the modern sector.

Why is such a large number of workers considered 'surplus'? In collieries of eastern India, unskilled labourers are employed, and by the time that they develop a level of occupational skill – usually within 10 to 12 years – their health deteriorates and they become regarded as redundant labour. Such a rapid deterioration of health is related to poor conditions in the working environment, and ECL has so far not initiated any organised enquiry into this matter. Besides being considered surplus on health grounds, a worker may be retrenched when the specific skills he has acquired become redundant due to the introduction of new technology. Finally, ECL has increasingly shown a gender bias against women workers by immediately classifying them as 'surplus.' Many of these women were originally employed on their own merit, but such cases of direct employment of women are becoming rare, and now ECL actively discourages their employment even on compensation grounds, that is, when a male working member of the family dies while at work.

Mechanisation has been initiated to improve productivity, particularly in the OCPs. A high degree of mechanisation is not possible in older underground mines as the cost of production far exceeds returns. However, much of this technology had to be imported from coal-producing developed countries. Total dependence on imported technology has not been wholly successful in ECL, where the strata conditions differ from those in European and Western countries. Kottadih, for example, was a flagship project of ECL assisted by the French technology of underground longwall support. In this kind of operation, 6–8 huge steel pillars hold the roof while giant shearers raze the coal seams, the entire process being operated from the surface by computers. Three factors are of great importance while selecting this technology: the thickness and weight of the overburden, the nature of the surface land use – as the land is allowed to subside after mining, displacement of communities and loss of cultivable land is inevitable – and the number of steel pillars which support the roof. In early April 1997, the roof of the underground Kottadih mine collapsed over a stretch of 5,600 feet, killing three people and injuring six others, and destroying machinery worth millions of rupees. The project has been discontinued since then. Moreover, use of cutting and loading machines in underground mines using longwall technology generates an enormous amount of dust and noise, thereby posing safety and health hazards. Lastly, some technologies are indivisible and hence may in turn create other problems when adapted in part to new mining environments. For example, in an open cast mine there may be an eight-foot thick upper layer of coal and a 20-foot thick layer at a lower level. In Sangramgarh (Samdi), this eight-foot layer has been ignored and mechanised operations have begun to exploit the lower layer, which is better suited for heavy machines. The exposed layer near the ground surface has attracted the local people, who have excavated the coal to leave honeycomb-like holes that are likely to exacerbate subsidence in the area. Instead of depending on imported technology, a combination of indigenously available technologies could have been the answer in a case like Sangramgarh.

Dispossessed peasants have taken to illegal coal mining and its trade; sealed and abandoned collieries are broken open and coal from pillars and sidewalls is extracted. In open cast mines, coal is stolen openly and regularly from ECL collieries – often in broad daylight in connivance with industrial security guards. Informal mining has flourished in several locations in the region, on village or privately owned land. Interestingly, the entire region possesses no retail outlet for domestic consumers, a fact that further expands informal trade of coal that is mined and procured without license. To remedy matters, ECL expects the district administration to protect its property, because its large retinue of security guards is unable to police its mines, but the district administration prefers to turn a blind eye to the illegal mining and trading of coal.

Neglected realities

The ICICI divided the 15 mining areas under ECL into the following four groups based on their 1997–'98 economic performance:

> Group I: Profit-making, high-capacity open cast mines: Raj Mahal, Sonepur Bazari and Santhal Pargana.
> Group II: Areas with profit-earning potential: Kajora, Bankola, Kenda and Kunustoria.
> Group III: High-capacity, longwall mines: Jhanjra and Kottadih.
> Group IV: Mines with difficult working conditions: 64 mines covering Satgram, Sripur, Salanpur, Sodepur, Mugma, and Pandaveswar areas.

This classification was constructed on the basis of two indices: the 'competitive' index – profit margins before interest and depreciation/profit margins after interest and depreciation; and the 'attractiveness' index – assessed capacity per employee per year/production value per employee. The report made no mention of how these two indices were adopted as preferred choices.

The cost structure among the loss-making mines varied; the highest loss-making mine was Sripur, followed by Ningah, Nutandanga, Temehani, New Kenda, Kalipahara, Konardihi, Ratibati-R and Chinakuri 1. To produce 917 thousand Mt coal, the percentage share of cost in these mines was as follows: wages 66.08 per cent; explosives 1.2 per cent; timber 0.84 per cent; POL 0.4 per cent; others 3.5 per cent; and sand stowing 27.93 per cent. Clearly, sand stowing in underground mines comprised the second largest head of expenditure, but it was also one that is most intangible. For example, once an underground void is filled up, there is no way to check whether the amount of sand claimed to have been stowed in the colliery has actually been stowed. Allegedly, there have been incidents of subsidence in and around recent mine workings, pointing to negligence in back-filling underground cavities. One path of investigation would be to compare the amounts of coal raised and sand stowed. Table 6.2 shows the discrepancy between the amounts of coal raised and the amounts of sand stowed in one sample year.

Table 6.2 Difference (in Mt) between Sand Stowing and Coal Extraction during April 1995–November 1995 and April 1996–November 1996

	Area					
	Apr. 95–Nov. 95			Difference Apr. 96–Nov. 96		Difference
	Sand Stowed	Coal Raised	Sand Over Coal	Sand Stowed	Coal Raised	Sand Over Coal
East Division						
Kunustoria	1,530,173.7	348,821	1,181,352.7	1,546,749.6	353,507	1,193,242.6
Kajora	798,342.6	200,523	597,819.6	970,899.6	209,084	761,815.6
Kenda	300,105.3	44,667	255,438.3	317,219.1	49,370	267,849.1
Bankola	922,102.5	193,080	729,022.5	1,109,559	231,311	878,248
Pandaveswar	841,051.2	194,014	647,037.2	1,151,244.6	239,769	911,475.6
Total (E.D.)	4,391,775.3	981,105	3,410,670.3	5,095,671.9	1,083,041	4,012,630.9
West Division						
Sodepur	500,379	138,253	362,126	618,380.4	147,216	471,164.4
Sripur	498,319.8	87,966	410,353.8	460,623.9	92,863	367,760.9
Satgram	1,310,238.6	267,004	1,043,234.6	1,091,887.5	260,233	831,654.5
Total (W.D.)	2,308,937.4	493,223	1,815,714.4	2,170,891.8	500,312	1,670,579.8
Total (ECL)	6,700,712.7	1,474,328	5,226,384.7	7,266,563.7	1,583,353	5,683,210.7

Source: data collected from J. K. Ropeways and ECL

The data indicate that more sand has been stowed than coal. Even accommodating for the different material properties of sand and coal, this is quite impossible. The data demonstrate ECL has neglected to conduct the process of sand stowing in underground trenches properly, and reveal the hypocritical attitude of ECL towards environmental concerns. The sand-stowing business neither benefits the region nor the underground collieries, but rather increases the power of the contractor raj. However, the underground mines apparently lose economic viability in terms of purely financial cost–benefit analysis because of this added cost of sand stowing in underground mines. The ICICI report had completely ignored this aspect.

That ECL's claims of adequately stowing the underground voids are also questionable in view of the radically different accounts of the condition of mines pre- and post nationalisation: whereas a pre-nationalisation report identified 22 areas as 'unsafe,' a 1997 report by the Director General of Mines Safety (DGMS) has marked 171 areas as unstable and unsafe for human habitation. A physical survey of these areas would have been sufficient to determine how many of the mines are post-nationalisation workings.

Taking this logic further, if even a semblance of environmental care and protective measures for land reclamation and rehabilitation that are followed in other coal producing countries and are delineated in best practice mining were adopted in eastern Indian open cast mines, instead of leaving the dug-up lands and huge overburden dumps, the profitability of OCPs is bound to decrease. Since there is no effective environmental planning and supervision, the OCPs avoid these costs just as the underground mines are sealed without being properly filled. However, as mentioned earlier, the losses made by the underground mines are only apparent; once the long-term environmental and social costs are taken into consideration, the open cast mines would also not be regarded as economically successful to the current extent.

Finally, any decision to close down underground mines will also have to be placed in the perspective of future economic and environmental concerns. Mining operations would compulsorily turn to deeper mining projects because of the exhaustion of shallow and comparatively easily mineable deposits within the next two decades. Underground mining, therefore, is the inevitable future of the coal industry – in this regard ceding underground mines to private operators would be an extremely hasty decision.

Cess of the State

The reason for the chronic sickness of ECL involves several aspects, most important of them being the coal tax – referred to as a 'cess' – imposed by the state government of West Bengal. Whereas in other states only a royalty is imposed on coal amounting to an average of Rs 135 per tonne, the West Bengal government imposes a 47.5 per cent cess (ad valorem) on coal produced in the state. According to Entry 49 in the State List, a state can charge cess on the land being used for mining. Moreover, according to the Mines and Minerals Regulation Act, a state that owns a coal belt is entitled to royalty. Coal cess, almost like direct tax, is calculated as part of the annual value of the land where a mine is located. The revenue earned from cess, according to the West Bengal government, is spent on employment generation and welfare, and also on paying for the coal purchased. The burden of this cost is passed on to consumers. The state earned Rs 610.05 crores in 1997–98. However, the state government is yet to receive over Rs 1,000 crores on account of cess from the central government. Even the Provident Fund dues of the employees have not been paid. After public outrage over the CIL's decision to close mines, the state government reduced the coal cess from 45 to 25 per cent, acknowledging a loss of Rs 250 crores to the state exchequer. Yet, the amount remains higher than the average royalty of 17 per cent levied by other states and, hence, ECL coal remains the costliest in the country. So far, even the state government has not considered returning the income from cess to the region through innovative planning measures.

The grossest flaw in the ICICI report was its dependence on economic terms that rely on the hidden assumption that better technological inputs will result in improved economic performance. Only at one point of the report did the ICICI observe that from the 'socio-economic perspective, it is not advisable to close down the mines under group IV category.' At around the same time, German coal production also met only 20 per cent of its total demand, but was carried on with a state subsidy of more than 6 billion marks in the face of popular demand. The ICICI report included sweeping statements as: 'The predominance of labour intensive underground mining operations coupled with the resultant low productivity is clearly reflected in the low profitability of the operations at ECL.' The report did not comprehensively consider either the problem it needed to resolve, or the compounding effects that would be triggered by adopting the strategies it suggested. However, as an afterthought, the report mentioned that the 'option of phasing out these mines would not be practical from the implementation point of view.' Justification for this claim is not included in the report.

ECL's reliance on imported technology is evident in the way it conducts its mining operations. In the wake of liberalisation, there have been attempts to introduce private capital – both from within India and abroad – into the coal mining sector, whereas the ICICI report has instead sought to undertake a program of closing mines. The restructuring of the coal industry has shifted CIL's priorities; this is evident from CIL's importation of non-coking coal from countries such as Australia, Indonesia and South Africa – incidentally, it has been doing so without appropriate docking facilities, indicative of haste and poorly-considered responses to continuing social and environmental concerns.

Present Uncertain, Future Tense

Facing resistance from trade unions, ECL delayed its program to close mines and retrench labourers; a revival plan for the company also appeared on the anvil. However, since nationalization, more than INR 4,000 million has been invested in ECL and new or further capital infusion does not appear to be likely. All major trade unions in ECL and the Coal Mines Officers' Association, India (CMOA) were united against closure, but everyone trusted that at least some mines will be closed in course of time. It should be noted that even with a popularised VRS, the impact of the closures on the overall economy would be considerable. Surprisingly, the productivity of ECL has shown an upwardly turn – an unprecedented statistic for the company. Trade unions claim that this has been possible through improved motivation at all levels – from the general *majdoor* to the general manager – and the trade unions as well as officers' union have played a significant role in this turnaround. Such a statistic demonstrates that worker motivation and human resource development, if taken seriously, can change even ECL work ethos. Still, one can say that the threat of impending closure of mines had shaken the core of existence of the Raniganj coal belt which, with its industries, comprise one of the economically richer areas of West Bengal. The crisis brought into focus the problem of non-performance of the state sector and highlights multiple facets – some specific to the coal mining industry of India and some to the region. It illuminated that mine closure in India is fraught with extremely complex issues involving not just labour but the regional socio-ecological-historical context, and there cannot be a simplistic 'one-size fits all' solution. Most importantly, it showed the underbelly of liberalization; as the coal mining industry opens up to private capital, pure economic cost and benefits of mining becomes the justification to close mines to deprive workers of their livelihood rights. The fact that reports such as the one by ICICI can and do overlook the cumulative environmental and social costs of open cast mining, makes them as outdated as the dinosaurs.

To answer the question: 'what could the company actually do under the circumstances?' one wonders why, in its 40 years of existence, given its unique nature of an old mining region, ECL has never considered undertaking its own R&D and planning. It has at all times been lured by the sophistication of machines that are used for coal mining in other countries. The results have often been disastrous; many machines are left unused, and many rusted ones fail to perform when needed. In its focus on immediate economic concerns, ECL has allowed environmental problems of the region to reach crisis levels. On winter evenings, hundreds of thousands of knee-high stacks of stolen coal are burnt along the streets and in residential yards, making it difficult to breathe the air. It is not surprising, therefore, that human productivity is low because of the lack of skills and poor health, which comprise a grey area largely ignored by ECL. In context of the intricate environmental and human dimensions to the problems of the region, a simple cost–benefit analysis is not adequate. The emphasis on open cast mining at the cost of underground extraction needs closer examination; one might ask: 'to what extent is the viability of OCPs justified, and how much of it is only apparent because adequate measures of environmental protection are side-tracked?'

Finally, the episode raised a critical issue – does the state have full right to earn revenue on its natural resources at the cost of its citizens and the environment? As private foreign companies begin to operate in the region, one can only imagine the disasters looming overhead. Regulation and policing levels have to increase to protect the environment and the rights of disadvantaged groups, but in a country running on the steam of corruption and bribery, any such hope remains a distant dream. Given India's dismal track record in managing environmental matters by state initiative, the prospect of policy mismanagement of private/foreign companies and the consequent damage is great. Older inhabitants of the region have not yet forgotten the days just after India's independence, the *Company raj* when the main objective of the mining entrepreneurs was said to be 'More Hole, More Coal.' What is surprising is that in spite of being a state-owned enterprise, ECL has not been much different from its predecessors. With regard to environmental matters, instead of seeking to improve conditions, it now wishes to liquidate and surrender the region to a new breed of profit-seekers.

References

Lahiri-Dutt, Kuntala (2001) *Mining and Urbanization in the Raniganj Coalbelt,* Calcutta: The World Press.

Lahiri-Dutt, Kuntala (1999) 'The promised land: Coal mining and displacement of indigenous communities in the Raniganj coalbelt', *Asia Pacific Journal of Environment and Development,* 6(1): 31–46.

DRET (Department of Resources, Energy and Tourism) (2011) *A Guide to Leading Practice in Sustainable Development in Mining,* Canberra: Australian Government.

Directorate General of Mines Safety Report (1997) *Unstable Areas under the Raniganj Coalfield,* Directorate General of Mines Safety, Dhanbad.

PART II
Mining Displacement and Other Social Impacts

Chapter 7

World Bank, Coal and Indigenous Peoples: Lessons from Parej East, Jharkhand

Tony Herbert and Kuntala Lahiri-Dutt

Background

The development and expansion of India's coal mining sector is one the major planks in India's push for economic growth and industrial modernization, and in providing the fuel for electricity generation and coke for steel-making. The current emphasis is on the exploitation of shallow coal reserves using open-cast mining techniques in erstwhile forested areas, mostly in Jharkhand and West Bengal, but also elsewhere such as in Chhattisgarh. Open-cast (also known as open cut) mining, as it is carried out generally in India, removes soil and rock (which is commonly known as the 'overburden') from the top of the coal seams by blasting, followed by removal with large earth-moving equipment (draglines and dump trucks). The exposed coal is then broken by blasting or crushing and loaded on trucks or railway wagons to be transported away to the consumers. The excavations form a pit typically a few hundred metres long, 50 m wide and up to 80 m deep, depending on the depth of the coal and the thickness of the seam. Usually, once the coal is extracted, the pit then moves laterally and the overburden is then dumped into the previous pit. At the end a void is left behind in India. One can easily see that open-cast mining has a large footprint. A mine producing 40 million tonnes of coal in its lifetime (approximately 15 years, but often double that) will leave a scar about 25 sq km in area. Consequently, in a heavily populated country such as India, displacement of people is inevitable. Where coal occurs in lands held traditionally by indigenous peoples, mining these lands raises serious questions of social justice. When coal mines destroy and degrade forested tracts, devastating the local flora and fauna – and along with them the lives of local poor – then one begins to see through the politics of selective dispossession that hides within the official messages of development.

This chapter deals with this issue of displacement of *adivasi* and other local communities in the case of one particular mine, namely, Parej East in Jharkhand. Mining activities in Parej East mine were and are carried out by Central Coalfields Limited (CCL), a subsidiary of Coal India Limited (CIL), a public sector company. In 1996, CIL entered into a loan package worth nearly USD 600 million from the World Bank (WB) to upgrade and expand 25 select coal mines and improve their environmental performance. This was one of the largest loans received by the

Indian coal sector. One of these mines was the Parej East project. At the ground level, project-affected persons (PAPs) of Parej East reacted to its negative social impact on their living, and with the support of a local organisation took issue with the project, particularly with the way resettlement and rehabilitation was being handled by CCL in apparent non-compliance with the WB's guidelines. Being Bank-funded, the Parej East project was subject to stringent guidelines and criteria against which project performance could be more openly assessed. These were set out in the WB's operational directives (ODs) in force at that time, and they provided a framework within which its project implementation was to take place.

These concerns resulted in the establishment of a World Bank Inspection Panel to investigate its supervision of Parej East. The Inspection Panel found many flaws in the planning and implementation at the Parej East project. The report indicates important lessons to be learnt, pointing particularly to the need of high-level policy changes with regard to resource projects and indigenous populations. As we show through our analysis of this recent event, the entre experience also raises questions about the procedures followed by the World Bank, in this case in India.

Coal Sector Loans and the Inspection Panel

With the aim of supporting India's reform and expansion of the coal sector, the WB, in 1997, provided finance for expansion and modernization of 25 selected CIL mines. This was done under the Coal Sector Rehabilitation Project (CSRP), approved in September 1997, with an International Bank of Reconstruction and Development (IBRD) loan of over USD 530 million. Parallel with this, another loan was provided for the Coal Sector Environmental Social Mitigation Project (CSESMP). This was to assist in CIL's efforts to mitigate the environmental and social impacts of this mining expansion in the 25 mines. It was approved in May 1996, with a loan of USD 63 million from the International Development Association (IDA).[1] It was envisaged that after being tested and revised as necessary during the five-year time period financed by the Bank, CIL would apply its new environmental and social mitigation policies in its 495 mines.[2]

---

1   At the early planning stage, the CSESMP was initially conceived as a component of the CSRP, but in November 1995, the project was split into an environmental and social component, the CSESMP, and an investment component, the CSRP. Progress on mitigation activities was linked to the CSRP through a series of covenants in Schedule 9 of the CSRP loan agreement. This meant that disbursements under the CSRP for any particular mine would be contingent on timely and effective implementation of the mine specific RAPs, EAPs and IPDPs.

2   Due to unsatisfactory performance under the CSRP regarding coal sector reform and financial covenants, as well as unsatisfactory performance in the area of economic rehabilitation under the CSESMP, Management informed the Ministry of Coal and CIL on 20 January 2000 that it was moving towards suspension. On July 25 2000 Management cancelled the undisbursed balance of the CSRP Loan. However, Coal India Ltd. decided

One of the 25 mines was the Parej East Project in Jharkhand. On 21 June 2001 a formal complaint was made to the WB's Inspection Panel (henceforth the Panel or IP)[3] by project-affected persons (PAPs) of one of the villages of Parej East, through the Chotanagpur Adivasi Sewa Samiti (CASS), a local NGO. The complaint was that the Bank was in violation of its own policies in force at that time. These policies related to involuntary resettlement, indigenous peoples, environmental assessment, project supervision, disclosure of information and management of cultural property. Preliminary assessment of the complaint found that it had substance and an Inspection Panel was set up to enquire into the Bank's supervision of the Parej East mine.

The IP made two visits to Jharkhand for extensive on-the-spot inquiries and interviews with relevant people. The Panel released its 100-page report on 25 November 2002 (WB 2002) which lists over 30 violations of the Bank's own policies, with a further 10 issues of serious concern. The Bank Management (BM) are the project supervisors in this case and are obliged to submit to the Bank board a response to the IP report, with suggestions for remedial action. This it did in May 2003, five months beyond the stipulated time. The Bank's board of Executive Directors approved BM's response on 22 July 2003 (WB 2003).

It should be noted that in an IP process, the WB is technically investigating part of itself. Typically, throughout a project the BM teams consist of about three people who visit a particular project for which they have supervisory responsibility, for two or three days every six months or so. In the case of Parej East, the team – as well as a special Environmental and Social Review Panel – visited roughly once a quarter, making it one of the most heavily supervised WB projects. However, the BM is not hands-on, so the report is not merely damning of the WB management, but also of CIL and its subsidiary, CCL, which has created the substantive grounds for these complaints.

Although the IP reports are usually publicly available, they seldom circulate beyond those involved. The purpose of this chapter is to highlight some of the key issues raised by the IP (here quoted with paragraph references) to a wider audience. This is because often the mining companies claim that their activities are sensitive to the rights and needs of local communities. The Parej East mining operation, being WB financed, was subject to intensive review and one can assume, on the basis of this report, the pitiful situation of other coal mines that are not under such scrutiny.

---

to continue with mitigation programs started under the CSESMP. On 20 April 2001 it extended the CSESMP closing date for one year to 30 June 2002. At the time the extension was granted, about USD 24 million was undisbursed. The CSESMP project eventually closed on 30 June 2002, with approximately 79 per cent fund-use.

3   The IP is a quasi-independent body created by the WB as a mechanism for holding the Bank accountable for violation of its policies and procedures. The three-member Panel investigates claims brought by claimants for inspection.

On July 22, 2003, a WB Board Meeting discussed the findings of the IP, and in a press release said it 'endorsed the finding of the IP, while noting that the project had positively influenced Coal India's policies on environmental and social issues, and that nearly 90 per cent of the PAPs had improved or restored their incomes at the time of project completion' (WB News Release no. 2004/30/SAR July 25, 2003). At this meeting the Bank Management committed itself to 'continue to supervise the project and assist stakeholders to reach substantial resolution on issues raised by the IP,' with periodic reports to go to the Board (WB 2003a).

An extensive six-page brochure summary of the IP conclusions and Management response was made by FIAN International. This schematically lists the IP paragraphs, the relevant issues, the IP findings on those issues, Management Response, and their 'Action to be Taken' (FIAN International 2009). Of the 37 issues raised by the IP, 25 were classified by Management as 'no action to be taken' and nine were identified as Action items but which require as their action little more than 'continuing supervision' or 'reassessment.' It points out that 'the ongoing Management reports confine themselves to discussing only items for Action to be taken, and the discussion is centered on technicalities of programmes, number of eligible candidates, remaining (declining) number of people to come under the programmes.'

We first detail some of the report's key criticisms along with BM's responses and then identify the nature of the issues or problems that have been encountered. Due to the lack, and in some cases impossibility of compliance, it is difficult not to conclude that collusion must have occurred in initiating the CSESMP. This is a story of how *not* to do things.

Justice and Equity for Displaced People

Full rehabilitation of PAPs includes both income restoration and house resettlement – an aspiration that is notoriously difficult or, as many would like to say, nearly impossible. With regard to income restoration, the Bank's policy statement (OD 30, para 24) says that 'displaced persons are assisted to improve, or at least restore, their former living standards, income earning capacity, and production levels,' a statement echoed in CIL's resettlement and rehabilitation (R&R) policy in that 'affected people improve, or at least regain their former standard of living and earning capacity after a reasonable transition period' (CIR&R 1994).

The difficulty lies in the fact that not only do the village people lose their houses, they are also deprived of the land and natural resources that constituted their economic survival base. The village subsistence dweller may possess a small plot of land, but has access to natural resources such as springs and rivers for water, forest for fuel and furniture and the earth and wood for building houses – all critically important for survival. Because the natural resources are non-formal sources of income, they are rarely recognised or documented, and hence rarely compensated for. The wealth of these resources is transferred to the mainline economy, one in which the displaced communities have little or no place. The transfer of these natural resources, from one sector of the economy to another, is a major structural change, one that raises questions of inequalities within our society and can lead to further impoverishment of poor people if not carefully addressed.

The main issue raised by the petitioners to the IP was that of income restoration, sometimes called economic rehabilitation. This is a critically important factor that has to be taken into consideration, especially while dealing with people with few skills or assets.

On this issue of income restoration, the IP report squarely holds the management responsible:

> A major continuing problem is the failure of income restoration.... Because of the inadequacy of the income restoration programs, some of them have been forced to spend whatever remains of their compensation simply to survive. This is an extremely urgent matter. It should not happen in a Bank-financed project. Steps should be taken to ensure compensation of these PAPs, not only because they have spent their original compensation for their assets on survival, but also for the losses and harm suffered due to delays in restoring their income potential (WB 2002, para 478).

The BM response (WB 2003, Annex 1.11) questions the validity of the points raised by the Panel, saying that the latter did not have access to the latest data, and now: 'Management is pleased to note that recent data indicates that the situation continues to improve...income restoration had been achieved with respect to 87.1 per cent of all PAPs entitled...it may be still too early to draw a final conclusion regarding the status of income restoration.' Data are given (WB 2003, 25–8) to the effect that a large proportion of PAPs' income increased significantly, especially that of women.

There is no statement as to the source of these figures, nor their availability for cross-checking. It could also be noted that in many cases restored income derives elsewhere than from the provisions of the CSESMP. A significant number of PAPs have settled on relatives' land in other places, many live from head-loading in the adjacent local coal sale dump, and many live by running pilfered coal on cycles to local markets, an illegal activity.

> In principle, income restoration can be achieved either through providing jobs in the mine, or providing replacement land, or by promoting self-employment of PAPs, or by combinations of the same. In the following sections we consider at each of these options.

Jobs in the Company

Lost income has traditionally been restored by providing a job in the company (*naukari*). This has been the most realistic means of ensuring those give up their houses and lands 'share in the benefits of the project.' And CIL has always used this hope of *naukari* as a 'bargaining chip' as it has been called (WB 1996), to induce people to willingly transfer their land for mining. However, the offer of a *naukari* is no longer an option. A radical curtailment of the workforce in the CIL subsidiaries was one of the conditions of the coal sector loans offered by the Bank. The curtailment is now being implemented through voluntary retirement schemes, retrenchment, and radical reduction in new employment to make way for more mechanised processes.

While claiming to follow the WB directives, the actual implementation is restricted by criteria that filter out many genuine claimants. Such criteria include the need to possess three acres, non-recognition of *gair-majurwa* (village commons) land and following age-old *kathiyani* documents. Often, awards are based on incorrect data, forcing village people to run from pillar to post seeking to rectify them. This is because possession of the land often does not match with settlement records.

In spite of radical curtailment of jobs, it is clear that CCL personnel continue to use this as a bargaining chip to convince indigenous and local communities of their share of the benefits from mining. Indeed, CIL has a tradition of offering jobs to replace land lost to mining, and this tradition is banked upon. This artificial creation of job expectation was criticised by the IP:

> A very misleading message [was] being given to the PAPS (WB 2002, para 224)... the previous mine manager had given PAPs promises of jobs that were not available (WB 2002, para 225) ... and only during the updated census of PAPs carried out in 1997, were the majority of Parej East PAPs presented with the fact that they would not get a mine job, and must instead choose a self-employment income restoration scheme (WB 2002, para 226) ...it must have been a shock for them to discover [this] when finally presented with the reality of their situation in early 1997 (WB 2002, para 224–7).

Current reports from other new CCL mining areas indicate that it is still the practice of CIL personnel to build false hopes of *naukari* to persuade future displacees to relinquish their land. Such misleading messages are very persuasive to village people, and such blatant manipulation by officials is common practice. A similar ploy is to offer jobs to families of middle men as a trade-off for winning acquiescence of the people they claim to represent. Such practices cannot but have long term negative results when the deceit becomes clear. The BM response (WB 2003, Annex 1.12) is that 'despite the communications (of 6 consultative meetings with the PAPs), PAPs continued to press for provision of additional mine jobs because of the obvious economic security.'

The window-dressing nature of such consultations should be well-known to the Bank, as should the suggestion that the fault lies with the obtuseness of the PAPs and their desire for the comparatively high wages of CIL employment, not in the expectations which CIL has deliberately built over decades, and still continues to use as its bargaining chip.

Replacement Land

Another form of possible income restoration is replacement land, (land-for-land) of equivalent productive value, either in another place or from reclaimed mine land. The Panel observed that, contrary to Bank policy:

> this (land–for-land) option has never been offered in any of the subsidiaries visited. Partly as a result, the question of the adequacy of compensation paid for land is an important source of discontent with landowners ... and according to Management, CCL received no requests for such assistance. But in the Rehabilitation Action Plans (RAP) some 117 PAPs opted for this assistance and 115 qualified. Management also indicated that a large number of PAPs found replacement land, indicating that, with effort, it could be obtained (WB 2002, para 231–5).

The Panel, quoting a report of International Mining Consultants (IMC 2000), recommends that PAPs be allowed to select an area of similar size and productive capacity to that affected by the project, and that transitional costs, such as legal fees, moving allowance, and first harvest equivalent be included (WB 2002, para 234).

The BM response (WB 2003, Annex 1.13) claims that a certain number had purchased replacement land, and 54 per cent of the PAPs were landless anyway. Consequent action is that 'continuing supervision will follow up on the issue of reclamation of previously mined lands.'

Self-employment Option

With the virtual bypassing of the land-for-land option, and the effective exclusion of a company job, the project placed all its hopes on a new possibility, that of 'assisting project-affected people in developing opportunities for self-employment' (WB 1996, 4.6). This self-employment option was to become the 'central pillar' (WB 2002, para 255) of CIL's R&R policy, that on which the success of the Environmental Social Mitigation Project (CSESMP) was to rest. The aim was formidable; namely, to turn subsistence-living farmers, many of them members of scheduled tribes, into entrepreneurs. It might be successful for a few, but for several hundred, in a span of five years was, as NGOs predicted, well-nigh impossible. Early on, the project acknowledged that 'training in itself for self-employment is not enough, even when it is supported by loans or grants. The majority of project-affected persons are farmers or agricultural laborers, and the transition into a new profession requires a considerable amount of follow-up assistance' (WB 1996, 4.7).

Predictably, that central pillar was not too strong, and the IP's criticism is vigorous (WB 2002, para 244–67). In 1994, only 26 of 418 PAPs expressed an interest in self-employment. Yet, it is evident that three years later in 1997 most eligible PAPs had to choose a self-employment option to restore their former standard of living. There is nothing in the 1994 Baseline Survey of the Rehabilitation Action Plans to indicate that the eligible PAPs were counseled about the implication of the self-employment option, and nothing to suggest that the PAPs were aware of the implication of trying to become full-time entrepreneurs (WB 2002, para 238).

Further, the Panel says that it was misleading to advocate training/self-employment as the means to restore most eligible PAPs' standard of living in East Parej (WB 2002, para 252): at most it could only provide a supplementary source of income (WB 2002, para 253). Further, the original RAP did not reflect the actual situation (WB 2002, para 56): appraisal failed to ascertain the adequacy or feasibility of the self-employment strategy (WB 2002, para 243), relying almost entirely on non-farm jobs as strategy (WB 2002, para 258): 'the Panel could not find any report of a professional analysis of the pre- and post-relocation (casual) labour market' (WB 2002, para 102); implementation since 1998 failed to follow-up on market survey (WB 2002, para 240), after a market survey was finally conducted in March 1998, management failed to ensure that the recommended follow-up measures were taken (WB 2002, para 243), and as it was unrealistic to expect to become entrepreneurs in five years, feasibility should have been reviewed on appraisal (WB 2002, para 267).

The failure of 'self-employment opportunities' was recognised by Management (WB 2002, para 257), but when Management placed the onus for the failure back onto the PAPs, the IP retorted that it: 'is surprised that Management would accuse those, who never asked to be relocated, of "not making the necessary effort," to do something that was imposed upon them, by those who acknowledged that such schemes had mostly failed elsewhere' (WB 2002, para 249).

CIL's response to the failure has been to modify its R&R policy to introduce a one-time cash grant of Rs 50,000 for acquisition of home and land. The Panel retorts:

> ... presenting a poor oustee, whose previous source of survival included a small patch of land, with a check, probably more money than he or she has ever seen or expects to see in a lump sum, may be a legal way of getting them to move on, but it should not be confused with development (WB 2002, para 88).

The BM response (WB 2003, 1.4) denies the absence of casual labour opportunities, mentions an agreement that CCL has made with contractors to employ PAPs, and claims that income restoration has been achieved with respect to 87.1 per cent of the PAPs. Its action plan is that 'during supervision the Bank will seek additional information and statistics on the issue of casual labour provided for PAPs.' Again, both claims of agreement with contractors and of percentage achievement are un-sourced management claims.

Monetary Compensation for Lost Assets

The basic principle of the WB's OD 4.30 3b is that displaced persons should be compensated for their losses at full replacement cost (not market cost) prior to the actual move[4]. Such compensation applies to land, houses and other non-moveable properties such as wells.

Regarding land compensation, the Panel reviews and confirms the difficulties regarding compensation, namely 'that many PAPs have not been and are not being compensated at full replacement cost which would enable them to buy similar land, and hence are still suffering harm' (WB 2002, 66, 72). The system is well-known to provide inadequate compensation (WB 2002, 65, 68), as middle men take a share (WB 2002, 68, 69), and it is based on market rather than replacement value (WB 2002, para 72, 73). It involves under-reporting of sale prices (WB 2002, para 66) that even with the customary 30 per cent 'solatium' (compensation given for loss) it is still less than replacement cost (WB 2002, para 73), it is based on rates at the date of notification, not the date of payment (sometimes a 10-year difference) (WB 2002, para 71), and there is lack of transparency with regard to itemised details of the compensation (WB 2002, para 76).

Examining the question of acquiring land through direct negotiations, the Panel quotes Coal India officials as 'shying away from them...because there is always the risk of allegations of corruption,' but then asks why it is being done in other types of projects, and echoes the opinion of NGOs that 'this is nothing more

---

4   According to the World Bank Operational Manual, OP 4.12, 'replacement cost' is the method of valuation of assets which helps to determine the amount sufficient to replace lost assets and cover transaction costs. In applying this method of valuation, depreciation of structures and assets should not be taken into account. For example, for houses it is the market cost of the materials to build a replacement structure with an area and quality similar to or better than those of the affected structure, plus the cost of transporting building materials to the construction site, any labour and contractors' fees, and any registration and transfer taxes. In determining the replacement cost, depreciation of the asset and the value of salvage materials are not taken into account, nor is the value of benefits to be derived from the project deducted from the valuation of an affected asset (OP 4.12 fn 12, Annex fn 1).

than an excuse to avoid the perhaps higher costs of privately negotiated purchases in the coal projects' (WB 2002, para 70).

The BM response (WB 2003, 1.2) outlines the intricacies of determining the current value of land, and repeats CIL's present policy, including the claim of 15 per cent interest given for each year after acquisition. It believes that the method used 'was considered adequate.' It fails to mention that the PAPs never receive awards with itemised details – not only of the area, grade and rates of the land, but of the same interest and solatium – and never know how much they are getting for what. When a member of the NGO asked at CCL Ranchi headquarters for an itemised copy of the award, he was told he could not have one 'because there might be a mistake!' And the management's claim that regardless of this many receive enhanced compensation after tribunal appeals only serves to show further that the original rate is inadequate.

For house compensation, the Panel notes that the process and the basis of this compensation also lack transparency, are open to abuse and raise serious questions (WB 2002, para 82, 85). The BM response merely repeats CIL policy, but again fails to address the lack of transparency in its implementation and that the PAPs are at the mercy of officials and bureaucrats. Action to be taken is a mere 'continuing Bank supervision'. In the tribunal system (for reviewing compensation claims):

> it is not appropriate that PAPs should have to go through a lengthy and costly judicial process to get just compensation, especially since not all PAPs can afford the direct costs of an appeal process and, even if they could, they would end up losing unless the costs of the appeal were added to their award. Even then the delays and uncertainties associated with the process could result in tangible harm, especially since the awards are subject to further appeal by CCL. It is unfortunate that CCL is appealing (to the High Court) all these decisions (WB 2002, para 74).

The BM response (WB 2003, 1.2) claims that there 'is a functioning Government grievance redress mechanism' and fails to answer the Panel's judgment about the fact that the claimant has to give heavy lawyer's fees or why CCL then appeals against the enhanced award in the High Court.

Reclamation of Mined Land and Post-mining Land Use

A common sight in the coal mine areas is moon-like landscapes, large voids and mountains of overburden, the land scarred and irrevocably destroyed for any productive use. As worldwide mining practice shows, systematic topsoil preservation, on-going back-filling and re-vegetation can prevent this. The IP report states that the World Bank Project Agreement (Article II, Section 2.01.b) 'clearly spells out CCL's obligation in Parej East...that Coal India shall carry out the Environmental Action Plans [EAP], and...shall promptly inform the Borrower and the Association of any material deviation in respect of the implementation there' (WB 2002, para 354). The Panel reports further, 'the Environmental Action Plan (WB 2002, para 224) states that "it is proposed to remove the top soil from the quarry and overburden dump and conserve it for re-use during the biological reclamation stage"' (WB 2002, para 360). Further quoting the Environmental and Social Review Panel: 'The commitment to reclamation of mined land in CIL's Environmental Policy is clear and unambiguous. The policy includes a commitment to progressive reclamation to achieve a post-mine landform and use consistent with the Environmental Management Plan (EMP), maximising backfilling, preservation and re-use of topsoil' (WB 2002, para 367).

Despite these requirements, 'the Panel was not shown nor did it observe any topsoil conservation during its visit to the Parej East Open Pit' (WB 2002, para 363), and 'although requested at the site, no documentation or information on the five-year CSESMP mine reclamation programme was ever provided to the Panel team' (WB 2002, para 364). Besides, the 'staff were unable to provide the Panel with evidence that the eventual configuration and rehabilitation of mined areas were being planned' (WB 2002, para 365). This approach was best summed up by CCL: 'CCL's Senior Mine Management told the Panel that CCL had no intention of reclaiming the mined areas for post-mining use' (WB 2002, para 372). The Panel also quotes the earlier Environmental and Social Review Panel (ESRP 2000, 10): 'at present virtually no effort is being made to reclaim mined land... all the topsoil resources of the mined land are being destroyed through burial in overburden dumps...we have seen little evidence of any fundamental change in attitude to overburden management and reclamation since our first visit...' (Coal India 2000, 368).

Drawing on reports of IMC (IMC 2000, Section 2.1, paras 3, 4, 5), the IP points out that there are no legal requirements or financial incentives needed for such land reclamation (WB 2002, para 372). Hence, 'because present legal conditions prevent the transfer of land acquired under the CBA Act, the IMC recommends that Coal India Limited should lobby the Government to amend existing legislation to allow for the eventual transfer of reclaimed land...the implementation of the IMC recommendations is vital' (WB 2002, para 377).

The Panel does make concrete suggestions: 'for planning new mines, CIL explore the possibility of utilizing the available backfill to maximise the area restored to productive land uses' (WB 2002, para 373), and 'that each subsidiary of CIL be required to prepare and implement an Environmental Management Strategy' and further that 'Coal India improves planning systems for new mines, with particular reference to land use issues and reinstatement of mined areas of agricultural use' (WB 2002, para 377). It also notes that 'improving reclamation of mined land in the future...is...an issue...fundamental to CIL's future environmental and social performance' (WB 2002, para 376).

Such restoration of mined land could also be used as a base for income restoration (WB 2002, para 288–91). In December 1997, the management and CIL agreed on a land-based restoration scheme to be carried out on unused or reclaimed mine land (WB 2002, para 288). It did not materialise. CCL had first responded that there was presently no land available, then later told the IP that there was no financial incentive to undertake such land restoration (WB 2002, para 290). Yet 'this would have been the most promising possibility for restoring or improving the lives of PAPs...' (WB 2002, para 291). The Panel's indictment is strong, although it does conclude that 'at the same time, this does not constitute a formal violation of Annex C of OD 4.01 as far as land reclamation in Parej East is concerned' (WB 2002, para 375).

The BM's response (WB 2003, para 25) is merely to pick out this one last sentence, note the Panel's finding of compliance and dismiss the entire issue with 'no action to be taken.'

Wider Issues of Process

The Panel makes incisive comments and recommendations on many other issues, such as information sharing and consultation, the Indigenous Peoples' Development Plan, NGOs as implementing contractors, resettlement sites – their size and legal possession – land held under customary title, access to forest sources, and so on. Lack of space does not allow even a summary treatment of these issues here.

In general, the BM response claims that the Bank 'has made every effort (WB 2003, 11)...[,] remains committed...to the achievement of the objectives of the CSESMP (WB 2003, 41)...[and] intends to continue supervising the CSESMP project until all outstanding issues have been resolved' (WB 2003, 11, 41, 37). The Bank's press release is similarly soft on these purported efforts, stressing that 'the Panel commended the Bank supervision team's subsequent attempts.'

It is difficult to reconcile BM's commending itself for every effort with the IP's strident criticisms containing over 30 counts of non-compliance. There is an unresolved inconsistency here. Has the Bank heard its own IP? What of the IP's many recommendations of coal sector policy changes vital to its sustainability? It is difficult to see how the board of executive directors could approve the BM's response: which in 24 instances was constituted as 'No Action to be Taken.'

Moreover, the two main thrusts of BM's follow-up commitment have both been rejected by Government of India (GoI), and this was known by the board at the time. These were to be: first, 'to advise Government of India on apparent entitlements for subsistence allowances' as per the Parej East RAP, in the form of a lump sum payment made to the 121 eligible PAFs to the sum of USD 300,000, to be disbursed by 31 March 2004; the second was establishment of an Independent Monitoring Panel (IMP) to review the various issues. Even if not rejected, this would have been of doubtful value, as experience with an IMP in National Thermal Power Corporation's Singrauli area has shown. There the IMP made many recommendations, but these were not binding on the Bank or the borrower, NTPC. NTPC resisted the efforts of IMP for improvement, and failed to implement its recommendations. The Bank and IP remained silent spectators of this. As such, the IMP was of no benefit, and function exclusively as an escape strategy by which the Bank could exit the project.

Now it appears that we are left with the oft-repeated phrase 'ongoing supervision and monitoring.' The Panel recorded that this project was already one of the most-supervised WB projects ever. The Bank undertook 21 supervision missions between 1996 and 2001. However, the Panel found that 'the supervision team's knowledge of ground realities was limited, and for that reason, their efforts to resolve problems had virtually no impact on the ground' (WB 2002, para 458). A total of USD 1.6 million was spent on these supervision missions, compared with USD 300,000 that was being offered as follow-on compensation for the PAPs.

How, then, was the Bank continue to 'supervise and monitor'? The key issues are the commitments of GoI and CIL to meet the guidelines/ODs set by its financier of the mining projects. The question was: Would the Bank be satisfied with reports given by them, and use the reports as evidence that these issues have been sufficiently addressed?

In its response, the BM says it has learnt a number of lessons. These include the need for: realistic assumptions about organisational change; strengthening legislation; mechanisms for institutional coordination; critical issues to be resolved before implementation; obligations of implementing agencies to be clear; innovative approaches for income restoration to be explored; and analysis of resettlement options. These are lessons that should have been learned a priori, especially as many of them were pointed out by NGOs even before the project commenced (Bhengra 1996), and which in institutional review programs globally appear obvious. What of the Bank's claimed relations with civil society? One would like to see further reference to specific actions that could be taken to improve future performance, and, on humanitarian grounds, it would certainly be worthwhile if more references were made to the ways the PAPs are adjusting to or surviving the massive changes in their lives.

Policy Changes

The recommendations of the IP point out the urgent need for changes to both CIL policy and legal structure to protect rural land and the indigenous poor in Jharkhand. The Panel report makes many constructive recommendations for high-level policy changes. These include:

- Amendment of national legislation to ensure the concurrent reclamation of mined land, to create incentives for the surrender of this land after mining to GoI, with or without compensation, and for the use of reclaimed land for income restoration (WB 2002, para 372, 377).
- In early planning, to make an area plan that includes the social and environmental impact of surrounding mines and ancillary industries (already existing and future planned) so that an accurate Environmental

Impact Assessment (EIA) and EAP may be formulated (WB 2002, para 46); also to make the RAP's action specific to projects (WB 2002, para 56).

- As part of a Base Line Survey, to enumerate common property resources, value them, establish the income from them and provide a proper basis for compensation (WB 2002, para 194, 200).
- With regard to CIL's R&R policy and as a means of fulfilling its aims, to remove discretionary language which leaves loopholes – e.g., 'where *feasible*'; 'a *reasonable* transition period'; 'will *persuade* contractors' [emphasis added] – to give a definition of adult individuals; to include a sworn affidavit also as evidence that a person is a legitimate PAP; to spell out appeal mechanisms for PAP grievances; to increase the plot sites to be at least 200 sq m (WB 2002, para 45); and to make an evaluation of long-term results of the new cash settlement approaches (WB 2002, para 88).
- To create a legal framework for recognition of land formally in possession of people under customary tenure arrangements, not just registered *rayati* land (WB 2002, para 158, 170, 179); and to ensure that the PAPs are given legal titles to their resettlement plots (WB 2002, para 146).
- To establish public information centres at projects where information in a form and language meaningful to the PAPs is available (WB 2002, para 394).

There is little in the BM's response to indicate much acceptance of the need for the Panel's recommended policy changes. One professed objective of the CSESMP was to test the effectiveness of government policies (WB 2003). Such policies may be based in the *Coal Bearing Areas Act*, in directives of the Coal Ministry, in CIL's R&R policy, in the implementation of that policy by CIL's subsidiaries, and as concretised in project-specific RAPs.

Questions Regarding the Bank's Strategy

Commitments

Strategies to gain loan approval include commitments made a priori and then broken. For example, the Parej East EIA – on the basis of which the project is approved – gives an unambiguous commitment to reclamation of mined land (WB 2002, para 354). Yet CCL's senior mine management made it clear to the IP that they had no intention of reclaiming the land for post-mining use (WB 2002, para 372), nor was the IP able to find any evidence of such reclamation (WB 2002, para 363, 364). We are faced here with the hard reality of commitments made – by both the borrower and the Bank at the planning stage in order to obtain project approval – and yet apparently willfully ignored at the implementation stage. This raises questions about the credibility of the Bank's commitments.

Further, to what degree is the Bank committed to follow its own OD? The CSESMP provided a situation where it was clear that the ODs could not be implemented, for example, Indian land legislation that made it impossible to give the required recognition to customary-held land; nonetheless, the Bank proceeded with land transfers. Here, we are not referring merely to lapses in implementation of the ODs, but to predictable and foreseeable impossibility of their being implemented. To what degree are the Bank's own ODs binding it?

Exaggerated claims

The Bank's Staff Appraisal Report of April 1996 stated that for the 25 mines 'implementation of the Environmental and Social Mitigation Project will safeguard the rehabilitation of 9,260 people and the proper resettlement of about 10,000 people. The implementation of the Indigenous Peoples Development Plans will improve the lives of about 186,000 people, of which 56,900 would be tribals' (WB 1996, para 4.2). A basic defect in the project is that from the beginning there was heavy 'oversell' of its capabilities. Such oversell is found elsewhere in the planning documents (cf. WB 1996, paras 1.8, 1.9, 2.11, 2.24, 2.25, 4.40), appearing to exaggerate the value of the product in the hope that it would be approved by the Board. Such exaggeration, when accepted, has allowed the mining to proceed, with the goals of environmental and social rehabilitation virtually impossible to achieve.

Another example is that of post-mine reclamation in the EMP which was made available to the public when the mine was being considered. This suggested that only about half of the 253 ha of mine area would be reclaimed for agricultural land after mining, while the rest would be left to fill with water. This water, it was argued, would help the local population as a source for irrigation, drinking or industrial demand. The Panel countered this, however, by saying that the water would be inaccessible, as it would be 'tens of meters below the surrounding countryside and separated from it by vertical quarry rock faces' (WB 2002, para 357), further pointing out that it would be very costly to pump for irrigation, and impossible to use for drinking as it would be poisoned by contact with coal seams. Why are such claims made?

Misrepresentation

In the whole section on consultation (WB 2002, para 410–48), the IP makes it clear that requisite consultations with affected people have been either bypassed or unsatisfactorily met with mere disclosure of information (WB 2002, para 421). The BM had claimed that consultation requirements had been adequately fulfilled (WB 2002, para 412). Yet, these claims collapsed under the scrutiny of the IP – this is with reference to: the Indigenous Peoples Development Plan – IPDP – (WB 2002, para 28–30); the EAP (WB 2002, para 424); the RAP (WB 2002, para 429–34); the host community (WB 2002, para 430); and NGOs (WB 2002, para 441–

8). The issue here is not one of consultation, but of the Bank's misrepresentation in falsely claiming that these consultations had been completed satisfactorily.

Another example or misrepresentation is that of the IPDPs formulated by a reputed consultancy firm in Kolkata. For each of the different mining projects, there was to be a different development plan, each plan with information specific to the particular villages affected by each project. The IP notes that in the IPDPs of the different projects 'the whole of chapters 4 to 8 of each IPDP are all repeated verbatim,' and that the claimed 'felt needs' of widely divergent villages are all the same, with no location-specific information (WB 2002, para 311–13). All this was presented to the Bank, and accepted by it, as competent consultancy. The Bank had covered this up by firmly claiming that the IPDPs had been formulated with due consultation (WB 2002, para 328–30). This issue had been pointed out by local NGOs as far back as 1996 (CASS and Jharkhand Janadhikar Manch 1996).

Supervision   Early in this project, international NGOs[5] pointed out that the Bank's performance indicators were all 'input indicators' (money invested, persons appointed, time frame adhered to, and so on), but that they were not matched by 'output indicators' of the ultimate outcome on the ground of these inputs. Such 'check-box' appraisal of inputs would not have helped the project (WB 2002, para 464).

Continued Supervision

In its Implementation Completion Report (ICR), the Bank lists fourteen issues under 'Lessons Learned' (WB 2003a, 30). These include: overly high expectations in achieving the project's objectives; the need to establish compliance with national laws; issues requiring decisions by government ministries being made conditions of appraisal; that the cash grant was helpful in overcoming PAP resistance; and the need for an external monitoring and review panel to be factored into investment lending. The Bank committed itself to address the issues.

It still had, however, to respond to the IP report. Bank Management followed up with its commitment to continue to supervise the project and assist stakeholders. This it did until the final report of 2007 (WB 2007): 'Over three and a half years, a supervision team made nearly a dozen visits to the project area...the Country Director held five meetings with GOI...and the Bank has devoted US 650,000 and 170 staff weeks in service of this commitment.' In January 2007, the Management report informed the Board of its efforts to resolve the issues, and presented a case for bringing to closure the Bank's involvement in this IP case.

---

5   Letter of Peter Bosshard of Berne Declaration to Jean Francois Bauer of the World Bank, 12 June 1997. Berne Declaration, and also Minewatch, London, were supportive throughout in articulating the concerns of the PAPs and presenting them to the Bank and Borrower.

The activities which Bank Management committed itself to continue monitoring, and the stated outcome as reported in WB 2007 are outlined in the table below.

Table 7.1   Issues and Outcomes

	Issues	Concluding Outcome
1	Economic rehabilitation of 73 PAPs who reported a decrease in income.	'substantially resolved.'
2	Settlement of claims for PAPs cultivating land under customary tenure.	'substantially resolved.'
3	Provision of long-term leases to PAPs for their new house plots in resettlement site.	'a formal (legal) agreement is not achievable.'
4	Disbursement of subsistence allowance funds by GOI.	'CCL has demonstrated substantial compliance with the agreed measures.'
5	Delivery of compensation and relocation entitlements.	'this process is governed by well established procedures over which the Bank has no influence.'
6	Water quality in the resettlement sites.	'assessed as being satisfactorily addressed by Coal India's CMPDI.'
7	Reclamation of mine land for agricultural use.	'CCL's own need of land…lack of legal clarity regarding user rights ..existing legal complexities…have complicated a decision on this matter…but 95% of PAPs have been able to restore or increase their income (without using reclaimed land).'
8	Mechanisms for consultation dissemination of project related information PAPs-	'as stakeholders are communication directly with one another as a means of resolving their differences, Management finds that this issue has been substantially resolved.'
9	Constitution of an Independent Monitoring Panel (IMP).	Bank accepted GOI position that 'the majority of rehabilitation work had been completed and that there was no need for an IMP, as institutional arrangements in place were sufficient.'

With this the Bank Management concluded its involvement in the project.

Conclusion

The experience at Parej East indicates that the WB, despite its guidelines and ODs, was unable to supervise properly the coal mining projects it funded, particularly with regard to effectively dealing with peoples' displacement and environmental mitigation. This was primarily for the reason that in some instances there were major GOI legal blocks that could not be resolved. It was apparent at the outset that, owing to legal conflicts, some ODs could not be complied with. One could be forgiven for thinking that there was a degree of collusion between the borrower, anxious for finance, and the Bank, whose *raison d'être* is to provide loans. There was enthusiasm to initiate the project and commence coal production at the expense of social and environmental impacts. The professed aim of the CSRP – modernisation to meet India's high demand for coal – was later contradicted by the reasons for cancellation of its second phase on July 24, 2000: 'revisited demand for coal…overall demand for coal would not grow.' But other factors were also guessed at, namely, non-implementation of coal privatisation and other coal sector reforms, and the non-success of CSESMP (Philips 2007).

CCL, with its size and limited rehabilitation experience, has been an unfortunate partner in social and environmental efforts. What, for example, has happened to the 'time-bound action-plan' for CIL to extend these social and environmental mitigation measures beyond the 25 supported mines (WB 2002, para 299, 480)? Further – WB 2002, para 88, notwithstanding – it is now common practice, especially in the private sector, to give oustees a lump sum of money as a justified way of resolving the problem.

Finally, who, in the vast bureaucracy, takes responsibility for the instances of 'non-compliance,' the scared countryside, the families who have lost their land and income? The Bank? GOI as the borrower? CIL? Each of these institutions sees clearly that the local communities, often the indigenous and poor, carry the burden.

One is faced with the possibility, given the dire realities of livelihood constraints in rural India that continued expansion of coal mining cannot but dispossess and impoverish large numbers of people unless a rethinking is initiated. Under the current scenario, patchy environmental and social rehabilitation measures simply do not create a situation in which the rural people can benefit from new mining projects. For example, the recent Greenpeace report (2012) has observed that the general area where new open cut mines such as Parej East are being constructed have been corridors of elephant and tiger movement. How can there be any compensation for the loss of such invaluable biodiversity? The proactive rhetoric used by the state conceals the inequitable and shortsighted apportionment of costs – in environmental and in social terms – in order to meet the demands for coal by the unban-industrial nexus.

References

Bhengra, Ratanakar. 1996. 'Coal Mining Displacement.' *Economic and Political Weekly*, 16 March, 646–49.

Chotanagpur Adivasi Sewa Samiti (CASS) and Jharkhand Janadhikar Manch. 1996. *Report on the East Parej OCP*, submitted to CCL and the World Bank, 20 April.

Coal India Resettlement and Rehabilitation Policy (CIR&R). 1994. *Report no 7*, Calcutta: Coal India Limited.

Coal India 2000. Environmental and Social Review Panel (ESRP). 2000. *Report*, Note 73: 10, Calcutta: Coal India Limited.

Dias, Xavier. 2005. 'World Bank in Jharkhand: Accountability Mechanisms & Indigenous Peoples.' In *Law, Environment and Development Journal* vol. 1/1, 73–79. http://www.lead-journal.org/content/05071.pdf.

Fernandes, Ashish 2012. *How Coal Mining is Trashing the Tigerland*, New Delhi: Greenpeace.

FIAN International 2009. No Action to be Taken: Tracking Down the World Bank's Inspection Panel Report and Management Response, Coal Mining Project Parej – East India. http://www.fian.org/resource/documents/others/no-action-to-be-taken/pdf.

Herbert, Tony. 2010. 'Mining and the World Bank Inspection Panel.' In Michele Kelley and Deepika D'Souza, (eds), *The World Bank in India: Undermining Sovereignty, Distorting Development.* Hyderabad: Orient BlackSwan, 63–80.

IMC Group Consulting Ltd (IMC). 2000. *Technical and Operational and Management Guidelines, Volume 1: Strengthening of Coal India Ltd.* Social and Environmental Management Capability, Final Report. September.

World Bank. 1996. Staff Appraisal Report No 15405-IN, Annex 2.3, No. 3. 24 April 1996, Washington, DC: World Bank.

———. 2002. *Inspection Panel Report no. 24000*, 25 November 2002, Washington, DC: World Bank.

———. 2003. *Management Report and Recommendation in Response to the Inspection Panel Investigation Report No. 24000.* Report no. INSP/R2003/0002. 25 July, Washington, DC: World Bank.

———. 2003a. *Implementation Completion Report (IDA28620), Report No.25021.* 10 April. Washington, DC: World Bank.

———. 2007. *Management Report on Status of Outstanding Issues following the Inspection Panel Investigation Report and Management's Response.* Washington DC, World Bank.

Chapter 8

'Captive' Coal Mining in Jharkhand: Taking Land from Indigenous Communities

Kuntala Lahiri-Dutt, Radhika Krishnan and Nesar Ahmad

Losing Land to Mining

When Vishnu Ganju, a long-time resident of Hempur village of Balumath block in district Latehar[1] fell ill with tuberculosis, the representatives of a coal mining company took him to hospital for treatment. Vishnu was then asked to put his thumb impression on some papers, which he thought were related to his medical treatment. Later, it transpired that they were legal papers sealing the sale of his inherited land. Back in the village, the representatives opened a bank account for Vishnu to deposit the compensation money. While such blatant deception of illiterate or semi-literate tribal/*adivasi*/indigenous[2] or low-caste villagers to give away their land may be uncommon, it is not the only means by which land is being acquired in Jharkhand to accommodate the expansion of captive[3] coal mining for power and steel production. Middlemen pressure peasant farmers to sell their land; Puran Ganju of Hempur village was coerced into selling his land. Chedi Ganju, also from this village, claimed that he has neither received the receipt for the sale nor the full amount for the land he supposedly has sold.

India had legislated the *Chhotanagpur Tenancy Act* (CNTA) of 1908 to make such land inalienable. Why cannot this and other legislation such as the *Panchayat Extension to Scheduled Areas Act* (PESA) prevent acquisition of tribal land for coal mining and protect their livelihoods? What does such complicity of the state in allowing land transfers from poor peasants to mining-industrial corporations suggest about the nature of development being pursued in India? Lastly, how do Vishnu, Puran and Chedi's stories relate to the broad picture of the interconnections between state, people, resources and capital? We sought the answer through

---

1   Land in Hempur village is being acquired for a power plant, not for mining. Though our survey was about coal mining, Vishnu Ganju's experience illustrates the coercive practices used to acquire land.

2   Subsequently, the terms 'tribals,' '*adivasi*' and indigenous communities will be used interchangeably.

3   'Captive' implies mining of coal meant for an associated power station or steel-making, instead of the coal being sold in the market. In 1996, the MoC issued a notification allowing mining of coal for captive use for the production of cement.

investigating land transfers in North Karanpura collieries in Jharkhand. One of the authors (RK) undertook an exploratory field survey in late 2009. She visited Gumla, Latehar, and Hazaribagh districts. This fieldwork complemented KLD's and NA's ongoing research the in Hazaribagh (Ashoka-Piparwar mines) area. Various stakeholders were interviewed; these included company officials and representatives of the Jharkhand Pollution Control Board (PCB), a number of key resource persons, and villagers. Secondary data such as Environmental Impact Assessment (EIA) reports, literature published by the companies and distributed among the villagers, and pamphlets released by NGOs and activist groups were collected. Material from PCB and the Jharkhand government was also analysed. This research was supplemented with a literature review.

The state argues that in today's globalised world, rapid expansion of capital-intensive mining and industry can enhance economic growth and improve rural people's livelihoods. If globalisation is locally experienced, as suggested by Randeira (2003, 325), then asking these questions can unravel far more than what is detailed in Indian government statements and memoranda. Tribal land acquisition for captive coal mining demonstrates the violence of mining and reveals the formal and informal tactics deployed by the state and private companies to dispossess the poor. Although conducted in 2009, this research is still relevant in that it illuminates a neglected aspect of development and dispossession in India. The gaps in the national policy process and the implementation of mining-related policy are also considered. The investigation also shows that a solution to the problem of land acquisition for infrastructural purposes, as suggested by some economists, cannot be left to the market because, at the micro-level in rural India, the land market operates informally.[4] Coal holds a singular place within the overall mining boom because of the special status granted to it historically in postcolonial India.[5] Coal was associated with nationalism and the working-class movement, and has assumed the status of a national asset. Coal India Limited (CIL) symbolises public interest, and Acts such as the CBAA have been legislated to circumvent the existing protective legislation of tribal lands.

Jharkhand is not exceptional; other Indian states have also welcomed mining corporations to generate revenues. To assist mining-led growth they have also acted as land brokers, adopted policies and changed mining legislation. Jharkhand is unique in its continued existence as a resource periphery[6] and

---

4   Morris and Pandey (2009, 27) observe the need for a complete overhaul of Indian statutes, organisational structures and processes to develop a titling system that can ensure the security of tenure to land and property.

5   Coal is one of the major minerals as per the *Mines and Minerals Development and Regulation Act* (MMDR, originally passed in 1957, amended in 1988 and 2003), allowing only large-scale operators to carry out mining. A new amendment to MMDR has been proposed that aims to offer 26 per cent of profits to local communities.

6   Conventional texts describe Jharkhand as a 'museum of mineral resources' – it has been the major producer of a large number of minerals. Jharkhand accounts for 29 per

in its complicated ethno-national frame, which was the *raison d'être* for its creation in 2000. Stuligross (2008, 83–4) considers that tribal people have never comprised a majority of the Jharkhand population, but Upadhyay (2009) affirms that the demand for a separate state of Jharkhand symbolised tribal aspirations of regaining control over land and resources. Attachment to land is one of the aspects of tribal identity, though the extent of this is arguable. The state has experienced intense struggles over land, minerals, forests, agriculture and water; resource ownership is contested and its control determines substantial benefits in terms of economic and political power (Jewitt, 2008). Stuligross (2008, 87) is of the opinion that, even before the current mining boom, the government had become the major land alienator since the 1960s. Consequently, two groups of people supported statehood: those who wanted to restrict land sale by individual tribes people, and those who sought to broaden the property market under the supervision of Jharkhand state. The process of land acquisition for mining has neither been straightforward nor smooth, whether by CIL or privately owned companies.

Taking Land for Raking Coal

Throughout the world, interrelated crises in food, finance, energy and climate have been spurred by corporate-driven globalisation, neo-liberal policy regimes and natural resource exploitation. The global wave of land-grabbing has had serious consequences on the rural poor (see Land Research Action Network's 2011 report).The World Bank (2010) estimated that over 45 million hectares of land was bought from farmers in the developing world in 2009, a tenfold rise from the previous decade.[7]

Resource restructuring in post-liberalisation India involves securing land from farmers to allow for commercial and industrial uses, including mining. Coal dependence in overall energy supplies has led to acquisition of coal from overseas as well as the granting of mining blocks to private companies for captive coal mining. Forty-four captive coal blocks in Jharkhand have been allotted to private companies manufacturing sponge iron, integrated iron and steel, and cement (Ministry of Coal or MoC, 2009).[8] Between 2005 and 2009, 34 companies were allotted coal mining blocks in 210 villages in the Karanpura

---

cent of India's coal reserves and contains the largest number of collieries in the country, producing 21 per cent of Indian coal production in 2004–05 (Bhushan et al. 2008, 159).

7   See   http://siteresources.worldbank.org/INTARD/Resources/ESW_Sept7_final_ final.pdf viewed on 20 December, 2011.

8   See the website of MoC, available online at http://www.coal.nic.in/caplist070709a. pdf viewed on 18 January 2010.

Valley alone.[9] Some of the companies involved in captive coal mining in India have substantial foreign equity[10] and generally they operate open-cut mines that have a significant ecological footprint and serious social consequences (see Lahiri-Dutt 2007a). As long as CIL was the sole operator in coal extraction, the company–community–government relationship could be dictated by the eminent domain principle, which vests sovereign ownership of all land and natural resources embodied in the state. But when privately owned corporate entities enter the picture, it becomes problematic for the state to demonstrate that public interest needs are being met. A liberalised coal mining industry creates a dilemma: does the state protect its poor and weak citizens or does it assist private capital and corporations?

In seeking to avoid this moral quandary, the state equates coal with national development and energy security, and hence with strategic sovereignty. Coal is presented as indispensable to meet India's huge energy needs which, once met, will light up electric bulbs in rural homes and raise literacy levels to assist in the growth of India's soft power. As for energy security, coal mining becomes one of the keys to reinforce the sovereignty of India, and allowing private companies to conduct large-scale coal mining in order to expand production rapidly becomes imperative. The framing of this problem invokes only one solution: that is to expand coal extraction, and the need arises to overcome forcefully any obstacles that might obstruct the expansion of coal mines.

A major hurdle is the acquisition of land for coal mining. When a privately owned company acquires mineral-rich land, the eminent domain principle – and the notion of public interest – cannot, and should not, be invoked (see, e.g., Singh 2010 and Chaudhary 2009). Yet, old legal instruments of colonial vintage such as the *Land Acquisition Act* (LAA) of 1894 continue to be used to procure land from peasants. In Jharkhand, coal lies under forest tracts owned or customarily used by tribal peoples, land that is legally inalienable. The concept of eminent domain allows both the LAA and the *Coal Bearing Areas (Acquisition and Development) Act* (CBAA) to take precedence over any other Acts (Guha 2006, 157). The CBAA was passed in 1957 to 'establish greater public control over the coal mining industry and its development, provided for the acquisition by the state of unworked land containing coal deposits or of rights in or over such land.'[11] It thus fully reflects the older 'commanding heights' philosophy. But when it is

---

9   According to information provided in a pamphlet issued by the Karanpura Bachao Sangharsh Samiti, available online at http://www.firstpeoplesfirst.in/jmacc-alliance.php (viewed on 11 November 2010).

10   Privately-owned Indian companies setting up or operating power projects as well as coal or lignite mines for captive consumption are allowed FDI of up to 100 per cent. One hundred per cent FDI is also allowed for establishing coal processing plants, subject to some conditions (Singh and Kalirajan 2003, 145).

11   Full Act downloadable from http://coal.nic.in/cba-act.pdf viewed on 8 December, 2011.

used by private corporations to manipulate revenue-hungry state governments and to procure land at cheap or no cost at all, it can turn into a deadly weapon that undermines the very philosophy of social justice. The state's complicity in using this instrument is evident in its pro-industrialisation policies, policies that often benefit the richer classes at the cost of the poor and the environment, but which are adopted in the name of improving the lot of the poor.[12] More importantly, by virtue of the CBAA, the mining of coal generally supersedes community rights (Sharma 2003). Procurement of land, particularly tribal land, without free and prior consent can have devastating effects on small landholders. When the state acts as the facilitator of the land transfer process, its partiality towards large corporations is exposed (Levien 2011, 71). Such a land broker state differs from the old developmental state (Polanyi 2001[1944]). The unholy alliance between state and 'vulture capitalism' embodies a predatory growth (Walker 2011) that enables corporate access to land but dispossesses the poor. David Harvey (2003) places the dialectical relationship between the politics of state and capital accumulation in space and time at the centre of the analytical framework of 'capital bondage.' Such neoliberal developmentalism raises the question of social justice, and questions the very legitimacy of the state.

Regulatory Framework to Acquire Land for Coal

As is evident from the official Industrial Policy of Jharkhand (2001)[13] the state has adopted the role of facilitator in land acquisition processes to assist and encourage the rapid expansion of mining by private companies. The ability of corporate bodies to acquire and control land has been facilitated by a series of strategic amendments to the *Chhotanagpur Tenancy Act* (CNTA) of 1908 and the *Bihar Tenancy Act*[14] (Sharan 2009, 84–5).[15] The government intends not to be a neutral arbitrator in the land acquisition process; the blueprint of its Industrial Policy allows it to intervene actively on behalf of companies. It is contemplating setting up 'Land-Banks' that would contain information on all land available for industrial development. In addition, it is establishing 'Private Industrial Estates' for prospective entrepreneurs.

12　Such as energy security of the country. About 40 per cent of India's population do not have access to electricity as defined.

13　*Industrial Policy of Jharkhand* available at http://jharkhandindustry.gov.in/industry:policy.html viewed on 8 December, 2011.

14　This Act applies to Jharkhand as well. The Industrial Policy of Jharkhand refers to the amendment in the *Bihar Tenancy Act.*

15　CNTA was passed in 1908 to protect and recognise tribal rights over land. In 1996, mining and industrial purposes were included in Section 49 of the CNTA as reasons for which land could be transferred.

In the context of coal mining, the participation of the state in the land acquisition process for private companies is less straightforward given that coal mining is technically nationalised. The CBAA states that the Act can be invoked only to acquire land for coal mining by government companies[16] and similarly, the *Land Acquisition Act 1894* (LAA) can be invoked only 'to obtain land for the erection of dwelling houses for workmen employed by the Company or for the provision of amenities directly connected therewith.'[17] The LAA cannot be used by the state to acquire the entire amount of land required by private companies. LAA is, thus, being invoked in three major ways. First, land is acquired under Part II of the Act instead of Part VII, paying part of the compensation from the government exchequer.[18] Secondly, land is acquired by invoking the emergency clause (section 17) (Singh, 2010). Thirdly, since there is no definition of public purpose given in the Act, the Collector can define any industrial project as a public purpose (see the critique of land acquisition process in Choudhary 2009).

State's Commitment to Industry

The proposed *Land Acquisition and Resettlement and Rehabilitation Bill 2011* (LARR), currently before the Parliament, has provision to acquire land fully or partially on behalf of private companies only if prior consent is given by 80 per cent of the population affected by a project. However, it was found during fieldwork that the Jharkhand government informally committed itself to acquire 30 per cent of the total land required by the companies. The companies, however, submit requisitions for more land than is needed. The government acquires *gairmajurwa* or deedless land[19] that has been used for generations by communities but which is not recorded as such. Prem Prakash Soren, an activist with the local NGO *Adivasi-Moolvasi Astitva Raksha Manch* (AMARM), is working with the villagers in their protest against Arcelor Mittal's proposed 12-million tonne mega steel plant. Soren claimed that villagers have observed the company representatives quoting

---

16   *Coal Bearing Areas (Acquisition and Development) Act 1957*, 4. Available from http://coal.nic.in/cba-act.pdf viewed on 8 December, 2011.

17   Part VII, Section 41, Page 12 of the *Land Acquisition Act 1894*. Available from http://dolr.nic.in/hyperlink/acq.htm viewed on 9 December, 2011.

18   Under the LAA the land can be acquired under either Part II or Part VII. Part II is used when land is acquired by the government or the PSUs. Compensation is paid from government funds. Part VII makes provisions for acquisition for the companies for which an MOU is signed between the company and the state government and compensation is paid by the company.

19   *Gairmajurwa* land is a relic from the colonial land revenue surveys of the Chotanagpur region. Literally it is 'land without deed,' that is, land that does not have legal papers to prove an individual's ownership of it. Such lands generally include pastures, wasteland, roads and ponds, which have often been used or cultivated for generations and are hence owned de facto, or are frequently used community lands.

contradictory figures for the land requirement.[20] At the time of the survey, the government had identified a total of 1,025 acres to meet Arcelor Mittal's land requirement. Much of this is common land; in the Gumla district Kamdera block alone, the commons often villages has been handed over to the company. AMARM claims that this is a blatant contravention of community the rights clearly specified in Khatiyan (section) Part II of CNTA. Similar experiences were echoed in other parts of the state where the government is acquiring land for captive coal mining for private companies.[21]

Informal Tactics

Land acquisition by private companies in Jharkhand has also been characterised by the application of state force, and covert coercion by both company representatives and state agencies. During our focus-group discussions in Harla and Gondalpara villages in Hazaribagh district and in Hempur village in Latehar, villagers said that notices of land acquisition are issued without detailed field surveys, and sometimes acquisition is undertaken based on remotely sensed satellite data as both government and company representatives are wary of meeting angry villagers. *Gram sabhas* are often poorly informed regarding plans for land acquisition.[22] Middlemen who broker the sales deals often cheat villagers. Activists are prosecuted in law to discourage them from organising resistance against forced evictions or forced sale of land. Deepak Das, a worker of the *Karanpura Bachao Sangharsh Samiti* (KBSS) and the *Dalit Vikas Sangathan* (DVS), noted that the police prevent activists from campaigning near land acquisition sites.

The example of land acquisition for Chitarpur Coal and Power Ltd (owned by the Abhijeet Group of Industries) in the Balumath block of Latehar district is a case in point. At the time of this survey, the company had only acquired about 200 acres, due to opposition to the project, and on which the company has constructed its office – its mining operations had yet to begin. The prices paid for land can vary

---

20 According to Soren and Royan Kerkatta of AMARM, the company has sometimes quoted much higher figures – up to 12,000 hectares of land.

21 For instance, in the case of Nico Jaiswal's Moitra Coal Mining Project in Badkagaon block in Hazaribagh, the government has invoked Section 17 of the LAA in order to acquire land. Referred to as the Emergency Clause, Section 17 of the LAA vests Special Powers in the state in case of 'urgency.' Accordingly, if the collector so directs, the state can take possession of any land needed for a 'public purpose.' Once acquired, such land is completely in the control of the government, 'free from all encumbrances.'

22 *Gram sabha* is literally 'village gathering,' which all legal adults are members of and can take part in to make collective decisions pertaining to the village. The PESA Act recognises *gram sabhas* as the crucial decision-making unit under the *panchayati raj* system and gives them important powers for consultation in the matters of land acquisition and development projects in the area. However, the Jharkhand *Panchayati Raj* Act has diluted PESA in this regard.

widely. Abhijeet raised the offer as the scale and intensity of protests grew. Again, the purchase process is far from transparent; the papers of those who have sold their land show that the buyer is one S. K. Jha, the Assistant Manager of Chitarpur Coal and Power.[23] The court tends to refuse access when activists attempt to use the *Right to Information Act* to retrieve sale documents.

The New Refugees: Displaced by Coal

Mining not only leads to changes in land-use patterns and the creation of wastelands,[24] it also leads to the physical and occupational displacement of people. Over a million people have been displaced due to coal mining alone between 1950 and 1995 (Bhushan et al. 2008, 164), but these figures are not comprehensive because there are no official records of mining-induced displacement. With more open cut mines, the situation will only worsen.

Displacement figures provided in EIA reports by the companies are not entirely reliable. Villagers claim that numbers are greater than that declared by the companies. Being essentially focused on the environment, the EIA reports do not account for secondary/occupational displacement or displacement from mining-related activities. For instance, land requirement for the construction of a dam or waste disposal facilities is not mentioned; the EIA of Chakra Opencast Mines project in Latehar, intended to supply coal to the 2000MW Tori power station, notes that only 211 households (Table 8.1) will be displaced. However, Kalender Ganju, *gram pradhan* of Hempur village, claims that at least 12,000 people from five villages will be displaced by the project.

The existing mechanism for estimating population displacement is insufficient in gauging the social impact of the proposed projects. Even if the companies claim that the quoted figures include only legally-recognised project-affected people, excluding the landless and those living on *gairmajurwa* lands, the official assessments of impact should document their presence.

---

23   We were unable to procure documented proof of informal practices elsewhere wherein land is sold not to the company but to an individual representing the company.

24   Grazing grounds and pastures routinely appear as 'wastelands' in official records. In reality, the so-called wastelands are Common Property Resources (CPRs), serving as the livelihood-base for millions of the poor and marginalised.

Table 8.1 **Displacement Caused by Land Acquisition (as Declared by Mining Companies)**

Company	Annual capacity of coal (in million tonnes/year)	Land requirement (in hectares)	Estimated displacement in EIA (number of people/ families)
Essar Power Jharkhand Ltd (Chakra Opencast)	4.5	900	211 families
NeelanchalIspat Nigam Limited	1.3	383	1,737
Chitarpur Coal and Power Ltd (Abhijit Group)	0.68	1,378	149
Eastern Mineral Trading Agency	3	491	1,158
Nico Jaiswal	1	294	Not available

Source: Various EIA reports submitted by the companies to PCB, and MoEF, Government of India, New Delhi.

Compensation, Rehabilitation and Resettlement

As resistance to coal projects escalates, the issue of adequate compensation becomes politicised by companies upscaling their cash offers.[25] Following a higher offer by NTPC, most private companies with mining block allotments in Badkagaon also began to offer sums of anywhere between Rs 1 and 5 lakhs per acre. Such random decisions are dividing the village communities. In spite of better offers, many are unwilling to give up their lands as they realise that a one-off cash payment will not compensate for the permanent loss of livelihood. Continued opposition is also rooted in the perception that payment is not commensurate with the financial revenues from marketing the coal. This brings home the issue of seeking a more socially just means of benefit sharing.

In 2009 the Abhijeet Group of Industries proposed to offer company shares to those willing to transfer their land, beginning by recognising *malikanahuk* or local claims of ownership rights over mineral resources. The company offered 500 shares per acre; however, these offers were at a fixed price and not at market value.[26] The company has recently also offered community development projects which

---

25 An example of this is as follows: residents of villages in surveyed areas reported that the NTPC offered Rs 1.25 lakhs per acre. Over the years, this has increased to Rs 10 lakhs per acre – the most generous offer to date in the area.

26 According to a brochure issued by the company, each share can be sold three years later at the rate of Rs 200 per share.

include technical training and allowance, but these are somewhat haphazard. It has also adopted the old CIL policy of offering employment to one member of each displaced family losing >2 acres of land, depending on the competence of that potential employee. Many villagers, however, remain skeptical that the company will meet these promises, and are still unwilling to settle for an uncertain future. Abhijeet, for instance, will only employ 350 people in the mines as per EIA, of which only 55 jobs will be available to semi-skilled and unskilled labour, the only types of work available to the local communities. No compensation is paid for the land used for the transportation of coal, or for the air and noise pollution caused by truck movements. Villagers do not receive compensation for the water pollution caused by coal washeries or for the loss of natural water sources.

Toothless Safeguards: Environmental Impact Assessments

Environmental impacts of mining do not necessarily remain within the boundaries of the leasehold area; once watersheds and natural drainage systems have been tampered with and local streams have been diverted, the environmental consequences are far-reaching and difficult to redress. The pollution and depletion of water bodies also defy lease boundaries. The EIAs assess these impacts only within the leasehold areas. Almost all coal mining activity – opencast or underground – involves breaching the groundwater table. Though the water consumption of mines is not very high, the overall impact on local water regimes is immense. Air pollution from mining activity, as well as the storage of overburden, is another major issue.

The EIAs do not document possible impacts on local water availability, nor do they assess the ground-level impact of air pollution on surrounding communities and vegetation. The cumulative impacts of water and air pollution are rarely addressed in EIAs, which are generally insensitive to the multiple impacts on local communities. Generally, the EIAs present technical information that is devoid of the socio-economic and human context in which they exist, thus compromising their efficacy.

Consultation with the community and securing their prior and informed consent has been proposed globally (MMSD 2002) as the key for mining companies to obtain a social license to operate. The public hearing (PH) component of EIAs is the Indian version of 'Free, Prior and Informed Consent' (FPIC), but has no legal standing, and the Ministry of Environment and Forests (MoEF) is not bound by its decision. Moreover, it unrealistically presumes equal power between the company and the communities. A PH is seen as a one-off event rather than an ongoing interaction, and villagers do not have the right to refuse to sell land and stop a project. In addition, the company is not legally bound to address any issues raised at the PH to the satisfaction of those present. The regulatory mechanism to implement EIAs and ensure the follow-up action is generally weak and 'the regulating agencies in India within the OC [opencast] coal mines are not able to discharge their responsibilities of checking compliance effectively' (Jha-Thakur and Fischer 2008, 457). Consequently, the 'regulating agencies act as policing agents, but the power in the regulation process is tilted towards developers' (Ibid. 457). These limitations render the PH process impotent, yet the meetings are enthusiastically attended because it remains the only platform available for project-affected people to express their grievances. The resistance to projects naturally finds a voice at the PHs, to avoid which the companies often try to circumvent or manipulate the PH process in several ways: by changing the date of a scheduled PH without notice – as in case of Nico Jaiswal; by deferring it if a large crowd gathers – as in case of EMTA's Badkagaon project; or by changing the location without notice – as in case of Abhijeet Group's PH in Latehar. State complicity is evident in that the PCB holds PHs without widely publicising them to avoid the leading activists participating in them. Official communication from the PCB requires the announcements to be placed in only two newspapers.

Can Social Impact Assessments Deliver?

The *National Resettlement and Rehabilitation Policy* of 2007(NRRP), as well as the proposed LARR Bill of 2011, mandate that all projects must undergo a Social Impact Assessment (SIA) before they are approved.[27] Although it is better to have at least some social understanding rather than none at all, in absence of a strong, pro-human rights approach in mining legislation the SIA becomes a blunt tool. Currently, the SIA is a part of the EIA requirement, and is usually completed by environmental consultants as a supplementary section. As compared to other mining countries, where detailed ethnographic, cultural and social information on village communities are gathered by SIA experts, the SIA sections of Indian EIAs do not offer much more than names of affected villages, the estimated number of displaced people, and minimal census data on the socio-economic characteristics of affected villages. Primary survey reports giving a meticulous picture of the possible economic, social, cultural and emotional impacts on communities are non-existent. The definition of 'mine-affected villagers' is also limited; villages lying outside the lease area are generally excluded even though they often suffer from the impacts of the mines (Lahiri-Dutt and Ahmad 2012). Mere numbers and figures, however, are not sufficient to assess the social impact of a project. To determine the attachment of villagers to land, and the role of land in supporting local livelihoods, it is necessary to document land productivity and the manner in which people depend on it, not just the amount of land. If the use of forest land

---

27   The National R&R Policy has lost authority among NGOs; a pamphlet issued by KBSS claims that the policy is practically an 'investment and destruction' (*nivesh* and *vinash*) policy. The policy has also been criticised by others for its failure to address the concerns of the displaced and affected populations. See, e.g., a critique of the policy by the Asian Centre for Human Rights (2007).

is being altered, it is not enough to catalogue the varieties of trees and animals present in the forest, as is done in EIAs. To understand fully the impact of land-use diversion, the role of the forest – from firewood production through to minor forest produce – in supporting local livelihoods needs to be assessed. For this, a thorough assessment of the livelihood loss from mining should be mandatory.

Ethnicity Matters

Activists offer two important arguments why many tribal peoples desperately oppose land acquisition: The first set of reasons is rooted in culture, the close relationship with land, and the marked extent to which tribal identity and existence is entwined with land and with a long history, not just of suffering and exploitation, but also of resistance. The second set of reasons is rooted in the dependence on land-based resources for daily subsistence. One villager in Harl, Hazaribagh, observed during our field survey: 'We have neither education, nor political clout. All we have is this land. If we give up our land today, then can we hope that our children would become *babus* in these companies tomorrow?'

The close relationship with the land is not shared by absentee landlords living in the cities, or by shopkeepers and businessmen, who readily give up their land for cash. Father Tony Herbert, a missionary teacher in Hazaribagh for decades, points out that when large landowners sell land, the *bataidars* or sharecroppers – who are inevitably *dalits*, OBCs or tribal members – lose out. Social conflicts over land acquisition by a company, NeelanchalIspat, support this statement; while the *dalits* and *adivasis* are opposing the project, upper-caste Hindus and Muslims are favouring it. The struggle to protect tribal customs and the resistance against land acquisition are linked with, and feed into each other. The desire to preserve tribal culture partly arises from the perception that it is by circumventing traditional tribal customs and practices that celebrate community ownership and control of land that the state alienates tribal members from their land. This is one of the reasons why the tribal peoples are holding tenaciously to CNTA and PESA.

Need for a New Paradigm: Mineral Ownership for People

In the early days of mineral-based industrialisation in Jharkhand, local communities were not entirely excluded and at least, as mine labourers, some 'earned incomes in excess of anything they could have earned in their villages (Corbridge 2004, 185). Today, they are not much better than captives on their own land, as jobs in machine-driven open-cut mines are few. Coal mining by private companies points to a complex political power game played in the name of energy security. In this process the powerful invoke the poor for whose development the energy is needed. After proclaiming India's arrival on the world-stage, the powerful then neglect the peripheries as resource-enclaves. The poor are forced to sell their livelihoods and become indebted bonded labourers; they may become illegal miners, or move to cities as beggars and crowd the urban slums. The rich and powerful avoid the issues of slave wages and class injustice, preferring the similarly unfulfilled rhetoric of climate change and global justice.

Only a convoluted logic can equate coal mining by private companies with public purpose or national interest. If globalisation and liberalisation aim to integrate India with the global economy, then our industries should follow the sustainability principles proposed by the International Council of Mining and Metals or keep up to date with contemporary global enterprises such as the Extractive Industries Transparency Initiative (EITI). Even then, doubts remain as to whether these global solution templates would be useful for land acquisition, since private coal mining in Jharkhand not only involves the violation of constitutional and legal provisions, it also illuminates the inadequacies of the regulatory framework to safeguard the poor. A cosmetic reform of mining-related laws, such as those under consideration, is not the panacea. Land grabs for captive coal mining occur within a paradigm in which the contravention and manipulation of the law and the regulatory framework may be allowed with formal or tacit support of the state machinery. When companies face resistance and opposition in the process of acquiring land, they resort to informal strategies and methods which the regulatory framework is unable or unwilling to detect and control.

The critical issue is that the existing regulatory framework takes no account of, and makes no provisions to mitigate, the social impacts before and after the mining license is granted. Greater awareness of the need to protect environmental integrity has ushered in the requirement of environmental clearance, although the extent of compliance to regulations remains low.[28] With regard to the society, however, the laws presume a citizenry equal in every respect, while ignoring material and cultural impacts as part of the sustainability agenda, which are seen as amorphous and unscientific. In the circumstances, one's identity is poised between that of an anti-mining activist or a pro-mining sycophant. Such generalisations lead to the production of a set of simplistic policy instruments such as the gamut of short, medium, and long-term recommendations that were offered by a World Bank visiting team (2007).

It is not enough to reform the laws – such as those currently underway[29] – in absence of the state's capacity to act as an arbiter in relation to the rights and interests at stake. Decision-making with regard to the acquisition of land needs to be the result of negotiation. A negotiated justice regime might greatly improve mining governance compared to current products of adjudication or administrative

---

28　See the recent study released by the CSE that shows how easily environmental clearances are given in India to divert forest land to non-forest useshttp://www.cseindia.org/userfiles/Overview.pdf viewed on 20 Dec 2011.

29　See http://www.prsindia.org/uploads/media/minesamendment.pdf for the new Mining Bill and http://rural.nic.in/sites/downloads/general/LS%20Version%20of%20LARR%20%20Bill.pdf for the amendments in LAA and R&R Bill.

decision-making, or purely private individual decision-making. These negotiations to secure social justice enshrine a liberal idea of self-determination that is embodied in FPIC/PH (Szablowski 2010, 112–3). Communities have not enjoyed personal autonomy or state and judicial representation to the extent that might make FPIC a useful instrument. Mining companies know that the poor are never ceded the same bargaining power, and dissuade civil society organisations from participation in PHs. Although their introduction is needed, one must remember that neither the addition of SIAs, nor continuous engagement processes can fully address the gross imbalance of power.

We propose an overhaul of mineral governance in India. The state claims the sovereign ownership of all mineral resources is a result of the colonial past. It was during colonial times that dispossession of the poor located in resource peripheries began. In the neoliberal present, this dispossession has been exacerbated and has assumed more blatant forms with state assistance. Therefore, modifications or legal reforms of mining are not enough; in order to reshape the economic development of India's mineral belts there is an urgent need to think about radical and fundamental changes in ownership.

The gross injustices effected by corporatised enterprises exemplified in this paper have led to activists calling for a moratorium on mining. It is impractical to want to stop one of the most ancient human endeavours that have formed the backbone of human civilisation. We need to develop a new mineral extractive paradigm, one that does not inflict deep wounds on nature and people, and one that is neither state nor corporate-dependent. Towards this goal, we propose an overhaul of mineral resource governance before further mining leases are granted. For this, possible alternatives to ownership by the state of all minerals and the use by corporate entities of these resources need to be raised. This is not impossible; India has examples of community-led management in other areas of natural resource extraction and use. India's water users' groups, for example, were emulated by many other countries. People's rights to forest lands and forest resources are in the process of being recognized and enshrined in laws. Various modes of participatory management of natural uses are now being applied throughout the country.

To ensure that poor communities receive full benefits from the mineral resources of their land so that their livelihoods are transformed for the better, royalties should be paid to a tribal or local council on the amount of resource extracted. Those directly affected by mining would receive a proportion of the royalties, while the remainder would benefit the lot of those who used the land and its resources, sharecroppers, and living in the surrounding area. The proposed MMDR Bill 2011 provides for sharing 26 per cent of the profit for coal and lignite mining and amounts equal to the royalty in case of other major minerals.[30] Instead of an easy one-stop shop,developing this elaborate framework would require extended time for agreement-making and consent-seeking processes. No mining

---

30 Draft   MMDR   is   available   athttp://pib.nic.in/archieve/others/2011/sep/ d2011093002.pdf viewed on 20 December, 2011. See sections 43 and 56.

project should proceed without the consent of tribal representatives and poor landowners; they should have the right to refuse a project. Unless they have this pivotal right, no situation could be truly just. The proposed LARR Bill 2011 makes it mandatory to obtain the consent of the affected population only in case of land being acquired for private companies and for Public-Private Partnership projects. When the government acquires land for its own projects or private companies are buying land from farmers for their projects, the need to obtain consent has not been deemed necessary. Currently, it is necessary only to consult the gram sabha before initiating a development project in the Fifth Schedule areas, under the PESA Act. We suggest that it should be amended to make it mandatory to obtain the consent of the gram sabha. This requirement should not be restricted only to the Fifth Scheduled areas but be extended to the entire country. Micro-level understanding of the flow of money into and changing power relations within the community would be essential to optimise the level of royalties and consequent benefits and continuing impacts on the communities. Civil society organisations could manage the scheme with regular inspections from the Central government and in consultation with the states.

References

Bhushan, Chandra and Monali Zeya Hazra. 2008. 'Mining in the sates: Jharkhand and West Bengal', in *Rich Lands, Poor People: Is 'Sustainable' Mining Possible? State of India's Environment: Sixth Citizens' Report*. New Delhi: Centre for Science and Environment.

Choudhary, Tarun. 2009. 'Use of eminent domain: Process and its critique', in *India Infrastructure Report 2009: Land – A Critical Resource for Infrastructure*, 3iNetwork, OUP, pgs. 75–6, also available online on http://www.iitk. ac.in/3inetwork/html/reports/IIR2009/IIR_2009_Final_July%2009.pdf

Corbridge, Stuart. 2004. 'Competing inequalities: The scheduled tribes and the reservation system in India's Jharkhand', in S. Corbridge, S. Jewitt and S. Kumar (eds) *Jharkhand: Environment, Development and Ethnicity*, New Delhi: Oxford University Press, pp. 175–202. Originally published in *Journal of Asian Studies*, 59(1).

Guha, A. 2006. 'Dispossession of peasants by industrial projects', in S. Jain and M. Bala (eds) *The Economics and Politics of Resettlement in India*, Delhi: Pearson Longman, pp. 155–73.

Harvey, David. 2003. *The New Imperialism*, Oxford: Oxford University Press.

Jewitt, Sarah 2008. 'Political ecology of Jharkhand conflicts', *Asia Pacific Viewpoint*, 49(1): 68–82.

Jha-Thakur, Urmila and Fischer, Thomas. 2008. 'Are open-cast coal mines casting a shadow on the Indian environment', *International Development Planning Review* 30(4): 441–59.

Lahiri-Dutt, Kuntala. 2007a. 'Illegal coal mining in Eastern India: Rethinking legitimacy and limits of justice', *Economic and Political Weekly*, XLII (49, December 8–14): 57–67.

Lahiri-Dutt, Kuntala and Nesar Ahmad, 2012. Considering gender in Social Impact Assessments, in Frank Vanclay and Ana Maria Esteves (eds) *New Directions in Social Impact Assessments: Conceptual and Methodological Advances*, Cheltenham: Edward Elgar.

Land Action Research Network 2011. 'Global land grabs: Investments, risks and dangerous legacies', *Development*, 54(1): 5–11.

Levien, M. 2011. 'Rationalising dispossession', *Economic and Political Weekly*, Vol XLVI, No. 11, March 11, 2011, pp. 66–71.

Morris, Sebastian and Ajay Pandey 2009. Land markets in India: Distortions and issues, in *India Infrastructure Report, 2009: Land – A Critical Resource for Infrastructure*, IDFC, New Delhi: Oxford University Press, pp. 13–19.

Polanyi, Karl 2001 [1944] *The Great Transformation: The Political Economic Origins of our Time*, Boston: Beacon Press.

Randeira, Shalini. 2003. 'Glocalization of law: Environmental justice, World Bank, NGOs and the cunning state in India', *Current Sociology*, 51(3–4): 305–28.

Sharan, Ramesh. 2009. 'Alienation and restoration of tribal land in Jharkhand', in Nandini Sundar (ed.), *Legal Grounds: Natural Resources, Identity, and the Law in Jharkhand*, pp. 82–113.New Delhi: Oxford University Press.

Sharma, M. 2003. *Implications of the CBA Act and Coal India R & R Policy on DPs and PAPs of Rajmahal Opencast Coal Mine Expansion Project, with focus on Necessity for Gender-Sensitive Rahabilitation Programme Development*, Report Submitted as a Project, Indira Gandhi National Open University, New Delhi.

Singh, Kanhaiya and Kaliappa Kalirajan 2003.'A decade of economic reforms in India: The mining sector', *Resources Policy*, 29: 139–51.

Singh, Ram 2010. Ending misuse of Land Acquisition Act in the *Economic Times* dated 18 Nov. 2010, available on http://economictimes.indiatimes. com/opinion/comments--analysis/Ending-misuse-of-land-acquisition-laws/ articleshow/6944908.cms). Viewed on 8 December, 2010.

Stuligross, David 2008. 'Resources, representation, and authority in Jharkhand,India', *Asia Pacific Viewpoint*, 49(1): 83–97.

Szablowski, David 2010. 'Operationalising Free, Prior, and Informed Consent in the extractive industry sector? Examining the challenges of a negotiated model of justice', *Canadian Journal of Development Studies*, 30(1–2): 111–30.

Upadhyay, Carol. 2009. 'Law, custom and adivasi identity: Politics of land rights in Chhotanagpur', in Nandini Sundar (ed.), *Legal Grounds: Natural Resources, Identity, and the Law in Jharkhand*, 30–55. New Delhi: Oxford University Press.

Walker, Kathy le Mons 2011. 'Neoliberalism on the ground in rural India: Predatory growth, agrarian crisis, internal colonization, and the intensification of class struggle', *Journal of Peasant Studies*, 35(4): 557–620.

World Bank 2007. *Jharkhand: Addressing the Challenges of Inclusive Development*, Washington, DC: World Bank.

Reports and other Documents Consulted

Environment Impact Assessment Report, submitted by Essar Power Jharkhand Limited (nd).

Environment Impact Assessment Report, submitted by NeelanchalIspat Nigam Limited (nd).

Environment Impact Assessment Report, submitted by Chitarpur Coal and Power Ltd (nd).

Environment Impact Assessment Report, submitted by Eastern Mineral Trading Agency (nd).

Environment Impact Assessment Report, submitted by Nico Jaiswal (nd).

Pamphlet, Karanpura Bachao Sangharsh Samiti (nd).

Brochure, Abhijit Group (nd).

News Articles from *Prabhat Khabar*, Hindi Newspaper published from Ranchi (various dates).

Chapter 9

Coal Mining in Northeastern India in the Age of Globalisation[1]

Walter Fernandes and Gita Bharali

Introduction

In spite of being the largest producer of mica, the third of coal, lignite and barytes, fourth of iron ore, sixth of manganese and bauxite, and eleventh of crude oil in the world, the Indian mining sector is facing difficult challenges. Many of these challenges are relatively new in nature, and are connected to the rapid liberalisation and globalisation of the economy which has taken place in recent decades. To reduce its dependence on coal as the primary source of energy, the Indian state has been laying increasing stress on increased uranium mining. An ever widening gap between power production and demand has brought the country to a critical juncture where mining and related policies are being changed to attract foreign investment and to facilitate higher mineral production. As a consequence, coal production is expected to be raised from 400 million metric tonnes (Mt) at present to 1,400 Mt in the near future. Large blocks of mineralised land are being leased for coal and uranium mining, and petroleum and natural gas exploration. This has led private mining companies to eye mineral-rich land in the tribal areas of East and Northeast India.

In this gigantic effort to increase power production, what receives inadequate attention are the ill effects of ruthless and indiscriminate mineral exploitation, particularly the negative impacts on the sustainable livelihoods and fundamental rights of the people who relied on that land that has been acquired for mines. The newly promulgated National Mineral Policy 2008, for example, focuses on the need to change the laws and build the infrastructure required to speed up mineral production through private investment. It does not mention that villages, settlements and forests have disappeared and mining regions experience environmental degradation and ecological imbalance (Ministry of Mines 2008b). Dispossession, displacement and disorganisation of local – especially tribal – communities have led to people's marginalisation. The natural rights of the people

---

1 A different version of the article was published online on the India Environment Portal, http://indiaenvironmentportal.org.in/files/Dr%20walter%20fernandes.doc [Accessed 21 May 2013].

to land and forest resources and other means of sustenance are violated. Mining activities have made people homeless, landless, jobless and insecure.

The policy changes thus mentioned as well as the impacts of mineral exploration and exploitation are important for Northeast India. This region produces coal, petroleum, some metal ores, and also has uranium deposits. This chapter will look at mining in general, with special emphasis on coal. Although data on different types of mines have not been disaggregated, most of the information pertaining to Central, Eastern and Northeastern India concerns coal. The Northeast also possesses uranium mines and the region as a whole has gained importance because of its uranium reserves. Coal, however, remains equally important, particularly for Assam and Meghalaya in the North-eastern region.

The Extent of Mining in India

India possesses significant mineral resources. Until 1951 India mined 24 minerals with a total value of Rs 173 million. In 2000, India produced 89 minerals: 6 of them for fuel; 11 metallic; 52 non-metallic and 22 minor valued at Rs 568 billion. The value of minerals other than petroleum and natural gas was Rs 306,751 million. The main metals produced are iron-ore, copper-ore, chromite and zinc concentrates, gold, manganese, bauxite and lead concentrates. More than 90 per cent of the value of non-metallic minerals comes from limestone, magnesite, dolomite, barytes, kaolin, gypsum, apatite and phosphorite, steatite and fluorite. Eighty per cent of the minerals extracted in India comprises coal and 80 per cent of the mines are privately owned, but the public sector accounts for 91 per cent of the mineral value (Ministry of Mines 2008a).

In the 1980s the mining sector employed 800,000 persons, accounted for 3 per cent of the gross domestic product (GDP) and 11.5 per cent of industrial sector production. Today, mines employ only 560,000 persons and this number is declining because of mechanisation (Moody 2007, 19). The central government has enacted many laws and bylaws to support mining. First enacted in 1957, and subsequently amended almost every four years until 1999, the *Mines and Minerals (Development and Regulation) Act* (MMDR) is meant to govern the mining sector. The MMDR classifies minerals as minor and major, lays down procedures for granting reconnaissance permits, prospecting licences and mining leases, and prescribes violations. States control minor minerals such as clay and sand, while the major minerals such as iron ore and coal are under the centre. Apart from the MMDR, mining is also subject to the *Mines Act* of 1952, the *National Mineral Policy* and the land acquisition and environment protection Acts (Sethi 2007, 68). Mining may be underground or open cast. The former employs more workers, but has a higher toll in fatalities and occupational disease. The latter occupies a much larger acreage of fertile land. Other than coal, hardly any mineral is extracted from underground mines. That raises the issue of livelihood loss and the health hazards of mining. Mining operations in India have displaced thousands of people from their land.

This section will consider these complex issues. It places the *National Mining Policy* at the forefront, which acknowledges that indigenous (tribal) people's territories host most minerals and are consequently targeted by both private companies and governments. Many of the mineral reserves also occur in protected areas and biosphere reserves. This explains why a very large proportion of the hundreds of thousands of persons displaced by mining in India belong to the tribal communities that consider themselves indigenous to the land.

Impacts of Mining on Local Communities

In India development projects have displaced an estimated 60 million persons between 1947 and 2000 (Fernandes et al. 2012, 17). Of this number, two major categories are recognised, namely, displaced persons (DPs); or people deprived of their livelihoods without relocation, known as project-affected persons (PAPs). Accurate information or estimates on mining-induced DP-PAPs exist in some states, but no aggregated national data is available. Mining is concentrated in Andhra Pradesh, Tamil Nadu, Orissa, West Bengal, Jharkhand, Meghalaya, Assam, Maharashtra, Chhattisgarh and Madhya Pradesh. From 1951 to 1995, mining in Andhra Pradesh caused 100,541 DP-PAPs from 37,368.82 acres (Fernandes et al. 2001, 59); in Orissa 300,000 persons from 166,047 acres (Fernandes and Asif 1997, 84); and in Jharkhand, 402,882 persons from 208,552 acres (Ekka and Asif 2000, 93). The limited information in Assam shows that 11,394.75 acres acquired for mining projects have deprived 41,200 of their livelihood between 1980 and 2000 (Fernandes and Bharali 2011, 137). In the same period in West Bengal, mining has displaced 418,061 persons (Fernandes et al. 2012, 146–7). In Goa, 4,740 persons have been displaced (Fernandes and Naik 2001, 39) during the period 1965–95.

One may add that among the states where studies of displacement have been conducted, only Goa does not have any DP-PAPs from coal mining. In Assam and Andhra Pradesh they are divided equally between coal and other minerals. In the remaining states, more than 80 per cent of the DP-PAPs are caused by coal mining. No study has been undertaken until recently in the areas of Chhattisgarh, Maharashtra, Madhya Pradesh and Uttar Pradesh, each of which holds coal reserves. Moreover, most figures that one receives from the various studies are gross underestimates. In Jharkhand, for example, accurate data could not be obtained for common land and common property resources (CPRs), which constitute the main type of land that is used for mining. In most states, very limited information is available for all types of mines until 1980 and in the Northeast until 1990. In Goa, most mining is undertaken through private leases and there is very little information on their DP-PAPs. Thus, the number of DP-PAPs since Independence is necessarily higher than the figures currently available. Moreover, states with high-intensity mining such as Chattisgarh, Madhya Pradesh and

Maharashtra are not included in the list. It is difficult to estimate the numbers of DP-PAPs; nonetheless, the total is likely to be near to 10 million, around seven million of which would be caused by coal mining.

Since much of the land used for mining is Common Property Resources (CPRs), very few of its DP-PAPs receive adequate compensation since the colonial land laws that remain in force in the country recognise only *patta* land, i.e., land with individual title. The CPRs are considered State property, although they are the main source of people's sustenance (Sethi 2007, 71). For example, all the Manjhi–Manjhaar tribal and dalit DP-PAPs of Bharat Aluminium Company (BALCO) in Mainpat, Chhattisgarh, were promised compensation, but only 50 families received it at the rate of Rs 12,000 per acre in 1992. However, the government set the rate at Rs 50,000. The remaining 62 CPR-dependent families were ignored (Agrawal 2007, 105). The compensation paid for the little private land they own is low because the regions they inhabit are considered backward and land is compensated at "market value" which is interpreted as the average of three years' rent or the registered sale price, which is invariably far below market value, as it is commonly known that not more than 40 per cent of the price is registered in order to save on taxes. Since very little land is sold in the tribal areas (Fernandes et al. 19–20), even the DP-PAPs who are paid compensation in the coal mining areas are unable to buy any other land to replace that which they lose. For example, the Mahanadi Coal Field of Orissa took away 30 acres of Manglu Pradhan's land. With the compensation amount he could buy only 5 acres, in a distant area 25 km from his village (Panda 2007, 115). Those who depend only on the CPRs are dispossessed without any alternative.

The occurrence of resettlement is weak for all the projects, but is lower in mining than in other fields. Orissa resettled 35.27 per cent of all its DPs between 1951 and 1995 (Fernandes and Asif 1997, 135); Andhra Pradesh, 28.82 per cent (Fernandes et al. 2001, 87); Kerala, 13.8 per cent (Muricken et al. 2003, 185–9); Goa, 40.78 per cent during 1965–1995 (Fernandes and Naik 2001, 123–4); West Bengal, 9 per cent during the period 1947–2000 (Fernandes et al. 2012, 240–43); Gujarat, 19 per cent (Lobo and Kumar 2009, 239); and Assam, the DPs of only about 10 projects during 1947–2000 (Fernandes and Bharali 2011, 222–3). An alternative to resettlement is jobs in the project. In reality very few projects give jobs to displaced persons. Even many of those who gain employment lose those positions later. For example, Mahanadi Coalfields gave a number of unskilled jobs to its DPs and dismissed them after a short period of time since it did not consider them physically fit (Panda 2007, 116). The North Karanpura Coalfields in Jharkhand used to offer one job for three acres of land acquired. But about 50 per cent of the families owned less than three acres, thereby obviating the requirement for the company to provide jobs; most of those who were employed were non-tribal encroachers on tribal land (Roy 2007, 151–60). The DP-PAPs of BALCO were promised positions, but the promise was not kept (Agrawal 2007, 105). The Vedanta major bauxite sites spread across Chhatisgarh and Orissa appointed some DPs as labourers on a payment of Rs 60 per tonne of ore delivered (Moody 2007, 90).

Loss of Livelihoods

Although mining is a source of wealth for some segments of society, it also destroys the source of subsistence of a large number of poor people – the DPs and PAPs. The result of displacement, poor resettlement and meagre compensation results in a cycle of exploitation, poverty and violence (Helmut et al. 2006). It deprives most of the affected persons of their land and work and renders much of the remaining land uncultivable. As a consequence, even those who continue to work on their land are unable to live on it (Padel and Das 2007, 24). A high rate of landlessness is thus its first impact. Landlessness is typical of all DP-PAPs, but it is higher in mining than in other areas. For example, in Andhra Pradesh, landlessness grew by 41.61 per cent among all the DP-PAPs, but by 83.72 per cent in mining (Fernandes et al. 2001, 112–13). The impacts also have a caste-tribe bias. For example, in Orissa, the rise in landlessness was 16.7 per cent more among tribal and 13 per cent more among dalit DPs (Pandey 1998, 180) than among others. In the North Karanpura Coalfields in Jharkhand where most of the 1,520 displaced families are tribal or dalits, the area cultivated per family declined from an average of 4.41 acres before displacement to 0.57 acres after it (Roy 2007, 151–60). Also, support mechanisms such as ponds, wells, and livestock that supplement the family income have declined. For example, the DP-PAPs of Mahanadi Coal Fields lost all the *Mahua* trees from each of which they used to earn around Rs 1,000 a year (Panda 2007, 115).

Access to work generally tends to decline among the DP-PAPs. For example, in Assam it declined from 77.27 per cent to 56.41 per cent (Fernandes and Bharali 2011, 244–6) and in West Bengal, from 91.02 per cent to 53.18 per cent (Fernandes et al. 2012, 312). Downward occupational mobility is its immediate result. For example, in Assam 50 per cent of the cultivators became daily wage earners or domestic or other unskilled workers (Fernandes and Bharali 2011, 290–91). Due to lack of other work, many become daily wage earners in the mines on land that was once theirs (Panda 2007, 115). In most States studied, over 50 per cent of the displaced families have dropped below the poverty line. In Assam, 56 per cent and in West Bengal, 49 per cent of them have removed their children from schools and turned them into child labourers. The proportion of land losers and child labourers is around 70 per cent in the coal mining areas because much of the land was comprised of CPRs and inhabited by the tribes who were deprived of it without alternatives (Fernandes and Bharali 2011, 317–19).

Liberalisation and Mining Impacts

The situation is expected to deteriorate with liberalisation: first, because of loss of jobs as the enumerated data demonstrate. The main reason is open cast mining and mechanisation. Because of the status given to open-cast mining, land acquisition is higher than in the past (Rao 1990, 62), and as a consequence environmental degradation is increasing. Because of mechanisation mines are larger in area than in the past. The average size of a mine grew from 150 acres in the 1970s to around 800 in the 1980s (Fernandes and Asif 1997, 74–5) and is around 1,500 acres today. This results in greater displacement than in the past. Because fewer jobs are created, the DP-PAPs may not even obtain exploitative low-wage work, but instead displaced and ignored (Fernandes 2008, 93).

Because of mechanisation, the situation of the mining DP-PAPs is to some extent worse than that of those displaced from other activities. As stated earlier, the number of jobs has declined from 800,000 to 560,000. Moreover, most DP-PAPs do not have the skills required for mechanised jobs, which go to outsiders while the DP-PAPs remain unemployed. For example, in the 1980s, because of mechanisation the National Aluminium Company Limited (NALCO) mines in the Koraput district, Orissa, created a little over 300 jobs for the transport of bauxite to the smelter. All these jobs went to outsiders since the local tribal peoples did not have the skills required. Had they used traditional transport, they would have created about 10,000 jobs which would have gone to the DP-PAP of the mines, the smelter plant and the Upper Kolab dam that were built around the same time (Pattanaik and Panda 1992). Even in this case, women feel the negative impact more than men do. For example, until the 1980s, coal mines gave unskilled jobs to women: their proportion in the sector was 30–40 per cent. At the turn of the millennium, it had decreased to 12 per cent of a smaller number of workers (Bhanumathi 2002, 21).

Environment and Health

Studies show that loss of culture and the destruction of social structures because of land loss is worse in mining than in other areas because a very large number of the DP-PAPs are in the forest and other remote areas where the dependence on CPR is high. Displacement from this livelihood results in complete marginalisation of the communities depending on them, and most of them are tribal (Padel and Das 2007, 24). Environmental degradation has increased in mining projects; open cast mining involves removal of vegetation and topsoil, displacement of fauna, release of pollutants and production of mine overburden. Land excavation, discharge of mine pit and waste water, dumping of waste rock, tailings or slag, and the discharge of metallic smoke and dust into the atmosphere cause pollution. The disposal of water from the refining units and metallurgical industries causes environmental problems. Air pollution includes dust from mining and acidic gases, carbon dioxide and other greenhouse gases from smelting, refining and other kiln operations. A tonne of aluminium is estimated to produce 4–8 tonnes of toxic red mud and 13.1 tonnes of carbon dioxide. Bauxite mining reduces a mountain's water-retaining capacity (Padel and Das 2007, 33). For example, the Jharia mines in Jharkhand produce the best-quality coking coal in India, but the area, mostly tribal, has been smouldering with underground mine fires for several decades. The fires emit huge quantities of carbon monoxide, carbon dioxide, sulphur dioxide and methane, leading to air pollution, breathing problems and skin diseases (Pradhan 2006). In Mainpat, once known as the only hill station in Chhattisgarh, black dust fills the air because of bauxite mining (Agrawal 2007, 103). As a result, health hazards are considerable.

Despite these hazards, the *National Mining Policy* (NMP), accords very little importance to the people and the environment. It is obvious that much thought has gone into its formulation. It discusses every aspect of mining, such as the preliminary survey, prospecting, exploration, mining and improvement of methods. Its focus is on private capital and the state and the optimal use of resources to ensure these ends: 'In future the core functions of the state in mining will be facilitation and regulation of exploration and mining activities of investors and entrepreneurs, provision of infrastructure and tax collection' (No. 4). 'It will encourage technology required for cost-reduction, human power development and economic aspects' (No. 7). The people to be affected by mining are mentioned only in the context of exploiting small deposits. Its social and environmental impact does not receive the same attention as its production does. Thus, the policy is thorough from the perspective of the investor, but the people who will be deleteriously affected by it are ignored.

Coal, Petroleum and Minor Metals in the Northeast

Compared to other Indian states, the Northeast has very few mines. However, the region is attaining importance because of its petroleum, uranium and coal reserves. The following section will examine the implications of mining in Northeast India.

The Gazette data show that only 105.47 acres have been acquired in Assam for mining. Apart from it, mining leases in Assam occupied 5,378.07 acres in 1991. Thus, the total used for mining was 5,483.54 acres in 1991, most of it for coal, followed by limestone. A total of 3,126.98 of these acres were on lease for 24 coal mines. The Northeastern Council website[2] reveals that 6,397.3 acres – 26 square kilometers – were on lease in Dibrugarh district for coal and more in Sibsagar in 2002, but it does not provide the exact area in the latter. It states that 4,940 acres – 20 sq km – are being explored in the Dibrugarh–Sibsagar belt and that coal is available in the Tinsukia, Karbi Anglong and North Cachar Hills districts. But exploration does not appear to have started. Thus, the land under lease for coal was around 8,000 acres in 2005. Assam follows the pattern of the remaining states of having fewer mines than in the 1980s, but the area covered is nonetheless greater than in the past (Fernandes and Bharali 2011, 136–7).

Apart from today's main minerals, coal and limestone, Assam has eight leases for crude oil, each of them of over 1,000 acres. In addition to these leases is the Nazira township, which has been built for this purpose. Thus, the total area of

---

2   Available at www.nerdatabank.nic.in [Accessed May 15 2013].

land under petroleum prospecting is not less than 10,000 acres. Together, all the minerals, including oil exploration, have occupied not less than 20,251 acres in Assam alone (Director of Economics and Statistics 2003, 128; Indian Bureau of Mines 1991, 17). Much information regarding other leases was unavailable for this study, so the reality is certainly higher than this figure, which errs on the conservative side. Moreover, the little information available is only from the 1990s, though mining had begun in the state long before that. Thus, both the land used and DP-PAPs are likely to constitute a figure at least double the amount cited.

Coal is also available in Meghalaya and Arunachal Pradesh, and petroleum prospecting has commenced in Nagaland. Much land is being taken over for natural gas exploration in Tripura. There are reasons to believe that a considerable proportion of inter-state border disputes concern rights and access to mineral-rich land. For example, the border dispute between Nagaland and Assam is mainly for the coal and oil-rich land in the Merapani region of Assam and the neighbouring Naga areas (Kikon 2009, 69). Anonymous sources in Nagaland informed the researchers that prospecting for oil began in that state within three months of the ceasefire agreement between the Government of India and the Nationalist Socialist Council of Nagalim being signed in 1997. Information on coal mining in Meghalaya is difficult to obtain partly because it is undertaken by private individuals since the land comes under the sixth schedule and much of the coal mined is smuggled to Bangladesh.

Data from the traffic control police indicate that more than 1,000 trucks transport coal from Meghalaya to the surrounding areas, including Bangladesh, every day during the mining season. Though they are supposed to contain a maximum of 16 tonnes, coal trucks typically transport not less than 20 tonnes each. In other words, not less than 20,000 tonnes of coal is transported to the rest of India during the eight month dry season when coal can be mined. An equal amount of coal is estimated to be sent to Bangladesh by waterways and other routes. This large-scale transport of coal altogether indicates that several thousand acres are being mined.

Of importance is the prevalence of child labour and overall dangerous conditions of work under exploitative conditions. Understandably, the poor take risks to work in the coal mines because the wages they receive are higher than what they are able to earn in other occupations. North Eastern Social Research Centre surveys show that child labour is prevalent in the coal mines of Meghalaya because of the nature of the mining processes which are being used. Much of the mining is conducted underground through very small access shafts referred to as rat holes in which children will find it easier to operate. It involves the workers lying on their backs in the narrow tunnel for hours at a time. Such mining practices have serious health hazards, as demonstrated by recent deaths due to accidents in these mines.

Uranium Mining in Meghalaya

Though this chapter predominantly concerns coal mining, the picture in the Northeast in general and Meghalaya in particular is bound to remain incomplete without summarising the circumstances surrounding uranium mining, particularly as a major conflict is in progress in the Khasi Hills of Meghalaya. Until 2003, uranium mining in India was confined to Jadugoda in East Singhbhum district of Jharkhand, where the Uranium Corporation of India Limited (UCIL) owns three underground uranium mines. These mines became the foundation of India's nuclear fuel stocks. UCIL has however discovered uranium reserves at a number of locations across the country including at Domiasiat in the Khasi Hills and in Garo of Meghalaya, Karbi Anglong in Assam, and the Kameng district in Arunachal Pradesh, the Bhima Basin in the Gulbarga district of Karnataka, the Yellapur–Peddagattu area of Nalgonda district and Lambapur near Hyderabad, Andhra Pradesh, and Bandohoranga near Jamshedpur, Jharkhand. The ore in these mines appears to be of better quality than that being mined at Jadugoda. The company is in the process of acquiring land at Bandohoranga even though the Meghalaya and Andhra Pradesh governments are yet to approve mining within their respective states. UCIL plans to invest Rs 31 billion in developing the mines at Domiasiat in Meghalaya (Dubey 2007; Subbarao 2007). In addition, UCIL is exploring uranium deposits in Chhattisgarh and Rajasthan (PTI 2006).

Uranium mining affects people in two ways: first, it deprives them of their livelihood by displacing them from their land; and second, it destroys their environment. Displacement rates will increase, since the justification given for the Indo–US Nuclear treaty is that nuclear power is a source of sustainable energy (Dubey 2007). Studies of miners in Canada, Sweden, Czechoslovakia and India (Venkataraman 2007) show that uranium mining can release radioactive material into the environment, and plutonium – a by-product – can cause cancer. Problems such as severe birth deformities have been found at the Jadugoda mine area in Jharkhand (Dias 2007, 143–5).

We now focus on the mines in the Kylleng–Pyndeng–Sohing area of Domiasiat in the West Khasi Hills. This mineral-rich district is the largest in area in Meghalaya with significant reserves of coal, limestone, uranium and granite. In 1972, the Atomic Mineral Directorate for Exploration and Research (AMDER) declared Domiasiat a potential area for uranium mining. This was confirmed in 1986 and subsequently drilling was conducted from 1992 to 1996. However, AMDER abandoned work temporarily because of protests from local people and social-developmental organisations (Das 2003). The growing nuclear arms race in South Asia has renewed pressure to exploit the Domiasiat yellowcake[3] reserves. In December 2007, the Union Ministry of Environment and Forest gave conditional clearance to UCIL to proceed with uranium mining, based on the meeting minutes

---

3 Yellowcake is a uranium concentrate powder, an intermediary product made from uranium ore.

from the environmental public hearing (Meghalaya State Pollution Control Board 2007). Some say that the minutes of the Meghalaya State Pollution Control Board are contrary to what was stated at the public hearing.

The Khasi Students' Union (KSU) held a week-long anti-uranium campaign around the time of the hearing. Their activists picketed state government offices on 4–5 June 2007 and called a 36-hour *bandh* beginning 11 June. Earlier, KSU had reportedly given an ultimatum to the state government to stop the hearing or face agitation, but the Pollution Control Board went ahead with it on 12 June 2007 (Dwaipayan 2008). KSU also organised a night road blockade from 19 June 2007. This agitation is one of many signs that the project has met with significant local opposition which has polarised Meghalaya, with the former Khasi Hills Autonomous District Council Chief Executive Member H. S. Shylla in favour of the project and the student body and many civil society groups advocating against it (*The Shillong Times* 2007). In a recent announcement the local Congress government stated that it would not take any hasty decision without properly studying planning documents and consulting with experts.

The main reason for the opposition is displacement and environmental hazards. As many as 30,000 people are likely to be displaced (Das 2003). There has been debate on the issue in the Meghalaya Assembly and Central Parliament. Students' associations, human rights and environmental groups, and health experts opposing the mine feel that Domiasiat is treading in the footsteps of Jadugoda in Jharkhand, where cancer is on the rise along with other inexplicable health problems believed to be caused by excessive radiation from the mines. Officially, cancer-related deaths of UCIL employees in Jadugoda have been debied. And the company told people in Domiasiat that no such risk exists, but the picture does not seem to be as positive as UCIL wants to make it appear (*The Shillong Times* 2007).

Amid this opposition, some villagers are reportedly welcoming the project because the area is under-developed. They are ready to be displaced and face environmental degradation in order to obtain the employment, better education and transport infrastructure which UCIL is promising them (PTI 2006). Anti-mining campaigners speak of the problems around mining and claim that since mining began, the Hills have witnessed a rise in instances of cancer, many villagers are suffering from mysterious diseases, and that miscarriages are increasing (Das 2003). Some studies point to 32 deaths from mining-related accidents (Ahmed 2005), but UCIL claimed these reports as well as the claim that children born to people living near the site would suffer from congenital defects as false (Das 2003). Claims and counter claims remain irreconcilable to date leaving entrenched positions either for or against the mining with little space in between.

The KSU, the Meghalaya People's Human Rights Council, and the Hynniewtrep Environment Status Preservation Organisation organised an open session in July 2007 in Shillong where UCIL was invited. For the first time, senior officials of the mining agency came with a team of geologists and medical experts to seek to persuade the audience that uranium mining was not as dangerous as was popularly believed. Representatives of the Atomic Minerals Division claimed that the uranium found in Domiasiat was of 'very low grade' and would be used for generating power and not to make nuclear weapons. They thus hoped to convince people of the benefits of mining. But as yet no verified studies have been undertaken which can provide authoritative and independent information on uranium mining and the social, environmental and health risks that people face.

A number of social organisations query the need for development in Domiasiat. They argue this in relation to a lack of development in the Northeast in general, and in this area in particular. They feel that the promise of improved infrastructure and educational facilities is only meant to convince the people to leave their land. They feel that these promises will not be kept once the land has been surrendered. They thus focus on the right of people to development that is not subordinated to the nuclear race, maintaining that the dialogue should shift to actual development which benefits people in the area and that all projects should be situated in relation to this.

Conclusion

The discussion in this chapter of mining in India in general, and in the Northeast in particular, raises questions not merely in relation to the mining sector, but also concerning the paradigm of development which is being pursued at present. Mining appears certain to become more important in the coming years than it has been in the past, and mineral exploitation is presented as development. Some feel that this is an example of the development of mineral resources at the cost of poor people. And yet the new national mining policy is proceeding with the promotion of more mining. Globalisation appears to add to the woes of those paying the price of mining-based development. The Indo–US nuclear deal and the arms race that has engulfed South Asia are additional high-level issues that compound the problems of dispossession and environmental degradation, each of which comprises the misery of thousands suffering the effects of uranium and of other forms of mining.

References

Agrawal, Neeraj. 2007. 'How Green was My Mountain.' *Caterpillar and the Mahua Flower*, edited by Rakesh Kalshian, 103–23. New Delhi: Panos South Asia.

Ahmed, Kazimuddin. 2005. 'What Lies Beneath and Above.' *The Telegraph, Northeast*, 15 January.

Bhanumathi, K. 2002. 'The Status of Women Affected by Mining in India.' In *Women and Mining: A Resource Kit*, 20–24. New Delhi: Delhi Forum.

Das, Snehasis. 2003. 'National Security or Development? Uranium Mining in Meghalaya.' Available online at www.thenakedeyefilms.com accessed on 29 July 2009.

Dias, Xavier. 2007. 'Never Say DAE.' In *Caterpillar and the Mahua Flower*, edited by Rakesh Kalshian, 140–50. New Delhi: Panos South Asia.

Director of Economic and Statistics 2003. *Statistical Handbook of Assam*. Guwahati: Directorate of Economic and Statistics, Government of Assam.

Dubey, Sunita. 2007. 'India's Uranium Nightmare.' *New America Media*, 19 March. Available online at www.newamericamedia.org accessed on 29 July 2009.

Dwaipayan. 2007. 'Uproar over Uranium Mining in Meghalaya.' *The Assam Tribune*, Guwahati, 22 February.

Ekka, Alex and Mohammad Asif. 2000. *Development-Induced Displacement and Rehabilitation in Jharkhand 1951–1995: A Database on Its Extent and Nature*. New Delhi: Indian Social Institute.

Fernandes, Walter. 2008. 'Sixty Years of Development-Induced Displacement in India: Impacts and the Search for Alternatives.' In *India: Social Development Report 2008: Development and Displacement*, edited by Hari Mohan Mathur, 89–102. New Delhi: Council for Social Development and Oxford University Press.

Fernandes, Walter and Mohammed Asif. 1997. *Development-Induced Displacement in Orissa 1951 to 1995: A Database on Its Extent and Nature*. New Delhi: Indian Social Institute.

Fernandes, Walter, Nafisa Goga D'Souza, Arundhati Roy Choudhury and Mohammed Asif. 2001. *Development-Induced Displacement in Andhra Pradesh 1951–1995: A Quantitative and Qualitative Study of its Extent and Nature*. New Delhi: Indian Social Institute, Guwahati: North Eastern Social Research Centre.

Fernandes, Walter and Niraj Naik. 2001. *Development-Induced Displacement in Goa 1965–1995: A Study on its Extent and Nature*. New Delhi and Panjim: Indian Social Institute and Indian National Social Action Forum.

Fernandes, Walter and Gita Bharali. 2011. 'Uprooted for Whose Benefit?' *Development-Induced Displacement and Deprivation in Assam 1947–2000*. Guwahati: North Eastern Social Research Centre.

Fernandes, Walter, Shanti Chhetry, Sherry Joseph and Satyen Lama. 2012. 'Progress: At Whose Cost?' *Development-Induced Displacement and Deprivation in West Bengal 1947–2000*. Guwahati: North Eastern Social Research Centre.

Helmut, Haberl, Helga Weizz and Heinz Schandle. 2006. 'Ecological Embeddedness – 1700–2000.' *Economic and Political Weekly*, 25 November, 4896–906.

Indian Bureau of Mines. 1991. 'Mining Leases in India 1991.' Nagpur.: Indian Bureau of Mines.

Kikon, Dolly. 2008. 'Ethnography of the Nagaland—Assam Foothills in Northeast India.' In *Land, People and Politics: Contest over Tribal Land in Northeast India*, edited by Walter Fernandes and Sanjay Barbora. Guwahati: North Eastern Social Research Centre and IWGIA.

Kikon, Dolly. 2009. 'Ethnography of the Nagaland-Assam Foothills in Northeast India.' In *Land, People and Politics Contest over Tribal Land in Northeast India*, edited by Walter Fernandes and Sanjay Barbora, 58–87. Guwahati: North Eastern Social Research Centre and IWGIA.

Lobo, Lancy and Shashikant Kumar. 2009. *Land Acquisition, Displacement and Resettlement in Gujarat 1947–2004.* Sage Publications.

Meghalaya State Pollution Control Board. 2007. *Proceedings of the Environmental Public Hearing held on 12th June 2007 at Nongbah Jynrin in West Khasi Hills District in Respect of the Proposed Kylleng–Pyndengsohiong Uranium Ore Mining & Processing Project of the Uranium Corporation of India Limited.* Shillong: Government of Meghalaya.

Ministry of Mines. 2008a. Indian Mineral Scenario. Available at: http://mines.nic.in/imsene.html [Accessed May 24, 2013].

Ministry of Mines. 2008b. *National Mineral Policy 2008*. New Delhi: Government of India.

Moody, Roger. 2007. 'Iron in the Soul.' In *Caterpillar and the Mahua Flower*, edited by Rakesh Kalshian, 11–23. New Delhi: Panos South Asia.

Padel, Felix and Samarendra Das. 2007. 'Agya, What Do You Mean by Development.' In *Caterpillar and the Mahua Flower*, edited by Rakesh Kalshian, 24–46. New Delhi: Panos South Asia.

Panda, Ranjan Kumar. 2007. 'Under a Black Sky.' In *Caterpillar and the Mahua Flower*, edited by Rakesh Kalshian, 114–23. New Delhi: Panos South Asia.

Pandey, Balaji. 1998. *Depriving the Underprivileged for Development.* Bhubaneshwar: Institute of Socio-Economic Development.

Pattanaik, Gopabandu and Damodar Panda. 1992. 'The New Economic Policy and the Poor.' *Social Action* 42(2): 201–12.

Pradhan, Kalpana 2006. 'In the line of Fire', Available at http://www.indiatogether.org/2006/nov/hlt-mining.htm [Accessed November 13 2009].

Press Trust of India (PTI). 2006. 'BARC Report Rules out Health Hazards from Proposed Uranium Mining in Meghalaya.' Guwahati, 27 June.

PTI 2006. 'BARC report rules out health hazards from proposed uranium mining in Meghalaya' New Delhi: Government of India, 27 June.

Rao, Gururaja A. V. 1990. 'Sardar Sarovar Project in Gujarat State: The Project and Rehabilitation Policy.' In *Workshop on Persons Displaced by Development Projects*, edited by Aloysius P. Fernandez, (June 27), 60–67. Banglore: ISECS and MYRADA.

Roy, Pradipto. 2007. 'Social Implications of Mining in the Northern Karampura Valley.' In *Managing the Social and Environmental Consequences of Coal Mining in India*, edited by Gurdeep Singh, David Laurence and Kuntala Lahiri Dutt, 151–60. Dhanbad: Indian School of Mines and University of South Wales: The Australian National University.

Sethi, Aman. 2007. 'Road to Perdition.' In *Caterpillar and the Mahua Flower*, edited by Rakesh Kalshian, 65–82. New Delhi: Panos South Asia.

Subbarao, Buddhi Kota. 2007. 'Killing them Slowly.' *India Together*, 19 November.

The Shillong Times. 2007. 'Police kills five militants in Meghalaya.' Shillong, 31 October.

The Shillong Times. 13 June 2007. 'Majority at Public Hearing Says No to Uranium Mining.' Shillong, 13 June.

Venkataraman, S. H. 2007. 'Nuclear Energy is Toxic.' *The Times of India*, Internet Edition, 21 November.

Chapter 10

Marginalising People on Marginal Commons: The Political Ecology of Coal in Andhra Pradesh

Patrik Oskarsson

Introduction

Major efforts are currently underway to bridge – or at the very least to slow down – the continued divergence between supply and demand of power in the south Indian state of Andhra Pradesh. As in most of the rest of India, the key resource in these efforts is coal, available in large quantities in the northern parts of the state along the Godavari river valley. At face value, the present coal mining expansion and the construction of new private sector thermal power plants is only weakly related to one another; coal mines are providing fuel to already existing power plants, while the new private plants planned along the eastern sea coast will use imported coal. Being among the few industrialised states in India possessing large coal deposits, Andhra Pradesh serves as an indicator of equity outcomes in terms of the land required for coal-powered industrialisation.

Controversies over land and land use have become intense in recent years across India, with highly debated and publicised protests, especially against Special Economic Zones (SEZs) (Banerjee-Guha 2008; Ramachandraiah and Srinivasan 2011). This is not surprising, since land and natural resource-based activities continue to support a majority of rural people (Mearns 1999). The Andhra Pradesh government has been among those most active in promoting SEZs and the state has also seen a number of other land-related controversies when irrigation dams and mines have been proposed (Balagopal 2007a; Balagopal 2007b; Balagopal 2007c). Dozens of open cast coal mines are planned over more than 16,000 hectares and each of the 23 approved thermal power plants are expected to require at least 1,000 ha. It is clear that the pursuit of coal-based power will require significant areas of land (Oskarsson 2011).[1]

While it is well-known that coal mining overwhelmingly takes place on land officially settled as forest on which the rights to individual agricultural and

---

1 Expanded coal mining is naturally in addition to already existing land use. Between 1995 and 2005 8,644 ha of forest land was diverted for coal mining in Andhra Pradesh (Ministry of Environment and Forests 2005).

common property livelihoods are denied (Bhushan and Zeya Hazra 2008), this chapter considers the wider political ecology of coal at state level by including its connection with thermal power plants. The expanding use of coal in Andhra Pradesh, including the power plants located in coastal wetlands, expropriates supposedly uninhabited commons despite the evidence that these are vital for the livelihoods of some of the state's poorest inhabitants. The political ecology of coal in the state thus comes with significant distributional concerns about who should be able to benefit from scarce land resources.

The first section of the chapter situates the political ecology at state level in the larger context of India's policies for power generation. This is followed by two sections which examine specific cases studies of a coal mine and a thermal power plant respectively, after which conclusions are drawn.

Major Coal Ventures on Marginal Lands

Across India a large gap exists between demand and supply of coal contributing to widespread power shortages. Despite a plan to mine 630 million tonnes of coal in India in 2011–12, a shortfall of 83 million tonnes will have to be imported (Ministry of Coal 2011). At the same time power generation, overwhelmingly dependent on coal-fuelled power plants, is another problematic area which has continuously failed to meet planned increases in output. Recent data detail large-scale plans with approval for an additional 200 thermal power plants across the country producing 220,000 MW of capacity (Kasturi 2011). Significant pressure thus exists to increase both coal and thermal power output.

Public sector Singareni Collieries is the sole coal mining company in Andhra Pradesh. At present it is working on 34 expansion projects at a cost of 10,000 crore (10 billion) rupees. If completed, these will add 40 million tonnes of additional coal, representing close to a doubling of present production rates. However, the expansion will also permanently remove large areas of land presently supporting local livelihoods. It is becoming increasingly apparent that although India has significant coal deposits, it is going to be politically as well as practically impossible to increase coal output to match expected demand (Ministry of Power 2012). As a densely populated country with increasingly strong social movements, India is likely to experience strong objections to further expansion of mining. The continued monopoly of the public sector in mining makes it unable to modernise and therefore unable to improve the quality and quantity of its production, a situation that results in private power producers beginning to search abroad for fuel alternatives.

Part of the solution to insufficient domestic quantities of coal is the use of imported coal. All along the Indian coast thermal power plants are planned to be established near to ports, ready to burn imported coal. One additional benefit for these plants is the use of salt water, rather than precious fresh water, for plant cooling. Coastal power plants are planned in Tamil Nadu, Maharashtra and

Gujarat, but nowhere are there as many as in Andhra Pradesh. Between 2006 and April 2012 23 new projects in Andhra Pradesh alone, totalling 28,553 MW were granted environmental approval by the Ministry of Environment and Forests (MoEF). Of these, 15 plants with 25,000 MW capacity are coastal plants, and all but two private sector projects. In addition to these already approved investments, another 18 coastal power plants, representing an additional 34,300 MW of power generation, with only one being half-owned by the public sector, are at various stages of the environmental approval process. Away from the coast an expansion of only about 6,000 MW of power is proposed at present, indicating the emphasis on coastal power plants for overall increased power production (Centre for Science and Environment 2011; Ministry of Environment and Forests 2012). The planned expansion of coal mining and thermal power generation in Andhra Pradesh is on a scale never previously seen in the state. An expansion over tens of thousands of hectares will affect significant areas in the densely populated state. A political ecology study of this expansion would be useful to develop an understanding of how the rearranged uses of natural resources will affect different groups of people when large-scale coal ventures take the place of local resource-based livelihoods.

Provision of land for private companies is a sensitive topic in India; frequent reversals in policy decisions have occurred over the years regarding whether or not governments should intervene in the process, and if so on whose behalf they should act (Fernandes 2009). Although it has come to be accepted that displacement should lead to rehabilitation, including land-for-land compensation, implementation leading to satisfactory outcomes often never occurs (Iyer 2007). Continued contestation over land is to be expected across the country.

The land most likely to be mined for coal as well as other minerals is in hilly terrain which is typically forest, regardless of its uses which in many cases are agricultural or involve a variety of common property livelihoods (Bhushan and Zeya Hazra 2008). Though a much more recent phenomenon, similar land use choices are being made in the areas where thermal power plants are being proposed on the coast, including in wetlands, salt pans, and mudflats, rather than on private agricultural land (Kasturi 2011; Ramani 2010; Sarma 2011b).

The use of commons is a strong connection between coal mines and thermal power plants. Commons are officially owned by the state, but it is well known that such areas tend to have many resource users. Those who depend on commons are frequently more vulnerable than settled farmers with private land, which has tended to be the main focus in policy circles (Guha 2001; Baviskar 1994; Springate-Baginski et al. 2007). The new *Forest Rights Act 2005* offers some scope for undoing this injustice by allowing forest-dwelling groups the opportunity to make claims over both individual and community land. A similar law does not exist for coastal commons, however, leaving affected groups without vital legal support.

Political ecology is concerned with how political economic forces determine access to, and use of, natural resources and the environment. Where regular environmental inquiry sees nature as a given, political ecology examines the possibilities of achieving social justice over natural resource use (Springate-

Baginski et al. 2007). It has been shown in many studies that local, natural resource-based livelihoods tend to be lost when dominant interests move into marginal lands (Ariza-Montobbio et al. 2010). During land acquisition the use of land with different forms of legal protection, and resettlement policies which vary depending on the group of people as well as the type of land-holding, work to split potential opposition. The intricate details of how a project site is carved out of the countryside can determine who is officially seen as affected and who is not. Frequent last-minute changes depending on the exertion of interest group pressures have been known to be common. The expected equity outcomes when land-hungry projects are planned on such land are meagre (Oskarsson 2013). This inequity is further heightened when particularly vulnerable groups such as tribal peoples are involved. A lack of access to formal education has tended to make such peoples easier to displace since they are less likely to be able to demand proper compensation compared to other groups.

The marginalisation of local livelihoods is likely to continue, since expert forums and planners are inaccessible to those affected by displacement, pollution and the loss of natural resource support when large industrial projects such as coal ventures are planned. Control over knowledge and technology enables powerful elites to implement agendas that suit their preferences (Martinez-Alier 2002). The complex system of regulation and control over industrialisation in India, despite reform, remains impossible for any administrator to manage despite efforts in recent years to simplify it. Lack of information, split responsibilities in the federal system, multiple laws with overlapping and often contradictory content, unclear implementation procedures, and frequent changes make it impossible to stay up-to-date with regulatory demands. Finding and demarcating these supposedly empty areas can be seen as a process of constructing empty spaces in which displacement occurs, thereby allowing large-scale projects to be introduced without proper controls. This process occurs at the particular proposed site but also at policy level in various government forums in charge of mediating between different interests across federal India.

However, natural resources are not merely about material interests. As for other fuels, with coal comes notions of a crucial energy resource which can power overall economic growth, modernisation and progress for the nation (Baviskar 2008). Many industrialists, the urban middle class, and even many farmers, who are often dependent on electric pumps for irrigation, are currently hampered by a severe power deficiency, with lengthy power cuts in many parts of India. Even though many will agree that the existing deficiencies are at least partially man-made, due to the apportionment of free power to farmers and to some industries, and that power losses are significant, enough support exists for the idea that there is simply not enough power for everyone and that more needs to be generated. Continuing inequities in power usage and the negative consequences of mining are not going to deter those in favour of increased coal use. Nor are climate change concerns likely to lessen India's appetite for coal-generated power.

However, the cultural politics of natural resources can also provide counter-narratives to that of modernization, thereby offering protection for traditional livelihoods. Land is one such resource which has come to be seen as belonging to farmers. In Andhra Pradesh there is particularly strong public sentiment for protection of land belonging to India's Scheduled Tribes minority, based on a history of displacement from dams and commercial forestry, and consolidation by more influential non-tribal farmers (Oskarsson 2010). Another powerful narrative is that of the Telengana struggle for an independent state in the North-Western part of Andhra Pradesh. For decades Telengana has been denied representation in state politics and many have come to see people from other parts of the state as exploiters of valuable resources such as farm land and water, but also minerals including coal, and thereby retarding wealth development of the region's inhabitants. These counter-narratives continue to offer symbolic support for protests against displacement and environmental destruction in civil society including, at times, support from some factions of government.

Opposition to Expanding Coal Mines in the Godavari Valley

This section presents a case study of the social consequences of open cast coal mining at Manuguru town in Khammam district, based on fieldwork carried out in April and August 2011. In this location an existing mine is currently expanding to the immediate North-West of the town, in an area where adivasi and Dalit[2] groups are subsisting on agricultural activity on forest land. Immediately to the northeast of the town is a new open cast mine planned on the site of an exhausted underground mine. This is where mainly non-tribal groups have turned the Godavari river plains into a significant farming area.

Coal in Andhra Pradesh exists along the length of the Godavari River valley. The significant deposits which constitute eight per cent of India's total coal deposits stretch for more than 350 km along the river in the districts of Warangal, Adilabad, Karimnagar, Khammam and West Godavari. The Godavari can also be said to form the southern limits of India's central tribal region where, other than large mineral reserves, a large part of the country's remaining forests exist, as well as some of the country's poorest and most excluded citizens, the Scheduled Tribes. The area protected under the Fifth Schedule of the Constitution for the benefit of adivasis is thus crucial for the energy supply of the state, creating a considerable development dilemma.

The present coal mining expansion represents a dramatic shift in the more than a century old coal operations in the Godavari valley. Where earlier adivasi groups, as well as other more recent migrants from the rest of the state, had been

---

2   Dalit, meaning the oppressed, is a common term in Andhra Pradesh for people of the lowest part of the caste hierarchy. Officially the term Scheduled Castes is used across India.

able largely to co-exist with underground mining, or even at times to benefit from employment in the mines, a new phase of expansion has begun in recent decades with the onset of more land-demanding open cast excavation methods. Not only does open cast mining displace more people and create additional environmental problems, it also allows for increased mechanisation. Consequently, employment in the mines has been cut drastically since the late 1980s despite a strong increase in the amount of coal mined.

According to the technical planners a much greater number of mines exist in the Manuguru area. What appears to be a continuous coal mine dissects those affected into different technically defined projects, which are extended over time, since land acquisition proceeds incrementally as the mines expand. The existing Manuguru Open Cast Mine I, II, III and IV are part of the same open pit of coal mining on the Western side of Manuguru town. II and III are currently operating, whereas IV is in the planning stage. An extension has been made to the number II mine, known as the Manuguru Open Cast Mine II Extension, which, combined with the existing II and III mines, requires 3,206 hectares of land in total, of which 278 ha is officially agricultural land and 2,674 ha is forest. The planned conversion of the PK II Underground mine into open cast on 486 hectare of largely fertile double-crop land is similarly named Manuguru Open Cast Mine (Oskarsson 2011). With multiple mines with remarkably similar names, it is small wonder that it was difficult to achieve clarity among villagers as to which mine they were affected by.

Manuguru Mandal is one of the more remote parts of the district immediately bordering Chhattisgarh to the north, yet the social composition, and thereby alienation of agricultural land to non-tribes, is significant. This is true not only for regular agricultural land but also for the forest. Thus we find tribal, Dalit and BC – the officially known Backward Caste[3] – communities living next to each other but with significantly varying possibilities to carry out agricultural livelihoods as well as making claims for compensation when displaced. Adding to the social mix are recent migrants who have moved to Manuguru town for jobs in, or related to, the coal mines.

Land holdings in Manuguru Mandal officially follow the AP *Tribal Land Transfer Regulations.*[4] These Regulations specify that only members of the Scheduled Tribes can own or lease land across the Scheduled Areas of the state. But many of the non-tribes in Manuguru, and elsewhere across the Scheduled Areas, continue to hold farm land due to wide loopholes in the implementation of the law. It has been estimated that 52 per cent of all agricultural land is in the possession of non-tribals in Khammam District, of which Manuguru Mandal is part (Rao et al. 2006). Public sector industry, such as the coal mines of Singareni

---

3   BC groups are further classified into sub-groups from 1–4 depending on their perceived need for government support. Many Muslim groups will fall within the BC definition.

4   The *Andhra Pradesh Scheduled Area Land Transfer Regulation 1959*, as amended to 1970.

Collieries, can continue to acquire land, since the sector has been interpreted as automatically acting in the public interest, as opposed to private companies which are banned from acquiring or even leasing Scheduled Area land in AP.[5] Whereas private companies have been able to mine coal for their own power generation, known as 'captive mining' in other parts of India, this has not happened in Andhra Pradesh (Herbert and Lahiri-Dutt 2004).

Another peculiarity with coal mining and land acquisition in the Scheduled Areas is that, unlike in the rest of the state, Panchayat approvals are not required. PESA, the adaptation of Panchayats to the Scheduled Areas, was left inoperative in AP from notification of the *State Panchayat Raj Act* in 1999 until 2011 – long after the mines studied here were proposed (Dandekar and Choudhury 2010; Reddy, Anil Kumar, et al. 2010). While new coal mines proposed by Singareni Colleries thus might have to take local views into account via PESA, this has not been the case until now, thereby allowing the mines to proceed without community consultation.[6]

For the westward expansion of coal mining in Manuguru there are two distinct effects of land loss; one is the tribal village Kondapuram, which was displaced and resettled 13 years ago as the Manuguru OC II mine expanded. The second case is the Dalit village Srirangapuram, now displaced by the Manuguru OC II Extension, but without any rights to land, and therefore not receiving any compensation other than for lost houses.

In the original Kondapuram village, now known as Old Kondapuram, Koya tribes lived together with a handful of BC immigrants. Land was plentiful: in a 2011 survey people recalled possessing as much as 7–10 acres of land for cultivation, and they could also rely on the forest for additional income via the collection of various forest products. A number of households made a living from cattle rearing, using the surrounding forest land. As the mine – then known as Manuguru OC II – expanded, the village became displaced around 1997. Since the village was tribal, the entire group was compensated with new agricultural land and houses (Oskarsson 2011).

New Kondapuram was constructed on land close to Manuguru town. Each displaced household received two acres of agricultural land for cultivation and a small plot of land for housing. In addition there was a sum of money paid for the old land and houses. The new agricultural land was spread out in various locations near and far from New Kondapuram. Some of it was as much as 5–6 km from the village and not irrigated. The land titles were *D-patta*, so called assigned land, giving less protection than the private pattas which had in many cases existed in Old Kondapuram. Compensation money could not pay entirely for new houses, payment for which instead came via a number of routes including bank and self-help group loans and from the AP Government's housing program.

---

5   This was affirmed in the 'Samatha Judgement' 1997 (Supreme Court of India 1997).

6   The AP PESA rules places the approval of land acquisition at the Mandal level rather than among those directly affected at the local Gram Sabha level.

All 20 households of the village surveyed in 2011[7] still claimed to be in possession of the land holdings they received as compensation 13 years earlier, though a few had leased it to other, presumably non-tribal, farmers. No one had been able to obtain employment at Singareni and only very few worked away from the fields. Education had somewhat improved, but only a handful of the younger people had been able to maintain their education beyond tenth grade. A number of new independent households had grown since the displacement. These had their own houses in the village but no land, leaving no other option than to work as agricultural labourers. The main concern for the residents of New Kondapuram is its proximity to the proposed Manuguru Open Cast mine. Though the village is currently safe, it is likely that some villagers will lose their agricultural land for waste storage of overburden from the mines. It is not known what will happen to the people in 15 years when, if things go according to plan, the mine extends to the resettled village; this leaves significant uncertainty about the future.

The second displaced village in the forest was Srirangapuram, a Dalit village about 1.5 km from the perimeter of the present Manuguru OC II Extension open cast mine, which was evacuated in 2010 when explosive blasting from the mine made it impossible to carry on living there. This was a village of migrants who over the last 20–40 years had arrived from Warangal and Nalgonda districts of AP to cultivate land in the forest.[8] 129 households[9] comprised the part of the village that was displaced due to expansion of the mine. Since only very few in the village had land titles, no compensation was forthcoming for this land. And since the villagers had migrated into the area only a few decades earlier there was no possibility of being compensated as forest-dwellers under the *Forest Rights Act*. Furthermore, because of the non-tribal status of the villagers, there would be no attempt by the government to resettle the community in a new location, as with the tribes of Kondapuram. For houses people received a uniform cash grant of 108,000 rupees. A number of additional grants meant that each family was entitled to a sum of up to 207,500 rupees in cash (Government of Andhra Pradesh 2009).

Despite the move away from Srirangapuram by some of the educated youth, most remained in agriculture. Although displaced, villagers held on to their agricultural plots for as many more years as possible out of necessity, constructing new houses or renting accommodation in the nearby, mainly Dalit village, predominantly of the mala sub-caste. While this was a practical solution in the short term, it neglected future displacement in perhaps as little as 6–8 years when the mine expands further. When this occurs two additional villages will become a storage area for overburden of the coal mine and thereby uninhabitable. There appeared to be no other solution available to them which would allow continued

---

7   Out of a total of 42 households.

8   Based on interviews carried out in August 2011.

9   The AP government R&R survey counted 76 households, which does not accord with the 129 households compensated.

cultivation of their fields, and possibly some compensation for lost land in the future.

The proposed coal mining on the north-eastern side of Manuguru town was in direct competition with fertile agricultural land. The conversion of an underground mine into open cast required 623 hectares of land (Ministry of Environment and Forests 2008b). While earlier coal mine expansions in the area had confiscated the land of two villages at most, the Manuguru OC represented something new when four large villages are displaced simultaneously. From this one mine, projected to operate only for 11 years, 1,248 households, or about 5,000 people, were affected by either losing their agricultural land (89 households), their houses (507 households) or both their agricultural land and their house plots (652 households) (Ministry of Environment and Forests 2008a).[10]

Perhaps not surprisingly, the larger number of displaced and the fertile agricultural land generated stronger opposition compared to the mining expansion in the forest discussed above. Over the years the villagers have been organising protest meetings, made representations to the District Collector and the Chief Minister, as well as trying to garner support from opposition parties and media.[11] However, it has proven difficult to be heard, with no success in reducing the land used by the mine or improved compensation. The slight delay, despite the central government approvals of mining, appears to be the only effect.

The plan for conversion into an open cast mine relies on the majority of cultivators being non-tribal in an area where only tribal people are legally allowed to cultivate land. Many of the non-tribal households now affected by the Manuguru open cast mine have lived in the area long enough to obtain land titles before the 1969 cut-off date. These farmers, though certainly better off than the tribes of the forest, are not of very great means. 40 Questionnaires were completed in two villages affected by the Manuguru OC mine. Details from these mainly non-tribal villages showed a lack of land holdings larger than 4–5 acres of land, although this land tended to be well-irrigated double-crop land. Many households were found to comprise landless agricultural workers.

The question remained as to what people would do with the cash compensation, whether 3 or 6 lakh per acre, in order to support themselves and their families in the future. Most of those who would receive compensation would not be allowed to buy new land in the area since they are not members of the tribes; and the money would not be enough to buy agricultural land outside of the Scheduled Areas.

---

10   According to the AP R&R policy a non-tribal person is displaced if they lose their house (Government of Andhra Pradesh, Irrigation and CAD Department 2005). Lost households from the Manuguru OC project totaled 1,159.

11   Interviews with leaders of opposition movements: Manuguru, April and August 2011. A number of local newspapers have covered the struggle against open cast mining and later the work in favour of improved compensation (Eenadu 2011; Andhra Prabha 2011c; Andhra Prabha 2011a; Andhra Jyoti 2011; Andhra Prabha 2011b; Eenadu 2008; Andhra Jyoti 2008).

The Manuguru OC mine avoids acquiring nearby degraded forest land, today mainly used for grazing but also to some extent cultivated. The forest could have been used as the dumping ground for OB rather than the agricultural land had this been in the interest of the mine planners. But it is part of Singareni's environmental policy not to use forest land for OB dumps irrespective of the quality of the forest, while agricultural land should be avoided only if possible (Singareni Colleries n.d.). However, the preference for agricultural land over forest areas might simply come down to money. The OC-II Extension EIA report mentions that the cost of acquiring agricultural land is 2 lakh per acre compared to forest land at 9 lakh per acre, or 4.5 times more (Environmental Protection Training and Research Institute 2007).

Originally an even larger open cast mine was planned in the area, called Sivalingapuram open cast mine. This mine would have produced 3 million tonnes of coal per year while displacing 10 villages, including those presently proposed to be displaced by the Manuguru OC project. The Sivalingapuram project would have required 6,000 houses to be acquired along with 6,500 acres of land, but after protests was turned into the present Manuguru OC. While still causing large-scale displacement and acquisition of fertile land, now four, rather than ten, villages are affected and the land requirement has come down to 1,200 acres.[12]

Other struggles have fared better in the area and elsewhere. A recent example is the resistance to the expansion of the Sattupally open cast mine, also proposed in Khammam District but on non-Scheduled Land, which caused the mine to be cancelled.[13] Immediately south of Manuguru town the Kunavaram open cast mine was proposed on land belonging to Kunavaram village, a fully ST village of the Koya tribe with 300 households. Villagers secured support from Tedum Debba, an organisation working across Andhra Pradesh on tribal welfare issues, and managed first to delay the project and then to have it cancelled.[14] The smaller size of the Kunavaram mine, but certainly also the struggle of the tribal activists along with villagers, seems to have been crucial in this struggle.

In sum, the two cases detailed in this section demonstrate that displacement by open cast coal mining projects worsens livelihoods. The tribes, somewhat surprisingly, negotiated and retained the best living conditions in this study because AP R&R policy mandates that new land should be provided for them, something which has also been implemented for one displaced village in Manuguru. The displaced tribes, however, remain almost completely dependent on the goodwill of the administration, putting them in a very precarious situation should the bureaucrats decide to interpret the R&R policy differently in the future. The better education and greater bargaining power of the displaced non-tribes have so far not led to anything close to sufficient compensation. Reasons for this can be found in the communities themselves being divided, having only half-hearted political

---

12 Interview CPI representative, Manuguru 24 August 2011.
13 Interview activist, Manuguru, August 2011.
14 Interview with representatives of Tedum Debba, Manuguru, 26 August 2011.

support and also finding themselves divided between those immediately displaced and those who later lose land as the mines expand over decades.

The discussion now moves from the adivasi areas of Telengana to the coast, specifically the northernmost District of Srikakulam. This is where large-scale plans to generate coal-based power have begun to take shape in highly controversial projects involving a different form of marginal land, a wetland used by fishing and agricultural groups, but again not officially recognised as either a conservation area or as private property.

Thermal Power in Coastal Wetlands

India's environmental decision making is leading to fatal consequences. Perhaps never before has so much blood flowed due to faulty decision making. Last month saw the death of three villagers in police violence during staged opposition to the construction of a thermal power plant in Sompeta, in the Srikakulam District of Andhra Pradesh. Agencies after agencies (sic.) have certified the area as a 'degraded barren and waste land' whereas in reality it was a marshy land with significant cultivation. […] Sompeta became an overnight national environmental issue after the death of three people but the neighboring wetland Naupada where a similar thermal power plant is under construction has escaped national attention. Naupada is no less important as a wetland than Sompeta, and the violation of law is no less serious. The only difference is that people have not yet been killed (Dutta 2010, 1–2).

In March 2011 the Naupada wetland also made it to the headlines of national newspapers after two protesters were killed in a confrontation with the police over the Bhavanapadu thermal power plant. This section is concerned with how despite many warnings a part of the Nuapada wetland was approved for construction of a thermal power plant, leading to tragic outcomes. Two particular aspects are analysed: the lack of recognition of a large number of livelihoods depending on the chosen site; and the particular environmental values of the wetland.

East Coast Energy's Bhavanapadu thermal power plant was originally planned on 2,450 acres of land in the Santhabommali Mandal (also known as Block) of Srikakulam. The plant, apart from its core power generation facilities, included waste disposal areas, a jetty and transportation infrastructure. The Bhavanapadu plant would use imported technologically advanced equipment which 'will have higher efficiency resulting in reduced coal consumption and reduced emission of greenhouse gases' (East Coast Energy Private Ltd n.d., 2). Cooling water was intended to be taken from the sea, thereby reducing the need for scarce fresh water (Ibid). Even the use of imported coal was presented as a matter of improved operations due to its higher quality compared to Indian coal. At the same time, there was no restriction from using Indian coal in the plant should it become available in the future.

Despite the focus on technological efficiency, industrialisation plans in Andhra Pradesh including the coastal power plants have been known to exhibit a bias towards businesses run by Telugus. With strong political connections, or even being promoted by politicians, these companies have often started as contractors on infrastructure projects and later expanded into other industries when new opportunities have appeared. East Coast Energy is an unknown company seemingly created for the purpose of building a thermal power plant along the AP coastline. No official documents provide clarity on who is behind the venture; indeed, there are no such legal requirements. It appears as if one of the key supporters is T. Subbarami Reddy, a major contractor and Rajya Sabha MP for the Congress Party, which is in power at the state and the national government levels. News stories have also implicated other top AP Congress leaders in East Coast Energy (Syed and S. Dutta 2012).

Little is known about the origins of the site selected for the power plant. It appears as if it had already been identified by the state government as useful for various industrial projects in the 1970s but never used. Initial plans were for industrial-scale salt pans and later for a plantation, until finally the site was handed over to East Coast Energy (Ministry of Environment and Forests 2011a). The EIA report for the thermal power plant claimed that the 'site is barren, uninhabited, low lying and belongs to Government of Andhra Pradesh.... There is no rehabilitation and resettlement issue, since there is no habitation on the land' (East Coast Energy Private Ltd n.d., 3). The selection criteria notes that it is a site with perhaps no great merits to it for a thermal power plant but which is also not sensitive from a social or an environmental perspective. The site was described as being set apart from sensitive coastal areas and from the command area of the Vamsadhara River irrigation canal. It was also not on agricultural land as it was earlier used for salt pans, and thus would not result in displacement (Ibid).

Other sources detail the Naupada wetland as being located close to the coast and supported by a number of streams of water as well as a canal flowing from the Vamsadhara River. The 'brackish swamps are rich in fish, crustaceans and other nutrients that are essential for the breeding and survival of the visiting pelicans and storks' (Environmental Resource and Response Centre 2009, 13). This has enabled a large number of fishing families to make a living. A large part of the swamp is also used for salt pans. 'About 20,000 people do salt farming on it, 5,000 fish in its ponds and another 5,000 do farming' (Mahapatra 2011). Sarma (2011b) claims three villages with 1,000 families in total have traditional fishing rights in the area, with another 4,000 families depending on the salt pans.

The wetland is surrounded by a large agricultural area which depends on it for water. If the land was raised even a few feet, then a large area of 20,000–30,000 acres of agricultural land would be at risk of flooding (Sarma 2011a). This is what happened when construction started on the power plant. Water which would have normally drained into the swamp now had nowhere to go other than onto the surrounding fields (Narain 2011). In this way a common cause was found between the displaced fishing groups and salt pan workers at the proposed site and the nearby farmers indirectly threatened with land inundation. Since displacement in Andhra Pradesh is defined as losing your house plot, and the land for the power plant remained defined as commons belonging to the state government, none of these groups would receive land compensation from the thermal power plant. A complete loss of livelihoods for the people of the area naturally created a groundswell of support for protests.

That these people depended on the swamp had been noted also by the local authorities before planning for the thermal power plant had commenced. A letter from the local Revenue Department to the District Collector discussed in detail the existence of fishing via ponds and tanks in the area as well as the presence of nets (Ramalaxmi 2005). 'As per the revenue records [the land] is classified in revenue records as Kakarapalli swamp lands. [It] is seasonally water logged and is neither suitable for cultivation nor for any commercial plantation' (Ministry of Environment and Forests 2010a, 18). A report made on behalf of the MoEF came to a similar conclusion, stating that the 'seasonally waterlogged area of the project site can be considered as a wet land' (Reddy, Murali, et al. 2010, 27). The different conclusion reached in the EIA is likely to be due to the time of year the study was conducted. The EIA was made during the hot summer months from March to May 2007, when the region is at its driest.

The second main concern was whether there were areas in need of environmental conservation at or close to the proposed site. The EIA report pointed out that there are no significant conservation areas other than 'a minor bird-breeding site' (East Coast Energy Private Ltd n.d., 3) four km away. But this 'minor' site became the focus of several reports by the Bombay Natural History Society in which a large number of rare migratory birds were detailed (Rahmani and Rajvanshi 2009; Bombay Natural History Society 2008). The bio-diverse wetland and the many people who depended on it were also detailed in local newspapers, and on national television (Sudhir 2010; Benjamin 2008).

Planning for the Bhavanapadu power plant started in 2007 when the project Terms of Reference (TOR) were issued by the MoEF. According to this document forthcoming plans had to specify the land-use, including the part under coastal regulation zone protection, mangroves, and the status of important environmental protection areas (Ministry of Environment and Forests 2007). An EIA was quickly drafted in 2007. Once submitted, the project was discussed in MoEF meetings in June, November and December 2008. Committee members made a site visit and concluded that the '[t]he proposed site is an ecological entity with incomparable value requiring conservation and protection' (Ministry of Environment and Forests 2008c). The MoEF Committee thus 'recommended that the proponent should shift their [sic.] site upland sufficiently away from the marshy area and submit the details for further consideration of the proposal' (Ministry of Environment and Forests 2008c).

A modified proposal was presented at a MoEF meeting in February 2009. The plant site had now been reduced from 2,450 to 1,995 acres and the project was approved even though the intention of the earlier meeting appear to be for the plant to shift to an entirely new location rather than removing a part of the proposed site. In the environmental approval it was noted that '[t]here are no notified ecologically sensitive areas in the vicinity of the proposed project site' (Ministry of Environment and Forests 2009a). The approval letter did not mention lost land or livelihoods as a result of the project, but demanded that a Corporate Social Responsibility fund of 10 crore rupees be created for unspecified beneficiaries (Ibid).

The project was further discussed by another MoEF Committee, the Coastal Regulation Committee,[15] in October 2009. This committee was satisfied by the Government of AP verifying that the proposed location was not part of the Naupada swamp and that there were no notified conservation areas nearby (Ministry of Environment and Forests 2009b). Yet another committee, the Wildlife Committee[16] of the same MoEF, continued the discussion. The State Forest Department verified to this committee that there was no official protected reserve or migratory bird route at the site despite the very same department having spent funds to establish a bird-watching tower and keeping records of migratory birds next to the wetland since 1992 (Rahmani and Rajvanshi 2009).

The wildlife committee was not convinced, however, since one of its members was from the Bombay Natural History Society, which had been active on the issue. The committee continued to discuss the project in five subsequent meetings from July 2009 to May 2010. One such meeting concluded that '[t]he entire area within the bounds of the project starting from the approach road to the project site is a marshland' (Ministry of Environment and Forests 2009c), and recommended that it be regarded as a conservation reserve, an unclear definition. The following meeting noted that creating a reserve was not possible since 'a large number of families were staying in that area and were having [*sic*] their livelihood rights for fishing in these areas' (Ministry of Environment and Forests 2010b). The people who were not seen as displaced by the thermal power plant could thus prevent it from attracting the official label needed for conservation, and little further could be done to resolve the matter. In the end the Wildlife Committee appears to have been unable to prevent approval according to existing rules.

When it was not possible for the various high-level committees to listen to complaints, protests – as noted earlier – turned violent. On 1 March, 2011 tensions at the site resulted in a clash between police and protesters (Appala Naidu 2011; Srinivasa Rao 2011). This was the direct result of escalating concerns and desperation on the part of local residents; as one NGO noted: 'Time and again over the last three years, human rights organisations, environmentalists, political parties and concerned citizens have pointed out the undesirability of constructing a thermal power plant in the wetland at Kakarapalli' (Human Rights Forum 2011). Once the violent clashes were reported in national newspapers the MoEF was

---

15   'Expert Appraisal Committee for Coastal Regulation Zone, Infrastructure Development and Miscellaneous projects.'
16   Full name: 'Standing Committee of National Board for Wildlife.'

forced to address the environmental violations, including the commencement of construction before the environmental approval had been issued. The project was suspended one day after the clashes (*The Hindu* 2011).

The East Coast Energy model remains popular in coastal Andhra Pradesh. A very similar proposal made by Nagarjuna Constructions, another Andhra Pradesh-based company, also proposed for a swamp at Sompeta in Srikakulam District led to social protest with a violent police clampdown in July 2010 (National Environment Appellate Authority 2010). Nuclear power, pharmaceutical and chemicals plants are all planned on similar coastal land in the state (Narain 2011). The reduced size of the Bhavanapadu plant provided an opportunity for yet another smaller power plant, Meghavaram, on the land that had previously been discarded (Sarma 2011b). However, this proposal was cancelled by the authorities.

The central government expert committee which vets environmental approvals has rejected only one thermal power plant since 2009 in Andhra Pradesh. For a thermal power plant in the West Godavari District, 'the Committee…noted that the site proposed is primarily a highly eco-sensitive area with mudflats and salt pans all around' (Ministry of Environment and Forests 2011b). In this case the applicant was required to find a new location. In many other cases the site dimensions are reduced or shifted in attempts to diminish the negative impacts.

The current state of the East Coast Energy power plant appears to be on hold. The official MoEF site even lists the project as not yet having received a TOR, which could indicate the need for the project to begin its application afresh (Ministry of Environment and Forests 2012). It appears as if the project is waiting for yet another expert report concerning what caused violent conflict at the site (*The Hindu* 2011). This is a common tactic when policymakers do not know how to proceed in arbitrating the commitments to social justice for local residents and the promotion of industrial projects. For yet another report to resolve the underlying issues in this case or change how land use planning is carried out would be very unlikely.

Governance is expected to continue to operate on a case-by-case basis, with little accountability of the company – which submitted a fraudulent EIA and started construction before approval – or of the expert committees of the MoEF, which did not demand that construction cease despite information being available that the proposed site was a wetland. Although the present location may not see a power plant being built, government and corporate creativity in finding other forms of marginal land seems to imply that other marginal commons will continue to be used in the future.

Conclusion

In analysing how land was usurped for a coal mine and a thermal power plant, with the aim to understand the larger political ecology of power generation through coal-based plants in Andhra Pradesh, dramatic commonalities appear. Despite significant variations in histories, peoples and even legal settings between the sites, the shared element of the two coal ventures was the reliance on commons. These commons have long supported some of the state's poorest inhabitants, and as these lands are lost, the poor are increasingly finding themselves on the losing side of the present coal energy expansion. As a result, the growing inequity is further exacerbating already existing significant social and economic inequalities in the state.

Though the use of commons for large-scale coal projects raises equity concerns, these are further exacerbated by existing compensation policies. The acquisition of commons for coal ventures means that there will be no compensation to the displaced other than cash for houses.

In forests the recent Forest Rights Act offer some amount of hope that such livelihoods will finally be recognised, though implementation is still lacking in many locations. At the coal mine in Manuguru there were also additional difficulties with the Act when forest-living Dalits could not claim land since they had not resided in the area long enough. For coastal wetlands the only legal mechanism of protection is to declare it in need of environmental conservation, something which will make local livelihoods almost as marginalised as from the impact of a thermal power plant.

The marginalisation of peoples takes place not only at the proposed site but also in high-level policy forums. For the thermal power plant the expert committees of the MoEF repeatedly went along with the representation of the project proponents despite other credible evidence and even personal visits by some of the experts to the proposed site. This rendered common property livelihoods invisible and the land possible to acquire without including its residents in official plans. This behaviour of committees of the central government outside the immediate control of the AP government indicated a dispersed acceptance of present approaches to land-use planning. While more work is needed to understand how national expert committees operate and how decisions are made, this study provides insights into the inability of relevant committees to strike a balance between competing interests, even when clear evidence exists.

The use of marginal common land can continue in cases where the immediately affected themselves are split into many different groups with varying interests and capacities to object to plans. An example of this occurs at one of the Manuguru coal mines, where the slow expansion over many years displaced one village after the next, rather than many at the same time. Additionally, the different legislative support for the Scheduled Tribes compared to other groups seemed to prevent larger collective action. Once a project threatens many who are affected in similarly negative ways, then retaliation might occur which at least creates possibilities for the modification of plans. This was what happened over the thermal power plant in Srikakulam District, which is presently on hold. In the latter case the result was not a larger policy decision on how to proceed in the future, but simply an isolated statement in response to strong protest.

The approach of using commons for the present coal energy expansion can also be seen as recognition of the possibilities for social protest. It is only by making marginal peoples and their land and resource uses invisible, and thereby avoiding protests, that land can be made available for state use. Energy production in the state is likely to continue to increase, although it remains far below demand due to the many controversies which will continue to be generated in the environmentally and socially irresponsible processes of establishing and increasing power production.

References

Andhra Jyoti. 2011. 'No meeting with collector until sound rate will be announced: Manuguru OC affected.' (Original in Telugu: Collector samaavesaniki ram ram: manuguru oc prabhavithula jhalak, retu nachitene charchalaku). *Andhra Jyoti, Khammam edition*, 13.

Andhra Jyoti. 2008. 'We are ready to sacrifice our lives but won't give our lands: Eggadigudem farmers.' (Original in Telugu: pranaalu poina bhumulivvam: Eggadigudem rythula velladi). *Andhra Jyoti, Khammam edition.*

Andhra Prabha. 2011a. 'Justice should be done to OC displaced people.' (Original in Telugu: OC nirvasitulaku nyayam cheyali). *Andhra Prabha, Khammam edition*, 20.

Andhra Prabha. 2011b. 'Officials behaviour to displaced is not good: Complained Ayodhya to RDO, JC.' (Original in Telugu: Nirvasitula patla adhikarula theeru bagaledu: RDO, JC laku firyadu chesina Ayodhya). *Andhra Prabha, Khammam edition*, 11.

Andhra Prabha. 2011c. 'Provide livelihood, not compensation.' (Original in Telugu: Pariharam kadu upadhi choopandi). *Andhra Prabha, Khammam edition*, 20.

Appala Naidu, T. 2011. '2 killed, over 50 hurt in police firing.' *The New Indian Express.*

Ariza-Montobbio, et al. 2010. 'The political ecology of Jatropha plantations for biodiesel in Tamil Nadu, India.' *Journal of Peasant Studies* 37(4): 875–97.

Balagopal, K. 2007a. 'Land Unrest in Andhra Pradesh-I: Ceiling Surpluses and Public Lands.' *Economic and Political Weekly* 42(38): 3829–33.

Balagopal, K. 2007b. 'Land Unrest in Andhra Pradesh-II: Impact of Grants to Industries.' *Economic and Political Weekly* 42(39): 3906.

Balagopal, K. 2007c. 'Land Unrest in Andhra Pradesh-III: Illegal Acquisition in Tribal Areas.' *Economic and Political Weekly* 42(40): 4029.

Banerjee-Guha, S. 2008. 'Space relations of capital and significance of new economic enclaves: SEZs in India.' *Economic and Political Weekly* 51–9.

Baviskar, A. 1994. 'Fate of the forest: Conservation and tribal rights.' *Economic and Political Weekly* 2493–501.

214 *The Coal Nation*

Baviskar, A. 2008. 'Introduction.' In *Contested Grounds: Essays on Nature, Culture, and Power*, edited by A. Baviskar, 1–12. New Delhi: Oxford University Press.

Benjamin. 2008. 'Location of power plant near bird resort opposed.' *The Hindu*. Available at: http://www.hindu.com/2008/08/26/stories/2008082654030400. htm [Accessed May 14, 2012].

Bhushan, C. and Zeya Hazra, M. 2008. *Rich Lands Poor People: Is 'Sustainable' Mining Possible?* New Delhi: Centre for Science and Environment.

Bombay Natural History Society, 2008. 'Report on Violations of Environment Clearance Procedures by East Coast Energy Private Ltd. at the Proposed Site for the 2640 Mw Bhavanapadu Thermal Project in Srikakulam District.' Andhra Pradesh, Mumbai: Bombay Natural History Society.

Centre for Science and Environment, 2011. 'Public Watch: Thermal Power Plants.' Available at: http://cseindia.org/userfiles/Thermal%20power%20plant.pdf [Accessed May 14, 2012].

Dandekar, A. and Choudhury, C. 2010. *PESA, Left-Wing Extremism and Governance: Concerns and Challenges in India's Tribal Districts*, Anand: Institute of Rural Management. Available at: http://www.tehelka.com/ channels/News/2010/july/10/PESAchapter.pdf [Accessed July 4, 2010].

Dutta, R. 2010. 'Criminal Action for Uncivil Acts.' *Environmental Resource Centre Journal* (5): 1–3.

East Coast Energy Private Ltd. 'Executive Summary of 2640 MW Bhavanapadu Coal Based Thermal Power Project Proposed near Kakarapalli village, Santhabommali Mandal Srikakulam District.' Andhra Pradesh, Hyderabad.

Eenadu, 2011. 'Farmers denied meeting the collector: Fight against proposed Manuguru OC.' (Original in Telugu: Collector tho sammavesanni bahishkarinchina rythulu: pratipadita manuguru OC pai ragada). *Eenadu, Khammam edition*, 13.

Eenadu, 2008. 'Revoke the establishment of open cast: displaced.' (Original in Telugu: OC erpatu viraminchukovali: nirvasitulu). *Eenadu, Khammam edition*.

Environmental Protection Training and Research Institute, 2007. 'Environmental Impact Assessment and Environmental Management Plan for the Proposed Manuguru Open Cast-II Extension Project Manuguru Area, Khammam District.' Andhra Pradesh, Hyderabad.

Environmental Resource and Response Centre, 2009. 'EIA Report of the Issue: Thermal Power Plant of East Coast Energy Pvt Ltd in Srikakulam, Andhra Pradesh.' *Environmental Resource Centre Journal* (2): 13–14.

Fernandes, W. 2009. 'Displacement and alienation from common property resources.' In *Displaced by development: Confronting marginalisation and gender injustice*, edited by Mehta, 105–32. New Delhi: Sage Publications.

Government of Andhra Pradesh, 2009. 'Khammam District Gazette Extraordinary.' Published 3/10/2009.

Government of Andhra Pradesh, Irrigation and CAD Department, 2005. 'Resettlement and Rehabilitation Policy 2005: For Project Affected Families.'

Guha, R. 2001. 'The prehistory of community forestry in India.' *Environmental History* 6(2): 213–38.

Herbert, T., and Lahiri-Dutt, K. 2004. 'Coal Sector Loans and Displacement of Indigenous Populations: Lessons from Jharkhand.' *Economic and Political Weekly* 39(23): 2403–9.

Human Rights Forum, 2011. 'Press Release 28 Feb 2011 on the Police Firing at Kakarapalli in Srikakulam District.'

Jenkins, R. 1999. *Democratic Politics and Economic Reform in India*. Cambridge: Cambridge University Press.

Kasturi, K. 2011. 'New thermal power clusters.' *Economic and Political Weekly* 46(40): 10–13.

Mahapatra, R. 2011. 'The great wetland robbery in Kakarapalli.' *Down To Earth*. Available at: http://www.downtoearth.org.in/content/great-wetland-robbery-kakarapalli [Accessed May 29, 2012].

Martinez-Alier, J. 2002. *The Environmentalism of the Poor: A Study of Ecological Conflicts and Valuation*. Massachusetts: Edward Elgar Publishing.

Mearns, R. 1999. 'Access to Land in Rural India.' *World Bank Policy Research Working Paper 2123*.

Ministry of Coal, 2011. *Annual Report 2010–11*, New Delhi: Government of India. Available at: http://www.coal.nic.in/annrep1011.pdf [Accessed September 17, 2011].

Ministry of Environment and Forests, 2005. Lok Sabha Unstarred Question No 2770 Answered on 21.03.2005: Grant of Forest Land for Mining. Available at: http://164.100.24.208/lsq14/quest.asp?qref=9731 [Accessed September 16, 2011].

Ministry of Environment and Forests, 2007. 'TOR for Bhavanapadu Thermal Power Plant.' Available at: http://164.100.194.5:8081/ssdnl/showTorDetails. do?proiectCode=No:J-13012/102/ 2007-IA II(T) [Accessed April 8, 2011].

Ministry of Environment and Forests, 2008a. 'Environmental Clearance Letter for Manuguru Opencast Coal Mine Project.' Dated 24 Oct 2008.

Ministry of Environment and Forests, 2008b. 'Minutes of the 31st Expert Appraisal Committee Thermal and Coal Mining.' held 22–24 September 2008.

Ministry of Environment and Forests, 2008c. 'Summary Record of the 36th Meeting of Reconstituted Expert Appraisal Committee on Environmental Impact Assessment of Thermal Power and Coal Mine Projects.' Available at: http://164.100.194.5:8081/ssdn1/getAgendaMettingMinutesSchedule.do;jsess ionid=301DF27BAA7D907BC2D5DC0D80DF712B?indCode=THEDec%20 15,%202008 [Accessed April 8, 2011].

Ministry of Environment and Forests, 2009a. 'Environmental clearance letter for Bhavanapadu Thermal Power Project.' Available at: http://164.100.194.5:8081/ ssdn1/showApprovedletters1994.do?projectCode=No:J-%2013011/36/%20 2008-IA%20II(T) [Accessed April 9, 2011].

Ministry of Environment and Forests, 2009b. 'Minutes of the 78th Meeting of the Expert Appraisal Committee for Coastal Regulation Zone.' Held 20–22 July, 2009.

Available at: 164.100.194.5:8081/ssdn1/getAgendaMettingMinutesSchedule.
do;jsessionid=C74B47282B3FA58E6BBB79EE39757093?indCode=MISJ
ul%2020,%202009 [Accessed April 8, 2011].

Ministry of Environment and Forests. 2009c. 'Minutes of the 17th Meeting of
the Standing Committee of National Board for Wildlife (NBWL).' Held
22 December, 2009. Available at: http://moef.nic.in/downloads/public-
information/17MoM_WL.pdf [Accessed January 8, 2013].

Ministry of Environment and Forests, 2010a. 'Site Inspection Report on Thermal
Power Plants of Srikakulam District and Aluminium Refinery Project in
Vishakapatnam and Vizianagaram Districts.' Delhi: Government of India.

Ministry of Environment and Forests. 2010b. 'Minutes of the 18th Meeting of the
Standing Committee of National Board for Wildlife (NBWL).' Held 12 April,
2010. Available at: http://moef.nic.in/divisions/wildlife/18MoM_WL.pdf
[Accessed May 21, 2012].

Ministry of Environment and Forests, 2011a. 'Minutes of the 19th meeting of
reconstituted expert appraisal committee on environmental impact assessment
of thermal power and coal mine projects.' 7–8 March 2011. Available at:
http://164.100.194.5:8081/ssdn1/getAgendaMettingMinutesSchedule.do;jses
sionid=EA647BCF7E2C9773749F849A3FEBC396?indCode=THEMar%20
07,%202011 [Accessed November 12, 2011].

Ministry of Environment and Forests, 2011b. 'Minutes of the 1st Meeting
of Reconstituted Expert Appraisal Committee on Environmental Impact
Assessment of Thermal Power and Coal Mine Projects.' 7–8
February, 2011. Available at: http://164.100.194.5:8081/ssdn1/
getAgendaMettingMinutesSchedule.do;jsessionid=7E2E35D4DD97DD854
03D376228093099?indCode=THEFeb%2007,%202011 [Accessed April 7,
2012].

Ministry of Environment and Forests. 2012. 'Environment Clearance
Database.' Available at: http://164.100.194.5:8081/ssdn1/projectApproved
FinalEIAWithoutTORfinalapproved3.do [Accessed May 19, 2012].

Ministry of Power, 2012. *Report of the Working Group on Power for Twelfth
Plan (2012–17)*, New Delhi: Government of India. Available at: http://
planningcommission.nic.in/aboutus/committee/wrkgrp12/wg_power1904.pdf
[Accessed May 28, 2012].

Narain, S. 2011. 'Lessons from Kakarapalli.' *Down To Earth*. Available at: http://
www.downtoearth.org.in/content/lessons-kakarapalli [Accessed May 19,
2012].

National Environment Appellate Authority, 2010. Order on Sompeta Thermal
Power Plant.

Oskarsson, P. 2010. 'The law of the land contested: Bauxite mining in tribal,
central India in an age of economic reform.' Norwich, UK: University of East
Anglia.

Oskarsson, P. 2011. 'Jobless Openings: The Expansion of Open Cast Coal Mining at the Expense of Rural Livelihoods in the Godavari Valley of Andhra Pradesh.' Hyderabad, India: ActionAid India.

Oskarsson, P. 2013. 'Dispossession by confusion from mineral–rich lands in central India.' *South Asia: Journal of South Asian Studies* 35(2): 199–212.

Rahmani, A. R. and Rajvanshi, A. 2009. 'Report Based on the Visit to Naupada Swamp and the Project Site of the 2640 MW Bhavanapadu Thermal Power Project of M/S, East Coast Energy Pvt. Ltd.' Report Submitted to the Standing Committee of the National Board for Wildlife.

Ramachandraiah, C. and Srinivasan, R. 2011. 'Special Economic Zones as New Forms of Corporate Land Grab: Experiences from India.' *Development* 54(1): 59–63.

Ramalaxmi, C. 2005. 'Fishing lease rights in Lingudukotturu, Pitarikotturu and Karakapalli, Tekkali, Srikakulam.' Revenue Department.

Ramani, S. 2010. 'Development and Displacement: Resentment in the Kutch.' *Economic and Political Weekly* 45(8): 15–18.

Rao, S. L. Deshingkar, P. and Farrington, J. 2006. 'Tribal Land Alienation in Andhra Pradesh: Processes, Impacts and Policy Concerns.' *Economic and Political Weekly* 41(52): 5401–7.

Reddy, K.S. Murali, N.S. and Kaliyaperumal, C. 2010. 'Site Inspection Report of Thermal Plants of Srikakulam District and Aluminium Refinery Protect in Vizianagaram and Vishakapatnam Districts.' New Delhi: Ministry of Environment and Forests.

Reddy, M.G. Anil Kumar, K., et al. 2010. *The Making of Andhra's Forest Underclass: An Historical Institutional Analysis of Forest Rights Deprivations*, Manchester: IPPG Discussion Paper 42. Available at: http://www.ippg.org.uk [Accessed June 17, 2010].

Sarma, E.A.S. 2011a. 'Kakarapalli: Another Blot on India's Democratic Systems.' *Economic and Political Weekly* 46(11): 12–15.

Sarma, E.A.S. 2011b. *Will the Sparrow Ever Return?* Visakhapatnam.

Singareni Colleries. No date. Environmental Management in SCCL. Available at: http://www.scclmines.com/env/DOCS/Env%20mgt%20in%20SCCL.pdf [Accessed September 12, 2011].

Springate-Baginski, O. et al. 2007. 'Annexation, Struggle and Response: Forest, People and Power in India and Nepal.' In *Forests, People and Power: The Political Ecology of Reform in South Asia* 27.

Srinivasa Rao, K. 2011. '2 killed, 25 hurt in police firing in Kakarapalli.' Available at: http://www.hindu.com/2011/03/01/stories/2011030162630400.htm [Accessed August 11, 2011].

Sudhir, U. 2010. 'Wetlands presented as wastelands: Undoing the lies.' Available at: http://www.ndtv.com/article/india/the-big-green-lie-wetland-declared-wasteland-in-andhra-pradesh-38871?cp [Accessed May 19, 2012].

Supreme Court of India. 1997. 'Judgement in the case of "Samatha versus the State of Andhra Pradesh and others".'

Syed, A. and Dutta, S. 2012. 'Rs 20,000 crore Hydropower Scam in Sikkim; State Govt in Cahoots with Pvt Players.' *CurrentNews*. Available at: http://currentnews.in/?p=27932 [Accessed May 29, 2012].

The Hindu. 2011. 'Ministry suspends work on Srikakulam project.' Available at: http://www.thehindu.com/news/states/andhra-pradesh/article1501148.ece [Accessed August 11, 2011].

Chapter 11

Water Worries in a Coal Mining Community: Understanding the Problem from the Community Perspective

Prajna Paramita Mishra

Water Uses: Conflicts between Industry and Community

Mining induces water scarcities in the areas in which it operates. This chapter examines such water worries in the coalfield region of Odisha, where water problems comprise the two-fold issues of scarcity and pollution. The study selected 260 households from mining villages and 100 households from control villages. Result shows that mining villages face acute water problems not only during summer, but throughout the year, whereas this problem is totally absent in control villages. The study outcomes reject the general view that underground mines are less vulnerable than opencast mines, suggesting instead that with regard to water both types of mines are equally vulnerable. Solutions to this problem lie in finding new forms of technology and seeking to develop measures that ensure the social responsibility of mining companies. The state and central government, and the pollution control board, need to play a major role in resolving water quality and shortage problems. Finally, the study considers the urgent need for communities to take collective action in improving what amount to life-threatening circumstances.

The mining industry and water resources are related in many ways. Kemp et al. (2010) outline the ways in which the mining industry interacts and interferes with water. Mining is a heavy water user; the industry's most important uses are for processing and transportation of ore and waste, separation of minerals, suppression of dust, equipment washing and human consumption. By altering the local or regional water balance, each of these activities, causes a direct impact on human daily life and the ecosystem: extraction processes reduce the volume of ground water; poorly disposed waste and wastewater pollute waterways; and Acid Mine Drainage (AMD), which occurs throughout the industry, is one of the root causes of mining and water-related problems (Bebbington and Williams 2008; Ochienge, Seanego and Nkwonta 2010). Li's (2012) anthropological study in the mining area of Peru discusses common controversies over how water resources should be used and managed. Here mining companies and small farmers in the vicinity have experienced the landscape differently, resulting in disagreement over water quality and availability. Budds and Hinojosa's (2012) study in Peru offers a political ecologist's perspective to the same problem. They try to examine how the increased demand for water by the mining industry in Peru rescales water governance. Kemp et al. (2010) usefully connect issues of mining with those of water and human rights. According to their research, mining industries regard water prevailingly as a business asset. Though they have a strong commitment to human rights in their sustainability framework, the human rights implications of water are rarely considered.

Cote and July (2009) proposed a water accounting framework for the Australian mining industry, explaining that water accounting for the mining industry is related to two broad areas. The first area concerns the intersection of the site with the community, environment and other stakeholders, while the second relates to water use within the mine site. The proposed framework provides consistent language and metrics for quantification and communication of water use covering these two areas. Mudd (2008) has complained that the sustainability reports of mining companies do not include the main aspects required to facilitate water accounting. For example, the reports do not incorporate data concerning the amount of recycled water used, the quality of various waters, and the impacts on water resources.

Research around the world shows that water problems in the mining industry constitutes a very important issue. The main concerns are the quantity of water consumed, competition over water between mining and other sector such as agriculture, and impacts of mining on water quality (Younger and Wolkersdorfer 2004). This study of mining-induced water problems concentrates on the Ib Valley coalfield of Odisha. The two main objectives of this chapter are to outline mining-related water problems in general and those specific to the Ib Valley coalfield in particular; to develop an understanding of these problems and the solutions most likely to mitigate them.

The first section of this chapter introduces the research that will be considered; the second section discusses mining-related water problems, especially in the Indian context. A brief account of the study area is provided in the third section, which also includes a note on data collection and methodology. Results of the study, which comprise of the main water problems occurring in mining villages, are provided in the fourth section, which also seeks to develop an understanding of these problems and to identify appropriate solutions.

Demands that Mining Places on Water

With increasing industrial water use there has been a concomitant rise in conflict between local communities and the related industry on issues ranging from water pollution to water scarcity. In water-scarce areas, industries are under tremendous pressure from community and government alike to reduce the use of water. The two major water problems related to mining are depletion of ground water, and scarcity and pollution of surface water. The experiences of Tikamgarh in Madhya

Pradesh and Rajasamand in Rajasthan Gupta (2007) have outlined a number of negative impacts that mining causes on water availability and quantity. These are: reduction in water quality; contamination of local surface and ground water; significant changes in local hydrology; and destruction of the productive capacity of agricultural land and vegetation.

Large-scale coal mining operations in Jharkhand have adversely affected the ground water table, resulting in a drastic decrease in water from the wells of adjoining villages. Effluent discharged from mine sites has seriously polluted the streams and underground waters of the area. Acid mine drainage; liquid effluent from coal handling plants, colliery workshops and mine sites; and suspended solids from coal washeries have caused serious water pollution in the region, adversely affecting aquatic life (Areeparampil 1996).

A study by Singh (1985) shows that a single coal washery discharges about 40 tonnes of fine coal into the Damodar River every day. A major source of water in the region, Damodar is now perhaps the most polluted river in India. In the Mahanadi river basin of Odisha, several industrial sites – with many polluting industries including ten coalmines – discharge huge quantities of water containing heavy metal and sulphur compounds during the monsoon, posing serious health and environmental threats. The river basin of Brahmani in Odisha also has major industrial areas and a number of coal mines, with high levels of industrial pollutants and drainage from coalmines being discharged into its rivers (Das 2001). Tribal people, fishing villages, cattle and several species of birds and other wildlife previously existed with little difficulty in and around the river. Since the advent of coal mining the river has become so polluted that it can no longer support aquatic life. Several species of sea turtles have become endangered.

Coal mining, aluminium smelting, fertiliser and chemical production use the river Nandira into an industrial drain. The now black water of the river is poisoning and slowly killing people, animals, fish and plants as far as 50 miles downstream. Agricultural productivity of the farmers depending on this polluted water has dropped. Fishing communities have also been wiped out (Das 2001). These are the major water-related problems in mining areas. It is against this background that this chapter considers the major water-related problems in the Ib Valley coalfield of Odisha. The third section of the chapter discusses the methodology and data collection of this study area.

Study Area and Data Collection

The state of Odisha has two major coalfields: Talcher and the Ib Valley coalfield. Ib Valley coalfield is situated in the Jharsuguda and Sundargarh districts of Odisha, covering an area of 1375 square kilometres. The coalfield is named after the river Ib, a tributary of the river Mahanadi. This coalfield was discovered during construction of a bridge over the river for the Bombay–Nagpur railway.

The first mine to be excavated was the Himgir Rampur Colliery in 1909. Subsequently, the Orient Mine No. 1 was commenced in 1940; both of these were underground mines. After the nationalisation of coal mines, these mines came under the administrative aegis of Western Coalfield Limited (WCL). Later, they were subsumed by South Eastern Coalfiled Limited (SECL), and then by Mahanadi Coalfield Limited (MCL) on its formation in 1992. Currently this coalfield is divided into five areas comprising five underground and seven opencast mines. The five areas are: Orient; Ib Valley; Lakhanpur; and the relatively new areas of Basundhara and Garjanbahal. Basundhara started after 1997, and Garjanbahal is yet to begin operation. Because of the longevity of the first three mines, this study concentrates on these to examine the severity of the impacts.

Because coal mining began in the area nearly a century ago, it is extremely difficult to assess the extent of its effects on the environment by undertaking a temporal comparative study prior and subsequent to its advent. Therefore, this research relies on the geographical comparative method of studying areas with-and-without exposure to coal mining. The sample consists of five villages situated near the coal mines (mining villages are Ainlapali, Bundia, Chharla, Lajkura and Ubuda) and two villages that are situated away from the coal mines (control villages are Saletikra and Tangarpali), but which belong to the same district. Data was collected from January to May 2005, and information was gathered with the help of a structured questionnaire. The households were selected using the circular systematic sampling method. From the total of 360 samples, 260 households were from the mining villages and 100 from the control villages.

Water in the Ib Valley Coalfields

There is acute scarcity of water in each of the mining villages, which worsens during summer. The groundwater level has dropped substantially. People depend on open wells, but during the summer they also become dry. For bathing and washing purposes, all villagers depend on the village ponds. After mining, these ponds are filled with coal dust. The MCL provides water to the Lajkura village every day. To Chharla and Ubuda the company provides water only during summer, although the amount provided is not sufficient for the villagers. Since the water level in Lajkura and Ubuda villages is receding, the tube wells are not functioning. In the instance of the control villages, there is no water problem because of well-functioning tube wells established by the Panchayat. In addition to these, there are at least two ponds in both of the villages. In the mining villages, the numbers of ponds are less, as some ponds are now coming under the control of the MCL area (see Table 11.1).

It has been observed that in the control villages the percentage of households depending on tube wells is very high. However, there are no functioning tube wells in the mining villages except Ainlapali and only one in Chharla. People in Saletikra and Bundia villages use their personal open wells as a source of water.

The MCL provides water to only one village, namely Lajkura, throughout the year. Bundia is the only village where a few households use river water.

One major difference in both mining and control villages is the quality and quantity of drinking water. In the control villages people do not face water problems: their tube wells function properly; water levels in open wells are not low; they have more than two ponds in each village; and the quality of water is good. On the other hand, in the mining villages tube wells do not function. Rich households maintain their own wells, which they keep covered in order to protect them from the coal dust rising from mining activities. They also use water filters to improve water quality. It is the poor who suffer from a number of diseases due to the use of polluted water from the coal dust filling village ponds.

As women are burdened with all domestic responsibilities, they suffer the most due to water pollution. In Bundia village women complained that they have to walk miles to fetch water from a particular pond where water is comparatively fresh. In this village, the Scheduled Caste (SC) and Scheduled Tribe (ST) populations, considered untouchable, are not allowed to access the village well. The caste system prevails, even in twenty-first century rural Odisha. As a result, these people do not have any option but to purchase water from higher-caste people. One woman revealed that sometimes they stole water during the night, suggesting the acuteness of the water problem in the Ib Valley.

Table 11.1   Sources of Water in Sample Households (in %)

Villages	Drinking and Cooking			Bathing and Cleaning		
	Tube Well	Ground Well	MCL Pipe	Ground Well	Pond	River
Villages near open cast mines	15.23	84.76	0	5.71	94.28	0
Villages near underground mines	0	100	0	4	84	12
Villages near open cast and underground mines	29.52	24.76	45.71	3.80	96.19	0
Control villages	69	31	0	18	82	0

Source: Field study conducted by the author in coal mining villages of Jharsuguda District, Odisha, 2005

Understanding the Problem

According to research conducted by the Centre for Science and Environment (CSE 2004), there is a lack of effective government regulation of industrial water management, and there are little or no incentives for efficient water use. Water tariffs are generally ignored, and there is a rise in water conflicts between industry and local communities. The provision of water subsidies to Indian industry is one of the main causes of the poor pricing of natural resources, and one which leads to their exploitation. Since using more water does not affect Indian industry, it uses the resource without regulation or responsibility (CSE 2004).

Recently, the government of India revised the water charges for industries and amended the *Water (Prevention and Control of Pollution) Cess Act 1977*. In 2010 the Odisha government determined to increase the water cess by 100 per cent. The state government decided that water supply to industries would be charged according to the electricity consumed for drawing the water, at the rate of 3 paise and 5 paise per unit of electricity consumed. However, the government has yet to recover 500 crore outstanding dues from big industries in the state (*Financial Express* 2010).

Industrial water tariffs from public supplies in India are typically based on average cost pricing rather than marginal cost pricing. This ignores the opportunity cost of water, that is, the benefit foregone in alternative usage (CSE 2004). The pollution by industry of ground and surface water is not taken into account in determining water tariffs. As a result, charges such as pollution taxes and/or effluent charges to be paid by industrial polluters are absent. Excessive quantities of water are therefore used, and excessive pollution is caused (CSE 2004). Proper pricing of natural resources is important for the proper management of natural resources. Proper water pricing would encourage industries to conserve water and as a result pollution would be reduced to a large extent.

All developed countries of the world set water-consumption standards and targets for industries to achieve, and regularly revise them in a bid to control water usage. In India there is currently no law determining the exact amount of water required for consumption by the various industrial sectors (CSE 2004). India has the highest global rate of ground water extraction. It also retains outdated laws relating to groundwater extraction. In Indian law, the person who owns the land also owns the groundwater below it: this law may have some relevance for domestic use, but it has none for industrial and commercial use. As a result of this, groundwater usage is not covered by tax (CSE 2004).

Presently there is a global shift from concentration-based standards to pollution load-based standards. The latter determine the total amount of pollutant generated for per unit of production (CSE 2004). Introduction of the 'polluters pay principle' induces companies to reduce pollution loads. In India, where industrial water pollution standards are concentration-based, both these principles are absent. As a result, industry uses more freshwater and discharges more pollutants than in comparable nations. The concentration-based system measures the concentration of pollution in a given quantity of water. As the cost of water is very low, it is economical for industry users to dilute the effluent with clean water and then to treat this polluted water (CSE 2004). It is clear that in instances such as this the problem lies with the implementation and regulation of government policies. Invariably it is the poor villagers or the people in the vicinity of polluting industries who suffer most. The next section seeks to identify solutions to these water-related problems.

In Search of Solutions

The environmental impacts of mining have been a challenge throughout India. Steps are taken in modern mining processes to minimise these impacts, such as in improved planning and environmental management (WCI 2005). The key to the water problem lies in effective management of water resources. An integrated approach involving water treatment, source reduction, reuse of process water, effluent treatment, recycling of treated effluent and waste-minimisation is urgently required (CSE 2004).

For the control of environmental pollution three important entities need to be considered: the market, government, and the community. Generally, market forces fail to control environmental pollution, and thus require government intervention. The command-and-control approach is a non-market policy instrument operating through penalties and the threat of legal action. Market-based or economic instruments typically comprise the form of pollution taxes (Pigou 1932) and pollution permits (Dales 1968). The choice between command and economic instruments depends both on their efficacy in achieving the target level of emissions and the relative size of welfare losses they produce (Baumol and Oates 1988). To control environmental pollution, Murthy (2002) has suggested a mixture of both command-and-control and economic instruments.

Vijay (2003) studies the subject of industrialisation leading to water pollution in a village of Andhra Pradesh. According to him, solutions derive neither from education and technology nor from regulating institutions. According to Behera and Reddy (2002), collective action by victims is needed in order to establish bargaining capacity to force the regulatory bodies to respond. When market and policy fail, such collective action is the alternative method most likely to succeed in tackling environmental problems.

In such a situation, compensation is also not a solution. Compensation ultimately results in the institutionalisation of a system that enables payment for the right to pollute (Behera and Reddy 2002). A responsible compensation package should encapsulate the breadth of the costs of water pollution, including health costs. Although advanced process technologies can help industries to reduce their water demand, this solution is costly. A far more cost-effective solution requires that industry re-use poor-quality water in series process that recirculates and reprocesses water (CSE 2004), thereby indefinitely re-using the same water after treatment for the same purpose. This is a comparatively cheaper method than installing new process technology, and recent technological development has ensured its application to any type of water-intensive industry (CSE 2004).

Rain water harvesting is another option addressing environmental water-usage problems. Harvesting is capable of meeting a substantial part of annual industrial water requirements. There are many technologies capable of solving industrial water problems that are cost effective (CSE 2004). Indian industry can substantially reduce its water consumption and waste water discharge by incorporating efficient systems for recycling and re-use, but it can only be realized through enforcement by government policy (CSE 2004).

The notions of 'Corporate Social Responsibly' and 'Sustainable Development' have been gathering strength within the mineral industry in India. Each of these paradigms broadly defines the community obligations of companies and stresses the need to improve social as well as environmental performance. For example, the International Council on Mining and Metals' (ICMM) sustainable development framework includes an undertaking by signatories to 'contribute to the social, economic and institutional development of the communities in which we operate' (ICMM 2003). In a similar vein, the World Coal Institute (WCI), in its report to the 'Rio Plus 10 Earth Summit' held in Johannesburg in 2002, stated that a key 'action area' for the industry is to make a more effective contribution to the social and economic development of local communities' (WCI 2002).

In summary, this case study has facilitated an understanding of water-related problems, their extent, their causes, and their impacts on people's lives in a mining community. Not only does the problem of water pollution loom large, there is also water scarcity. Both water availability and water quality are matters of serious concern. Low water-levels affect tube wells closer to the mine, and coal dust affects closer villages more readily than distant villages. In all villages the burden of water collection is on the womenfolk, who are thus the worst-affected by water pollution. Along with other domestic responsibilities they also walk miles to collect water. A number of health problems arising out of this are reported by the villagers. As discussed earlier, caste continues to play a dominant role in determining access to water in the villages where the MCL does not provide an adequate quantity of water.

Three decades after the launch of the National Drinking Water Mission, villages in India continue to reel under water-related stresses (Krishnan et al. 2003). The National Drinking Water Mission has been set up by the Government of India under the Department of Rural Development, Ministry of Agriculture in 1986. The main objectives are to improve the performance and cost effectiveness of the on-going; programmes in the field of rural drinking water supply and to ensure the availability of an adequate quantity of drinking water of acceptable quality on a long term basis (Government of India 2010). The causes of water worries differ from place to place. In the case of the Ib Valley coalfield mining communities, collective efforts are needed from the government, non-government organisations (NGOs), industry and community to work together to mitigate this persistent problem.

References

Areeparampil, M. 1996. 'Displacement due to Mining in Jharkhand.' *Economic and Political Weekly* 31(24): 1524–28.

Baumol, W.J. and W. Oates. 1988. *The Theory of Environmental Policy*. (2nd edition). Cambridge: Cambridge University Press.

Bebbingotn, A. and M. Williams. 2008. 'Water and Mining Conflicts in Peru.' *Mountain Research and Development* 28 (3/4): 190–95.

Behera, B. and V.R. Reddy. 2002. 'Environment and Accountability: Impact of Industrial Pollution on Rural Communities.' *Economic and Political Weekly* 37(3): 257–65.

Budds, J. and L. Hinojosa. 2012. 'Restructuring and Rescaling Water Governance in Mining Contexts: The Co-Production of Waterscapes in Peru.' *Water Alternatives* 5(1): 119–37.

Centre for Science and Environment. 2004. 'Not a Non-issue, Water use in Industry.' *Down to Earth* (Supplement) 15 February: 1–8.

Cote, C. Moran and C.J. July. 2009. 'A water accounting framework for the Australian minerals industry.' In *Proceedings of the 4th International Conference on Sustainable Development Indicators in the Minerals Industry*. Publication Series No 5/2009, Australian Institute of Mining and Metallurgy, Gold Coast, Australia: 339–50.

Dales, J.H. 1968. *Pollution, Property and Prices: An Essay in Policy Making and Economics*. Toronto: University of Toronto Press.

Das, Prafulla. 2001. 'Orissa Government Blamed for Declining Quality of River Water.' *The Hindu*, Available at http://www.hindu.com/2001/08/06/stories/1406219e.htm accessed on 7 August, 2008.

Financial Express. 2010. 'Orissa hikes industrial water cess by 100%.' *The Financial Express* Available at http://www.financialexpress.com/news/orissa-hikes-industrial-water-cess-by-100-/654648 accessed on 6th April, 2012.

Government of India. 2010. 'National Rural Drinking Water Programme' Available at http://rural.nic.in/sites/downloads/pura/National%20Rural%20 Drinking%20Water%20Programme.pdf accessed on 20th May, 2013).

Gupta, P. 2007. 'Impact of Mining on Water Availability and Quality – Experiences.' Available at http://www.indiawaterportal.org/solex/6952 accessed on 9th May, 2012).

International Council on Mining and Metals. 2003. '10 Principles of Sustainable Development Framework.' Available at http://www.icmm.com/our-work/sustainable-development-framework/10-principles accessed on 10 July, 2008).

Kemp, D., C.J. Bond, D.M. Franks and C. Cote. 2010. 'Mining, water and human rights: making the connection.' *Journal of Cleaner Production* 18(15): 1553–62.

Krishnan R, S. Bhadwal, A. Javed, S. Singhal and S. Sreekesh. 2003. 'Water Stress in Indian Villages.' *Economic and Political Weekly* 38(37): 3879–84.

Li, F. 2012. 'Contesting Equivalences: Controversies over Water and Mining in Peru and Chile.' In *Social Life of Water in a Time of Crisis*, edited by John Wagner, 27–49. Berghahn Books.

Mudd, G.N. 2008. 'Sustainability Reporting and Water Resources: a Preliminary Assessment of Embodied Water and Sustainable Mining.' *Mine, Water and the Environment* 27(3): 136–44.

Murthy, M.N. 2002. 'Economic Instruments in Assimilative based Environment: A Case Study of Water Pollution Abatement by Industries in India.' No. 59/2002, Discussion paper, Delhi: Institute of Economic Growth.

Ochieng, G.M, E.S. Seanego and O.I Nkwonta. 2010. 'Impacts of mining on water resources in South Africa: A review.' *Scientific Research and Essays* 5(22): 3351–57.

Pigou, A. 1932. *The Economics of Welfare*. 4th edition, London: Macmillan.

Singh, J. 1985. *Upper Damodar Valley: A Study in Settlement Geography*. New Delhi: Inter-India Publications.

United Nations. 2006. 'Water, a shared Responsibility.' 2nd United Nations World Water Development Report, Nairobi: UN Habitat.

Vijay, G. 2003. '"Other Side" of New Industrialization.' *Economic and Political Weekly* 38(48): 5026–30.

World Coal Institute. 2002. 'Water Management at Coal Mining Operations in South Africa.' WCI Project Report, London: WCI.

———. 2005. *The Coal Resource, A Comprehensive Overview of Coal*. London: WCI.

Younger, P.L and Wolkersdorfer C. 2004. 'Mining Impact on the Fresh Water Environment: Technical and Managerial Guidelines for Catchments Scale Management.' *Mining, Water and the Environment* 23 Supplement 1: S2–S80.

Chapter 12

Gender in Coal Mining Induced Displacement and Rehabilitation in Jharkhand

Nesar Ahmad and Kuntala Lahiri-Dutt

Mining: Impacts on Communities and the Displacement of Women and men

Although coal is administered centrally in India by the Ministry of Coal, the government does not maintain reliable and comprehensive data on displacement caused by extraction of the resource. Most importantly, it is not possible to retrieve gender-segregated data from the meager records available. Often, mining displaced communities (DPs, also known as project-affected people or PAPs) are treated as a homogeneous and unitary group, without attention to the diversity within these communities. However, many new open-cut coal mining projects exclude indigenous people, especially women, from the social and economic benefits they produce, whether it is due to a lack of skills, to a lack of formal education, or to a lack of ownership and control over land and water. The social and gendered consequences of specific mining projects vary with local circumstances, but the common features in India include pauperization: the impoverishment of communities and the feminisation of that poverty. Physical removal from the original place of residence – primary dislocation – is the most critical among the various social impacts, and has complex outcomes and is immediately visible and measurable; displacement from access to local natural resources and resource-based traditional occupations occurs imperceptibly over a longer duration of time and leads to the complete destruction of livelihood bases (Basu 2000). Displacement has an undermining influence on social bonds and cultural roots of the entire community, with devastating and disruptive effects on the lives of women, who often operate at a subsistence level. The economic and socio-cultural consequences of displacement are traumatic. When communities are forced to leave the land that they have lived on for generations, they not only lose farming land, but are also deprived of the forests, waterways, springs and grazing lands on which their lives were dependent. The social bonds that existed between individuals are ruined, and there is often a lack of trust in rehabilitation colonies among those who have been displaced and resettled. However, there is yet another process at work – that of the inevitable secondary displacement, with its short and long-term impacts, such as the loss of commons, decay of agriculture, and displacement of peasantry, eroding the livelihood bases of the community. As communities are burdened with negative impacts such as the loss of traditional livelihoods, environmental degradation, and social and economic instability after mine development, the gender dimensions of livelihood reconstruction become an important area of inquiry for comparison and learning (Mallik and Chatterji 1997).

This chapter focuses on the impacts of coal mining on women as a means of considering the extent to which the impacts of mining are gendered due to the different roles women and men play in mining communities. Differences in gender roles mean that women and girls bear a disproportionate share of the unpaid work in almost all households in rural India; the inequities are exacerbated by shifting burdens: of procuring primary subsistence and livelihoods; and of providing for children. When mining projects lead to the restructure of agrarian relations, a decline in household assets, inequitable entry into labour markets and differential burdens of care, the capacities of women and men to seize the new opportunities and to cope with the risks and repercussions from such investments invariably differ. Most mining company and related professionals, for example, treat men as the head of the household in villages, although the role of such men within the household may be limited to earning cash income. Rural women bear the primary responsibilities of care for children and the elderly, as well as the livestock, and in poorer families also fetch water and collect food, fuel and fodder for the animals. Yet in assessing the social impacts of mining, experts readily perceive the community as a homogeneous unit, without taking into consideration the different roles, positions and situations of women and men. Such implicit assumptions result in the production of a kind of knowledge that claims to be universal and objective, but which is based exclusively on men's lives.

Guijt and Shah's (1998) influential work has contributed to breaking the myth of this imagined, homogeneous and equitable community. More recently, Razavi and Hassim (2006) have posed another powerful critique of social structure in development and have drawn attention to gendered structures within society that call for new and inclusive social policies. Research is emerging that differentiates social impacts along gender lines. However, from a gender perspective three characteristics of projects remain problematic. First, the impact assessment experts are generally yet to acknowledge and adopt the work by feminist scholars on the household. Impact assessment professionals need to incorporate greater specificity in the study of social impacts, and to explore the activities, roles and contributions made by individuals at a more personal scale than that of the household. Second, normative assumptions about men's and women's roles as breadwinners and mothers/carers respectively remain, even in the context of rural areas in developing countries where many women work as farmers or act in other ways as the primary subsistence providers. Even where women are engaged in paid work, universal stereotypes portray women primarily as homemakers. Third, women-headed households do not receive due consideration in the existing literature on social impacts, although it is generally known that many projects cause increased out-migration of males to cities and mines, and women therefore assume productive roles in many rural households. Male out-migration, along with desertion and the early death of many males due to poor health conditions and lack of medical facilities, mean that many rural households are now headed by women: such potential impacts on families should be specifically considered.

The global mining sector – that is, multinational mining companies; international associations; industry bodies; and regulatory and donor agencies – is becoming increasingly sensitive to circumstances in which social impacts are gender-differentiated. Nonetheless, it is still important to contemplate why gender impacts should be considered in order to avoid falling into a theory vacuum in which women are incorporated in the existing scheme as a supplementary measure. Such projects regarding the impacts of mining on women potentially invite inaccurate universalist conclusions by treating all women as the same irrespective of their race, economic class, ethnicity, caste or age. Without a gendered understanding, all women would be contrasted with all men, and women's interests would then become separate from and considered in isolation from those of men. Women would thus become regarded as a special category in a bureaucratic sense, the impacts on whom would constitute little more than an exercise in listing special interest problems. To avoid this, one must consider the ways in which gender is included in thinking about the impacts of mining and displacement literature from a theoretically-informed feminist perspective.

By using a feminist perspective we aim to create a sophisticated understanding that moves beyond the simplistic picture presented by the narratives of impacts of mining on women. Theoretically, the genre of impacts of mining on women literature is based on the approach of Women in Development, a 1970s perspective that has been discredited and is no longer in circulation in contemporary theory and practices of Gender and Development (GAD, see McIlwaine and Datta 2003 for a full literature on GAD). Nonetheless, such projects continue: for example, the World Bank's women in extractive industries action plan (Eftimie et al. 2009) uses an exclusive, 'women-only' focus. A similar approach was advocated in S. Sharma's (2010) article, which considers the effects of mining on women in Australia based on experiences in Asian countries, as though women from these contexts have similar interests and attributes, and the impacts on them would be comparable. There are some insightful research on impacts of mining on women that stand out because of their attention to politico-economic, and historical and cultural context. For example, Indira Rothermund singled out women in her exemplary study of coal mining in Jharkhand (Rothermund 1994), but placed her study in contexts of the region's unique cultural attributes and its complex history of involvement of indigenous men and women in mining. One can say that to connect theory with practical and empirical studies, researchers require a gender lens and a feminist perspective that renders visible not only the differences in roles and status between women and men, but also the heterogeneities that exist among women, for instance, between those from Australian coal mining areas and those in rural India. A feminist perspective is transdisciplinary, eclectic and does not rely on a standard methodological analytical framework (Harding 1987).

Feminist methodology is transformative and aimed at initiating positive change, foregrounding this as the purpose of academics and practitioners in researching social impacts. Gibson-Graham (1994, 206–7) has drawn attention to the analogies between mining and feminist social research, because, as she observes, both "intervene in and disturb a landscape by probing and digging for a rich lode of ore or layer of stratum that has hitherto lain covered, or unknown, perhaps until now unvalued." Similarly, in undertaking gender assessments the adoption of a feminist perspective assists in learning about unexplored areas. Unlike mining, however, there is no accepted single feminist method, which makes it difficult to translate feminist theory into analytical frameworks that fundamentally challenge the masculine biases in SIA. For example, the androcentric picture that one receives of mining and the social worlds around mines emerges from the process by which men test hypotheses generated by what men exclusively find problematic as subjects of enquiry. A feminist perspective would be useful to shift away from the current simplistic narratives of the negative impacts of mining on women. Available narratives generally portray women living in and around mines as victims, and neglect to assess their needs, interests and aspirations. They also ignore the differences economic class, race and age can make.

Gender: Ignored in Displacement Literature

The effects of displacement are not gender neutral, given that women and men have well-demarcated gender roles in indigenous communities, and that the impacts of mining on women and men are not the same. Wherever an entire community suffers from the losses of environmental resources, women suffer more because of their gender roles. For example, women in poor communities are primarily responsible for collecting the subsistence resources that maintain the family's overall livelihood. In indigenous communities the status of women in some productive systems is not unequal to that of men, but at the same time they remain responsible for collecting the essentials for subsistence: food, fodder, fuel and water (Agarwal 1992). This gender division of labour makes women still more dependent on and close to local resources offered by the local environment than men (Leach 2007). Often, these chores and responsibilities do not translate into better status or greater power in decision-making. Indian women, especially those living in villages, do not have legal rights over land, and are rarely the title holders. Compensation processes for land reclamation usually only accept an adult man as the head of a household, and fail to consider the needs and requirements of women. Compensatory jobs, if any, are usually awarded to men, and women are forced either to stay at home as housewives, or to look for wage labour work further afield (Fernandes 1998; Ganguly-Thukral 1996). In such circumstances, where only men are seen as legitimate wage earners, women who are left in the community to care for children while their husbands are away can find themselves overwhelmed with new responsibilities (Fouillard 2003). The resettlement and rehabilitation (R&R) process has continued to remain gender-blind. An instance of this is CIL R&R policy: until recently the policy devoted only one sentence to women, which concerned ensuring women's access to income-generating opportunities; however, the latest version of the policy (2012) does not even mention women. R&R interventionist measures look primarily at housing and other such practical needs of communities. As noted by Kumar (2005), the time has come for Indian R&R policy to incorporate recent feminist arguments and approaches arising out of empirical research and social activism.

An early study of a coal-displaced community in Jharkhand by one of the authors (Ahmad 2003) showed that women were more affected by displacement than men because physical amenities and services were rarely provided at relocation sites, and women were more dependent on them than men. Further, the facilities that were provided tended to fall into disrepair as the villagers often lack the skills needed to maintain the infrastructure. One example of this is the provision of hand pumps for drinking water supplies. The lack of drinking water facilities and health services was a common problem at all resettlement sites However, gender impacts go beyond a lack of physical assets. Although women participated fully in farming, they performed specific tasks in the field – such as planting, weeding and rice-thrashing – and in the home – such as fetching water, cooking, and caring for children. However, as customary rights are informal and rarely codified or supported by policy, these responsibilities were not supported with equal rights to land or equal input regarding decisions affecting home and village life. Rao (2006) has observed that while customary principles are flexible and carry social sanctions, they are socially embedded. Hence, in the context of growing competition over land, gender identities and a concern for gender justice tend to be subordinate to issues of ethnic identity and survival.

The gendered division of labour makes women more dependent than men on the resources offered by the local environment (Leach, 1992). Consequently, the impacts of mining and subsequent effects of displacement also become gendered. Whereas the entire community suffers from the loss of environmental resources, women suffer more as they are primarily responsible for food security of the family. The expropriation of subsistence resources – water, basic food, fuel and fodder – by the mining companies and the dispossession of these assets experienced by women are difficult to evaluate in economic terms and a gender analysis framework is necessary to reduce impoverishment risks for displaced communities. The social and economic impacts of mining are gendered in the sense that women are less likely to share in benefits such as employment, and more likely to bear the burden of risks such as loss of livelihood. Without a gender analysis integrated into the SIA process, people, especially women, experience poverty, and those already in poverty are likely to remain disadvantaged.

If provided against land loss, compensatory jobs usually go to men, who are the legal owners of land. Opportunities for women, particularly young working-age women, are almost exclusively limited to low-income activities. Fernandes (1998) and Ganguly-Thukral (1996) have observed that many women are compelled to look for wage work to supplement family incomes after displacement, particularly where male out-migration occurs.

One perspective in contemporary literature on displacement suggests that a gender assessment is a useful planning tool that can be incorporated into all aspects of project analysis, design, implementation and monitoring (Kolhoff 1996; Kuyek 2003; Overholt 1991). Legislation should make it mandatory to conduct impact assessment of the project from a gender perspective to reveal how the proscribed social roles for women and men are affected in a development project, and should also require the development of current professional-quality practices to guide the assessment (Archibald and Crncovich 1999).

Another view is that gender cannot be grafted onto existing planning traditions, and that gender concerns require their own separate tradition (Moser 1993). Young (1993) argues that incorporating a gender assessment into the current process may fail to acknowledge the complexity of gender inequalities within society. She proposes development of separate principles in combination with state, community and family level initiatives. Scholars feel that the inclusion of gender and women as factors to be examined in a social assessment may not result in any significant changes in the way companies and governments respond to the findings or to the specific problems that emerge for women and families (Overholt 1998, 25). Manchanda (2004, 4179) argues that in gendering displacement, one must consciously be wary of the infantilisation of women. For her, '[t]he contemporary image of the forcibly displaced, the refugee and the internally displaced, fleeing life and livelihood threatening situations, is a woman usually with small children clinging to her.' As a result, mine development policies are invariably insensitive to gender roles and responsibilities, discriminating against women either by completely ignoring their needs or by infantilising them.

Displacement, forcible eviction and dispossession are undeniable realities of life in the coal mining regions of India. Large-scale acquisition of land is the most important driver of this displacement; the Indian Constitution, courts and government justify the function of land appropriation as a public good, as evident from a statement of the Supreme Court:

> The power to acquire private property for public use is an attribute of sovereignty and is essential to the existence of a government. The power of eminent domain was recognized on the principle that the sovereign state can always acquire the property of a citizen for public good, without the owner's consent.... The right to acquire an interest in land compulsorily has assumed increasing importance as a result of requirement of such land more and more everyday, for different public purposes and to implement the promises made by the framers of the Constitution to the people of India.

The 'public' that is referenced in this statement does not clearly include women in displaced communities as valid and able citizens.

The *Coal Bearing Areas (Development and Acquisition) Act* (CBAA), which is used to acquire land for coal mining, supersedes the non-transferability of indigenous lands that is also enshrined as a legal instrument, e.g., the *Chhotanagpur Tenancy Act, 1908* (See Chapter 13, for a detailed discussion). Over the last three decades of extensive mining development, the power of the state has led not only to involuntary and forced displacement (Fernandes and Thukral 1989), but also to dispossession through the destruction of the livelihoods of entire communities, turning humans into faceless PAPs. The process of change has been aided by in-migration, through the erosion of local environmental resources, through dereliction of the lands that have traditionally provided subsistence resources, through poor implementation of national safeguards and the neglect of international guidelines of care, and through stripping communities of customary rights, many of which were not formalised in law.

Robinson's 2002 work put the question of development and displacement of indigenous peoples in the domain of migration, following some previous approaches on environmental refugees. Colchester raised questions of implications of indigenous people's rights in the context of sustainable resource use. He noted that tensions between local claims to own and control land and a national development policy, which excludes indigenous peoples from the process, have reached a critically high level: 'the three central claims are: the right to ownership and control of their territories, the right of self-determination, and the right to represent themselves through their own institutions' (Colchester 1995, 61). Political scientists such as Ivison et al. (2000) describe the political theory of rights: the local community must have rights over local natural resources, and this is the principle of 'primacy of rights.' A right gives the holder authorisation to use resources from a particular source and includes the particular social privileges and obligations associated with that right (see Zwarteveen and Meinzen-Dick 2001 in the context of water rights). The legitimacy of right holders' claims is linked to social relations of authority and power. However, there are significant differences between various kinds of rights, for example, user rights and ownership rights such as those customary uses of *gairmajurwa* (or 'deedless' or common) lands. Another strand of thought in this set is that of 'conflicts' of interests between those of the state/mining companies and local indigenous communities (Ali 2004).

One major characteristic of the displaced population is that most of them do not have significant amounts of property or assets; they are generally landless or marginal farmers coming from poor families of Scheduled Caste (SC) and Scheduled Tribe (ST) communities which form the lowest strata of the society. According to an estimate by the Commissioner of ST and SC, over 40 per cent of those displaced up to 1990 came from these communities, even though the tribal people make up only 7.5 per cent of India's total population (Kothari 1996, 1477).

The Government of India did not have any national R&R policy till 1994. Some of the state governments and selected public sector companies, whose operations displace people on a large scale, for example the CIL (1994) and National Thermal Power Corporation – NTPC – (1993), have their own rehabilitation policies, but these are barely implemented. However, the rehabilitation of DP in the country continues to remain gender-blind, and neglects to take up the welfare of women in the displaced communities as an essential part of the rehabilitation process. This is probably because rehabilitation is not perceived by these companies as an intrinsic human right of DPs. Pressures from external funding agencies like the World Bank (as in the case of CIL) or the concern to complete the project on time (as in the case of NTPC) remain greater incentives in designing R&R policies. The laws that are instrumental in land acquisition, such as the Land Acquisition Act and Coal Bearing Areas (Acquisition and Development) Act, have no provisions for rehabilitation, and only compensation for land is considered enough in those acts.

In India there is no comprehensive national R&R legislation. The current R&R policy draft initiated in 2007 is supposed to be supported by proper land acquisition and R&R legislation, but it is not yet effective. There is a bill proposed by the Indian government which is yet to be passed by the parliament. This bill will replace the currently used *Land Acquisition Act 1894*. The proposed bill has seen many changes since it was first introduced in 2008. Earlier the government introduced two separate bills: one for land acquisition and the other for R&R, which could not be passed. Then a unified *Land Acquisition and Rehabilitation and Resettlement Bill 2011* was tabled in the parliament and subsequently referred to the Parliamentary Standing Committee, which suggested a raft of changes. The government presented a revised LARR bill that ignored most of the suggestions of the committee. This revised bill, however, did not obtain the approval of the cabinet, as some cabinet members did not find the bill conducive for industry and private investment in the country. A group of ministers was constituted to look into the bill, which after very intense deliberations reached consensus and with further revisions and a new name – the *Right to Fair Compensation, Rehabilitation and Resettlement Bill 2012* – it was approved by the cabinet and is now finally passed by the parliament.

Studies in Jharkhand

As there is a lack of detailed gender-based information about the projects involving involuntary displacement and R&R we present here evidence from two field-based temporal studies conducted, amongst the displaced communities in one chosen area of eastern India to reveal how women are affected differently than men by displacement. One study was conducted in 2001 (Ahmad and Lahiri-Dutt 2006) in Hazaribagh and Chatra districts, and another in 2008 (Lahiri-Dutt and Ahmad 2010) in Benti village of Chatra district in Jharkhand, eastern India.

The industrialisation process in Jharkhand began in the mid-nineteenth century, depending heavily on mineral exploitation. The mines of Jharia, Bokaro and Karakpura coalfields started in 1856 and hired local people as coal mine workers. However, these mines generally were underground workings with little impact on surface land use. In recent decades the region has been one of the favourite destinations of mining, power, irrigation and other large industrial projects. Ekka and Asif (2000) estimate that in total 1.4 million acres of land were acquired for development projects during the period 1950–90, and that these projects have displaced 1.5 million people during the same period. Mining caused about one third of total land acquisition and 27 per cent of the total displacement of people, according to the same study. Similarly, a recent publication by Greenpeace has brought to light how the declining forest cover of the state is threatening the existence of local fauna (Fernandes and Kohli, 2012).

Four major mining projects were responsible for impacts on the lives of women and men included in the two studies; these are: Parej East Open Cut Project (OCP) of Central Coal Limited (CCL), a subsidiary of the public sector CIL and the private company Tata Iron and Steel Company (TISCO) in Hazaribagh district;and Ashoka OCP, Piparwar OCP in Chatra district of the CCL. The largest number of our research participants in the first study were affected by Parej East (24), followed by Piparwar (15), Ashoka (6) and TISCO (1), while the second study included project-affected families of Ashoka and Piparwar projects.

In 2001 we interviewed 46 randomly chosen women displaced by coal mines, through the help of two local NGOs, Prerana Resource Centre and Chotanagpur Adivasi Seva Samiti (CASS). Although our study population was only a small fraction of the total number of displaced women, we believe that they represent the suffering and trauma of mining-induced displacement for women in such communities in general. These participants, 46 in total, were from two districts, Hazaribagh and Chatra, belonging to SCs (26 or 57 per cent), STs (15 or 33 per cent), Other Backward Castes (OBCs –3 or 7 per cent) and Muslim (2 or 3 per cent). Before the advent of mining and displacement, the lives of these women were based on agriculture and the collection of the produce of the forests. All of the women except two were illiterate. Of the 46 women, 37 were married, eight were widows, and one unmarried. Their ages ranged from 16 to 60 years, with an average age of 35 years.

In 2008 we conducted a repeat survey of 50 women and 30 men separately through long personal interviews over a period of six months in Benti village of Chatra district, where families had lost their land, forest and water sources to Ashoka and Pirapwar projects. However, their homes are not wholly lost, as the entire village had not been relocated. All respondents belonged to either SC or ST communities. Around 80 per cent of the women were married and the rest were widows. They were aged between 30 and 65 years, with half being between 35 and 45. The vast majority were illiterate.

For both studies, we used structured questionnaire schedules to generate numerical data. The interviewees were from different *tolas* (neighbourhoods) of Benti village as well as from the three resettlement sites (New Malmahura, Chiraiyatand and Kalyanpur). Additionally, we conducted long unstructured interviews of some key informants such as the older residents and women, project officers and social activists working in the area. Three focus group discussions were conducted with affected women, and direct observation methods were also used. We returned to the village in 2009 to confirm our findings and consider any changes.

The material from the interviews and focus group discussions were analysed to assess whether the impacts of mining activities were gendered or not, that is, if they were different for women and men. The key social impacts were analysed according to gender, and categorised into six major social indicators: homelessness; loss of livelihood; marginalisation and food insecurity; ill health; psychological stresses; and social and cultural disintegration. The six indicators are based on Cernea's (2000) important work suggesting an eight-point risk model to understand the impact of involuntary displacement on people and our observations during the fieldwork.

Impacts of Displacement on Women

The foremost impact is the changes in gender relationships at home and in the communities. All families have lost their agricultural lands. At the moment, all families have somewhere to live. The cash compensation received for lost land has enabled some families to buy a vehicle, generally a motorbike. The families which have been able to secure employment also tend have financial reserves in the form of bank deposits and insurance policies. However, the ownership of these assets is highly skewed towards men. The bikes are owned and used by men. The financial assets tend also to be held by men. The enhancement of the economic power of men relative to women has turned women into being mere dependents of men. Such dependence is quite uncommon in tribal communities in India and it therefore has significantly affected the social relationships and women's sense of self-worth.

Homelessness

Almost all the women surveyed in the first study had lost their homes to the mining projects and, at the time of interview, about 30 per cent of them were living in temporary huts, some already for as long as ten years. The temporary homes included CCL's miners' quarters and guards' barracks, and relatives' homes. Many of the homes are built of raw materials that have to be procured from the market at a price, and hence have fallen into disrepair due to the lack of continued and adequate incomes to allow families to purchase goods from the market to repair the house. Toilets and other sanitary facilities are almost always absent from these houses. Taking shelter in relatives' houses has at times led to ill feeling amongst kin, leading to an overall loosening of the social fabric. As mentioned above, most of the women in the second Study in Benti had not lost houses. Only two of the Benti hamlets had been relocated to the rehabilitation sites.

Loss of livelihood

About 90 per cent of the women participating in the first study and all the women from the Benti study lost the agricultural lands owned by their families to the mining projects, although very few of them had accurate knowledge of the amount of land lost. This minimal awareness of family possessions illustrates the low status of women in the family and society. According to both studies, however, most of the land that was lost was deedless *gairmajurwa* land. Often people use these lands as community pastures or for farming as a matter of tradition, but these de facto circumstances are not recorded in government records. Consequently, no one in the villages was compensated for the lost *gairmajurwa* land. However, the importance of these commons to the women of local communities as a source of subsistence resources cannot be overstated. Similarly, natural water sources such as springs that do not belong to a particular individual but provide water to all the families in the communities are not compensated for.

Prior to displacement, most of the women were engaged in farming and also collected forest products. Both studies found that after displacement, women lost the earlier work on fields and in forests, and now most of the women are confined to household chores. The earlier forest-work undertaken by women was not paid work, but it provided these communities with produce. Leaves and flowers of different trees were used as vegetables and their fruits eaten; some roots and leaves were used for treating fevers and arthritis. Mahua is one of the most important forest products collected by the families. *Dori*, the seeds of Mahua trees, would be crushed for oil, which is used to make food and body lotion. The families had their own Mahua trees as well as common trees which were lost to the mines. The Mahua flower is dried in the sun for later consumption. Often collection involves the whole family, with women and children playing the major role. Mahua is used mainly for making liquor, but also as a food item. Families used to sell collected Mahua to earn money. The forest provided a range of other products used as food and medicine which were not available or available only in very small quantities after displacement. Firewood was also collected by almost all of the families and sold by some to earn money. Since relocation to places distant from the forest, it has not been possible for the women to collect these products.

Awarding jobs to the displaced families in its subsidiary companies is one of the major planks of the CIL's R&R policy. However, according to a Ministry of Home Affairs' study (Fernandes 1998) – which is the latest available data on CIL's compensatory jobs – CIL did not provide employment to people from about 30 per cent of displaced families during 1980–85. In this study 33 per cent of women reported that none of their family members was given any job with the company. The compensatory jobs, when provided by CCL, have gone mostly to men. None of the women from first study and only a few women from second study obtained compensatory employment with CCL. The Rehabilitation Policy of CIL does not restrict a woman from receiving a compensatory job, but patriarchal Indian society considers it self-apparent that only men should take employment. When we spoke with local men it was stated clearly that as long as men were around, women should not join the CCL. CIL's policy is not very encouraging for women. Technological upgrading and mechanisation of the mines have so far helped only in expelling women from the mine industry in India (Lahiri-Dutt 2012).

The management's bias towards men in awarding compensatory jobs is evident: many women reported that CCL preferred to hire males. CCL management discourages women from joining the mines and gives priority to men. Although it is CIL's stated policy that it will provide employment to a dependent of any of its workers after a workplace death or accident causing permanent disability, the rules are exercised in a discriminatory fashion with company officials, family and society generally discouraging women from taking up this entitlement. As one 55 year-old woman participating in the Benti study described, even though her husband died on the job, she was not given employment: 'The managers said: "Why do you want to take the job?... It would be better to let your son have the job."' Her son, however, has yet to be offered employment and the family is still fighting the case. Finding income-producing activities has become difficult for women simply because without forests and farmland, there is very little productive activity available. Some women informed us that it is difficult to acquire employment as the mine uses heavy machines and prefers skilled external labour. As a young woman of 22 years, interviewed in the second study, said:

> There is no work available now. Earlier there were some casual jobs. Earlier, there was *van-jhar* [forest] and *tandtapur* [upper land] so we could survive. Now nothing grows here. What can we women do now? There is no work. CCL is getting all its work like removing the bushes, cleaning, etc, done by vehicles. People are starving – just surviving somehow.

For those who cannot be provided with mining employment, the R&R policy recommends offering contract jobs if available, and also notes that the contractors be persuaded to employ the DPs on a preferential basis. The dump yards where coal is stored for sale by some authorised sellers – known as 'local sale depots' in the Ashoka-Piperwar area – had also for a while become a source of jobs for local people as loaders of trucks. However, this proved to be a temporary arrangement, as for the last year these dumping yards were empty in the Parej area. In Benti also, the local sale dump had discontinued.

Crisis of drinking water

Since the advent of mining, there has been a crisis of drinking water due to the depletion of groundwater. Open-cut mining has also led to the pollution of the surface and ground water. In the first phase of the study we found that one effect of displacement was the decline in women's access to various sources of drinking water. Before displacement, most of the families had access to more than one source. They could take water from a well – a *dadi* or a *jharna* – or a nearby river.

If in summer one or two sources dried up, they would continue to have access to water from other sources. But now many women have observed that they have problems in obtaining water in summer when their sole source of water vanishes. The falling water level due to mining causes this to occur more frequently and for longer durations of time than in the past.

Another issue of concern was the distance travelled to fetch water. For most women, safe drinking water was available in the village before displacement and usually they did not have to walk more than 1 km to access a fresh water supply. This situation was the same for most women even after displacement, except for a few, who had to walk about 2–3 kms to reach a water source. Women of Benti village, for instance have to walk about 2 km. Some women told us that in summer, when the local water sources run dry, they are forced to walk even longer distances.

In the second phase of the study also we found that all villagers in Benti experience a water shortage and that any available water is of poor quality. Water scarcity affects women more than men, as fetching water for drinking, carrying it to the home, and washing clothes and utensils are women's jobs in Benti. Many traditional water sources have disappeared. Women used to access local natural water sources such as the spring and the river, and now have to cover long distances on foot to retrieve water. Water shortages increase during the long summer. Women complained about having to use dirty and polluted water released by CCL. As one woman from Benti, interviewed for the second study, observed:

> For washing and bathing, we use water from the mines. It makes our skin itch, but what can we do? Our body becomes sticky after using the mine water. When there were no mines here, we used get water from the Dahenchawa River. We used the river water for drinking as well as for washing and cleaning. Now, even the little water that is left in that river is so dirty that we cannot use it.

Some women said that those who can afford it are now buying water from a local entrepreneur who extracts water from the Damodar River and pipes it, untreated, into their houses. A few better-off families, especially in Baniya *Tola*, have installed their own bore wells or deepened existing wells.

The differential access to safe and clean water sources has led to internal strife within the community and has widened social gaps. Women in Fulsi Tengri of Benti village complained during the focus group that lower caste families are bearing a greater burden of the water shortage. The shortage of water has fuelled conflicts between the community and company officials. At Piparwar, women have regularly demanded water by organising demonstrations and blockading project offices.

Marginalisation and food insecurity

Entire families were pauperised after displacement from their homes and consequent separation from the local environment. Their land and trees were lost, the number of animals they could rear decreased, and the number of families owning animals declined sharply. Their access to common land and forest trees was lost and forests often declined with the onset of mining. Compensation given for the loss of trees was nominal, with no provision of compensation for trees on common lands or the trees in forest. Thus, food security deteriorated significantly after displacement. According to the first study, families lost their land along with Mahua and other trees, resulting in 39 families (out of 46) reporting negative changes in their eating habits. Traditional agricultural produces like *madua*, corn, *bajra* and pulses are consumed less after displacement. The consumption of these staple cereals declined sharply for about two-thirds of the families. Rice and wheat replaced these cereals, as these food items were available in markets, clearly indicating the entrance of a cash-based economy in which indigenous women are at a disadvantage. Once women become jobless and remain housebound doing household chores, they have much less access to money. In other words, whatever cash flow there is after displacement due, for example, to the compensatory award of employment is invariably in the control of men. In the earlier system with lesser cash flow, women had access to and a level of control over the sources of food. This gave them a certain level of security in terms of food in spite of their unequal status in the society.

Families' consumption of vegetables, pulses, meat, eggs and fish also declined sharply, as evident from interviews in which 52, 54 and 26 per cent of the women reported that they would in the past consume pulses, vegetables and eggs/meat/fish respectively, seven days a week before displacement. After displacement, these percentages fell to 4, 13 and 0 percent respectively. In contrast, the percentages of those who said that they rarely consume these items increased from 22 to 24 percent, 28 to 33 per cent, and 23 to 39 percent respectively. Meat, fish and eggs were earlier available as most of the families used to possess animals such as goats, chickens, pigs, which provided meat, as well as cows, buffalos and oxen. The number of families owning these animals as well as the average number of animals per family declined considerably after displacement. Wild animals from the forest and fish from common water sources were also available earlier, access to which declined later as the commons were lost.

One woman observed: 'The forest is moving farther and farther away from us.' The increased dependence on bought commodities means the need for a cash income increases. Households with a CCL job holder are able to obtain a greater variety and quantity of food than before the mine. In contrast, two-thirds of the women felt the opposite: that their access to food and its variety had diminished. In general, the variety of food in tribal households was traditionally based on women's labour, a result of the collection of different edible food stuffs from the forest and crops grown. This affects women's loss of identity and self-worth. In New Malmahura, a resettled *tola*, one woman, interviewed for the first study, said:

> The company just threw us here. What would we do here? There is no land to farm. Those who have been able to get some kind of private job have survived. We are just hanging on to life scraping a livelihood. Earlier we had cows and oxen, goats and poultry....We used to say '*if you want to eat meat, just grab a chicken.*' Now, all that has become impossible. Even a chicken you have to buy now. Even our animals die rapidly here.

Families' access to the Public Distribution System (PDS) declined after displacement. Most of the displaced families were not given new ration cards after the displacement. Poor families' access to PDS shops is in any case very low.

Health and illnesses

Displacement also caused the health of families in general and women and children in particular to deteriorate. Families' access to government hospitals, traditional medicinal systems and CCL's health centres declined after displacement and their visits to health practitioners increased. Most of the families – about two-thirds – used to make use of services provided by the untrained or semi-trained private health practitioners, and this proportion increased after displacement to about three-quarters. Ironically, however, the share of families availing themselves of CCL's health services declined after displacement, although two-thirds of the families received jobs with CCL. This happened because the families have now been relocated to places away from the CCL's hospital. In effect, the families' access to government or company health centres declined, and their dependence on private health practitioners, who visit them on bicycles, increased.

More than 70 per cent of women felt that their husbands' health status had worsened after displacement, and 80 per cent of women expressed this view about their children. Thirteen women – 30 per cent – said that they were sick at the time of the interview. Twenty-five per cent of women observed an increase in the occurrence of malaria after displacement, especially for children. Overall, half of the women interviewed said that their access to health facilities had declined. Significantly, however, the other half felt that their access to health facilities had increased because they now visit private health centres for treatment. Previously many villagers used herbal medicines prescribed by a local healer (*ojha-guni*), whose treatment modality involved using forest herbs and appeasing the supernatural powers to cure illness. Not everyone can afford private medical treatment. According to one woman, (interviewed fin the second phase of the study): 'when one gets sick, treatment is now easy, but only if you have money. If you have money your life is saved, if you have no money you will certainly die.'

Psychological trauma

Many women reported mental stress after their displacement. Pressures were related to the uncertainty of residence, the responsibility of rebuilding the livelihoods, increased joblessness and resultant idleness, increased drinking by husbands, wife-beating and domestic violence including quarrels, and the shock and trauma of losing the few assets that they once possessed. The stress of adjusting to a new location was not easy to cope with; the frustrating impacts on women were evident in their reporting of sleeplessness, becoming short-tempered, feeling tense and depressed. About 30 per cent of sampled women were living in temporary homes for up to ten years. Almost all women whose husbands obtained jobs with CCL reported that domestic quarrels, violence and drinking by their husbands had increased. Though drinking is part of *adivasi* culture, liquor distillation being the occupation of women, the drinking habits of men changed as a money economy developed. Earlier, it was home-made liquor imbibed occasionally in moderation, usually in the home environment.

Social and cultural risks

As the displaced families become scattered in different rehabilitation colonies, seek refuge in relatives' homes, or obtain CCL's miners' quarters or coolie barracks, the process uproots people socially and culturally as age-old social and cultural networks are disrupted. Women's conventional support systems are lost, as are those of men. The traditional communal systems of conflict resolution and coping with emergencies also break down. In about 60 per cent of cases, women said that earlier conflicts would be resolved by the caste or *gram panchayat* (village council). However, many of these conventional systems have lost their powers and women are left to solve their problems themselves. The nuclear family has assumed greater importance, while the authority and influence of the traditional panchayats has declined.

As far as cultural impacts are concerned, 43 per cent of women noticed changes in the way they celebrated festivals such as the Karma and Sarhul. Women's role and status during those celebrations were not the same as before. In the new places, they were now supposed to remain only as spectators and not participants in these festivals. For children, orchestras, cinema and television had become more important than the traditional tribal dances and songs. More than 50 per cent of the women observed such changes in their children's behaviour.

One gendered effect of mining is changes in power relationships between men and women. The cash compensations, the consultations and the offers of employment empower men more than women, resulting in changed gender relations that are manifested in, amongst other things, the changed behaviour of men. Men who never drank become addicted to drinking; men who were closely involved in family become violent and beat their wives; men who never visited a prostitute now spend all their money on them. This is a direct result of increased mental stress resulting from uncertainty and frustration and from the experience of a sudden change in the institutional and social structures of life. The stress frequently leads to increased violence against women and children.

Women reported that men – both those with a job with CCL, who possessed disposable cash, and those who did not – are now drinking more alcohol and acting violently at home. One woman commented about her husband: 'He did not drink much when he was working on the farm, but after losing the land he started drinking a lot. He also started beating and abusing me.' This is one of the hallmarks of gender impacts – changes in the income sources of rural families, in the balance of profitability of various livelihood activities and the greater impoverishment of some households – that lead to gender conflicts within households. Gendered conflicts or tensions also arise in the rare cases in which the economic role of the woman is enhanced and accompanied by a reduction in that of the man. A failure to perform the desired economic role frustrates the man, who is supposed to be the head of the household and the provider, and enhances gender tensions within the household.

The mining projects have reduced women's mobility. Although one might expect that with the improved connectivity and roads made by the mines, mobility for men and women would increase, women pointed out that this was not the case for a number of reasons. First, as previously discussed, most are now confined to the house because they have no purpose to be out. Second, they tend not to have access to money. Third, for the majority of them, if they were to go anywhere, the local bus stop is now farther away. Finally, the mining projects regularly change the landscape and the locations of roads, creating confusion and uncertainty.

Another issue was that the mines have brought in many outsiders to work. The presence of a large number of strange men has created apprehension among the women, as one described: 'Earlier, if there was a community event we would go without fear. But it's not like that now. Now there is anxiety and fear. Everywhere there are outsiders and new people are coming into the area every day. Earlier, the forest provided us with some amount of privacy. Now there are strangers everywhere.' The influx of strangers has created difficulties for defecation and bathing, both of which used to be performed in the open in the village. Neither in the villages nor in the resettlement colonies are sanitation and sewage systems adequate. In general, there are no toilets in the huts and people are used to defecating in the relative privacy of the forests or fields. However, even in resettlement colonies, there are no toilets for the individual houses.

Compensation, Resettlement, Rehabilitation: Women Ignored

As mentioned earlier, the displacing agencies are not bound legally to rehabilitate displaced families. In this case, CCL had provided people with small cash compensations for the land and homes that they lost, and also gave jobs to about two-thirds of the families in accordance with its R&R policy. However, the situation of rehabilitation can hardly be considered satisfactory. Not all the families were living at the rehabilitation centres of CCL. Those who did were from Pindra rehabilitation centre, Premnagar rehabilitation centre (displaced by East Parej OCP), Chirayat and rehabilitation centre, and New Mangardaha (Benti Rehabilitation Centre—1 displaced by Ashoka and Pirarwar OCP). These people were not happy: the company had allotted them plots at the rehabilitation centre, but none of them were given *pattas* of the land they were living on. As a consequence they had no legal right over the lands they were living on at the rehabilitation centres. The rehabilitation centres lacked basic necessities such as drinking water, sanitation, and health and education services.

Practical gender needs were raised by women included in the study: bathing and defecating had become difficult. All women said that earlier there would usually be ponds or a natural spring (*jharna*) in the village. The traditional and natural sources of water also disappeared as the villages were mined, and about 55 per cent of women said that a pond or *jharna* in the village provided the source of water. A woman in the barracks said that she bathed once in about 10 days because for each bath she had to walk to the river. Men, she said, could go there more often, but for women, trekking such long distances became difficult in view of their numerous responsibilities at home. Similarly, about 60 per cent of women said that they had problems in finding a place to defecate, as forests are disappearing fast and becoming thinner due to mining. Some of them also noted the lack of privacy due to there being men onlookers everywhere.

Educational and medical facilities available at the rehabilitation centres were in a miserable state. Pindra rehabilitation centre, where families from Parej village were dumped forcibly by the police, had buildings for a school and a hospital, but even after four years there were neither doctors nor teachers. A woman noted that the CCL had originally intended to run the hospital and the school, but has now shifted the responsibility to the government. The Premnagar rehabilitation centre, where Turi families had been given plots, had no health facility; however there was one CCL health centre nearby. In the Chirayatand rehabilitation centre, where families displaced by Ashoka and Piperwar projects are rehabilitated, there was one health centre, where a doctor was said to be scheduled to arrive. But there was no medical centre at New Mangardaha (Benti Rehabilitation Centre—1), where families displaced by the same projects were rehabilitated, and a private practitioner had opened office.

Displacement and the Changing Status of Women

From these localised case studies it is clear that with displacement and relocation the place of women has changed in the family and the community. Yet households continue to be regarded by officials as units with joint-utility functions. Regardless of this, they are contested terrains, with women and men wielding differential power and control over household resources. Women's lower bargaining power is evident from the gender-based division of labour in which women take important but unpaid and informal roles in collecting and preparing foodstuffs, fuel and water, childcare, and care for the elderly. New gender roles dictate that women no longer remain productive and equal partners in the community; in most cases, they are meant to remain at home, taking little part in income-generating activities. At the same time, their household chores, including the collection of water and fuel from the now-distant forests, have increased. Men who now work for CCL or the local contractors in the dumps have become prominent as major breadwinners, with women being ascribed to the four walls of the home, where they faced increased violence. Overall, as gendered roles and functions have taken on different meanings there has been a devaluation of the roles and tasks that women traditionally perform and a relegation of women to a lower status.

This devaluation has to do directly with the way gender roles have been created and recreated during the process of compensation and rehabilitation established by the government agencies and the mining company. The visiting officials used men in the villages as legitimate representatives, voicing the apparent sentiments of the community. Women often remained uninformed or received information from the male members of the family. They generally did not receive details regarding monetary compensation for the land that the family had lost. As with the jobs, the cash for compensation was given to men. Intra-household inequities were not negligible in the indigenous gender roles, but at the same time women enjoyed some status, and support networks developed over time in their original locations. Displacement and the half-hearted R&R as practiced now in the mining areas is obviously insensitive to these gendered realities. Unless the R&R policy is made gender-sensitive, it will be difficult for any significant social and economic development to take place in the Indian mineral tracts.

Conclusion

Mining continues to remain a masculine domain in terms of the nature of its work in spite of much evidence to the contrary (Lahiri-Dutt and Macintyre 2006), and this has led to a masculine corporate culture that keeps women invisible in the workplace and in the community, permeating into the community development and R&R policies (Lahiri-Dutt 2006b). In view of differential access to land, means of production, income and property ownership, women's conditions of work remain dependent on survival strategies of households in specific relation to land and environmental resources. Evidence on gendered impacts of large mining projects is growing, and gender-differentiated impacts have been noted both in better-off countries as well as in less-developed countries (see Macdonald and Rowland 2002; Hill and Newell 2009).In their work among the indigenous populations of Canada, Hipwell et al. (2002) classified gender impacts into three broad categories: health and wellbeing; women's work and traditional roles; and gender inequalities in the economic benefits from mining activities. Our observations show that these apply in Benti too. The depleted agricultural subsistence base and environmental degradation create concerns for women who are primarily burdened with ensuring the family's food security. The changes to traditional culture have disproportionately affected women, particularly by devaluing their productive work at home, and by undermining their status as decision-makers and land-owners. The economic benefits of the mine go primarily to the men, which in turn exacerbates gender relations. The gender impacts of mining primarily derive from the shifting power relations within communities and families. Last but not the least, geographical distance from the mining operation also influences the winners and losers among women and men (Lahiri-Dutt and Mahy 2007). In general, physical proximity to the mines leads to the direct experience of excessive noise, and the visibility of gigantic machines arouses fear and a sense of insecurity, creating a heightened sense of the negative impacts among women. In Benti, which is caught between two large open-cut projects, geographical proximity to the large open-cut mine plays a key role.

As is evident from the above studies, there is no doubt that large open-cut coal mine projects negatively affect women more than men, and that poorer women, such as those from tribal communities as studied here, are especially vulnerable. Indeed, other scholars of India have pointed out that women bear most of the brunt when rural communities are under stress (Vepa 2010). This chapter raises broad issues concerning the possibility of creating an alternative, feminist model of development in India's mining areas capable of integrating gender-specific concerns regarding the social reproduction of displaced communities. By illuminating the gender-segregated impacts of coal mining, our research enhances the visibility of women in the development of social impact assessments, and in thinking about displacement and rehabilitation, and opens up dialogue towards empowering the women in local communities in ways of ending gender-based relations of domination, and in how to redefine development. Such redefinition can occur not only through resource redistribution, but also through non-distributive processes that can promote women's capabilities within the society with a view to empowering them.

From a policy perspective, environmental policy must be strengthened by gender-aware social sciences. Environmental Impact Analysis and clearance from the environmental departments of State Pollution Control Boards have now become mandatory before large-scale projects such as coal mines are initiated. However, environmental management should involve caring for people, and this awareness has not yet developed among these officials in India. Social impact analyses are yet to be treated as a prerequisite for such projects, with gender analysis as an important component. R&R programmes need to respond to the complex ways in which women and men use power and resources through their different identities, access and entitlements. This will make the notions of a 'displaced person' or a PAP gender-inclusive, and help to divert some of the detrimental interventions in women's and men's lives. This would begin with the recognition that women are as much economic and political citizens as men, with entitlement to the full realisation of their human rights. Women need to be made co-beneficiaries with men in any resettlement benefit packages, and in women-headed households special assistance should be given before, during and after displacement. The inclusion of gender issues and social equity in assessment, design, implementation and monitoring must be made a compulsory element for large-scale development projects such as coal mines.

Acknowledgments

The authors are grateful for funding support from the Indian Social Institute, New Delhi, to undertake the research, to Father Tony Herbert of Prerana Resource Centre, Hazaribagh, and to Chotanagpur Adivasi Seva Samiti (CASS), Hazaribagh for facilitating the research, and finally to Dr Gillian Vogl for helping out with directing us to important theoretical literature on displacement.

References

Agarwal, Bina. 1992. 'The Gender and Environment Debate: Lessons from India.' *Feminist Studies* 18(1): 119–58.
———. 1995. *A Field of One's Own? Gender and Land Rights in South Asia.* Cambridge: Cambridge University Press.
Ahmad, N. and Lahiri-Dutt, K. 2006. 'Engendering Mining Communities: Examining the Missing Gender Concerns in Coal Mining Displacement and Rehabilitation in India.' *Gender Technology and Development* 10(3): 313–39.
Ali, S.H. 2003. *Mining, the Environment and Indigenous Development Conflicts.* Arizona:University of Arizona Press.
Banerjee, Paula, S. Basu Ray Chaudhury and S.K. Das, eds. 2005. *Internal Displacement in South Asia: The Relevance of the UN Guiding Principles.* New Delhi: Sage Publications.
Basu, Malika. 2000. 'Locating Gender: Feminist Perspective and Gender Approach in Development-induced-displacement.' Unpublished manuscript submitted to Indian Social Institute.
Baviskar, Amita. 2002. *In the Belly of the River.* New Delhi: Penguin Books.
Cernea, Michel M. 2000. 'Risks, Safeguards and Reconstruction: A Model for Population Displacement and Resettlement.' *Economic and Political Weekly* 7 October.
Coal India Limited (CIL). 1994. 'Resettlement and Rehabilitation Policy of Coal India Ltd.' In *Rehabilitation Policy and Law in India: A Right to Livelihood,* edited by Walter Fernandes and Vijay Pranjpye, 345–54. Delhi: Indian Social Institute.
Colchester, M. 1995. 'Some Dilemmas in Asserting Indigenous Property Rights.' *Indigenous Affairs* 4: 5–7.

de Wet, Chris. 2001. 'Economic Development and Population Displacement: Can Everybody Win?' *Economic and Political Weekly* 15 December: 4637–46.

Dias, Xavier. 2005. 'World Bank in Jharkhand: Accountability Mechanisms and Indigenous Peoples.' *Law, Environment and Development Journal* 1(1): 73–9. Available online at http://www.lead-journal.org/content/05071.pdf (accessed on 27 July, 2009).

Eftimie, A., Heller, K. and Strongman, J. 2009. *Gender Dimensions of the Extractive Industries: Mining for Equity, Extractive Industries and Development.* Washington, DC: The World Bank.

Ekka, A. and M. Asif. 2000. *Mining and Displacement in Jharkhand.* Indian Social Institute, New Delhi.

Fernandes, A and K. Kohli 2012. *Trashing the Tigerland*, Greenpeace Report, Bengaluru: Greenpeace India Society. Available from http://www.greenpeace. org/india/Global/india/report/How-Coal-mining-is-Trashing-Tigerland.pdf Accessed on 17 January, 2013.

Fernandes, Walter. 1998. 'Development Induced Displacement in Eastern India.' In *Antiquity to Modernity in Tribal India*, Vol. 1, edited by S.C. Dubey, 217–301. Delhi: Inter-India Publications.

Fouillard, Camille. 2003. *Refugees and Forced Displacement: International Security, Human Vulnerability, and the State.* Tokyo: United Nations University Press.

Ganguly-Thukral, Enakshi. 1996. 'Development, Displacement and Rehabilitation Locating Gender', *Economic and Political Weekly*, 15 June 1996.

Gibson-Graham, J.K. 1994. '"Stuffed if I Know!": Reflections on Post-modern Feminist Social Research.' *Gender, Place and Culture* 1(2): 205–24.

Guijt, I. and Shah, M. 1998. *The Myth of the Community.* London: Intermediate Technology Publications.

Harding, S. 1987. 'Introduction: Is there a feminist method?' In *Feminism and Methodology: Social Science Issues*, edited by S. Harding, 1–15. Bloomington: Indiana University Press.

Hill, C. and Newell, K. 2009. *Women, Communities and Mining: The Gender Impacts of Mining and the role of Gender Impact Assessment.* Carlton: Oxfam Australia.

Hipwell, W., Mamen, K., Weitzner, V. and Whiteman, G. 2002. *Aboriginal Peoples and Mining in Canada: Consultation, Participation and Prospects for Change.* Working Discussion Paper. Ottawa: North-South Institute.

Ivison, D., P. Patton and W. Saunders. 2000. *Political Theory and the Rights of Indigenous Peoples.* Cambridge University Press, Cambridge.

Kolhoff, A. 1996. 'Integrating Gender Assessment Study into Environmental Impact Assessment.' *Project Appraisal* 11(4): 261–66.

Kothari, Smitu. 1996. 'Who's Nation? Displaced as Victims of Development?' *Economic and Political Weekly* 15 June.

Kumar, M. 2005. 'Incorporating gender issues in national responses.' Paper presented in a Regional Workshop on NHRIs and IDPs, 14–15 November, 2005, organised by APF, Colombo, October.

Kuyek, J. 2003. 'Rhetoric vs Reality: Investing as if Human Rights Mattered.' Presentation of Mining Watch Canada, 11 June. Available at www.miningwatch.ca

Lahiri-Dutt, K. and Mahy, P. 2007. *Impacts of Mining on Women and Youth in Indonesia: Two Mining Locations* (report to the World Bank). Canberra: Australian National University.

Lahiri-Dutt, Kuntala. 2001. 'When the mines come.' Report of an Oral Testimony Project Amongst Mining Displaced Indigenous People in Jharkhand, Panos Institute Ranchi and Hazaribagh.

———. 2006a. 'Kamins Building the Empire: Class, Caste and Gender Interface in Indian Collieries.' In *Mining Women: Gender in the Development of a Global Industry, 1670–2005*, edited by Jaclyn Gier and Laurie Mercier, 71–87. New York: Palgrave Macmillan.

———. 2006b. 'Mainstreaming Gender in the Mines: Results from an Indonesian Colliery.' *Development in Practice* 16(2): 215–21.

———. 2012. 'The Shifting Gender of Coal: Feminist Musings on Women's Work in Indian Collieries.' *South Asia: Journal of the South Asian Studies Association* 35(2): 456–76.

Lahiri-Dutt, Kuntala and Martha Macintyre, eds. 2006. *Women Miners in Developing Countries: Pit women and others.* Aldershot: Ashgate.

Lahiri-Dutt, Kuntala and Nesar Ahmad. 2012. 'Considering Gender in Social Impact Assessments.' In *New Directions in Social Impact Assessments: Conceptual and Methodological Advances*, edited by Frank Vanclay and Ana Maria Esteves, 117–37. Cheltenham: Edward Elgar.

Leach, M. 2007. 'Earth Mother Myths and other Ecofeminist Fables: How a Strategic Notion Rose and Fell.' *Development and Change* 38(1): 67–85.

Mallik, S. Basu and Samyadip Chatterji. 1997. *Alienation, Displacement and Rehabilitation.* New Delhi: Uppal Publishing House.

Manchanda, Rita. 2004. 'Gender Conflict and Displacement: Contesting "Infantilisation" of Forced Migrant Women.' *Economic and Political Weekly* 11 September: 4179–86.

Mathur, H.M. ed. 2006. *Managing Resettlement in India: Approaches, Issues and Experience.* New Delhi: Oxford University Press.

Macdonald, I. and Rowland, C. eds. 2002. *Tunnel Vision: Women, Mining and Communities.* Fitzroy: Oxfam Community Aid Abroad.

McIlwaine, C. and Datta, K. 2003. 'From Feminising to Engendering Development.' *Gender, Place and Culture* 10(4): 369–82.

Moser, Caroline. 1993. *Gender Planning and Development: Theory, Practice and Training.* New York and London: Routledge.

Muggah, Robert. 2003. 'A Tale of Two Solitudes: Comparing Conflict and Development-induced Internal Displacement and Involuntary Resettlement.' *International Migration* 41: 6–31.

Ota, A.B. 1996. 'Countering Impoverishment Risks: The Case of Rengali Dam Project.' In *Involuntary Displacement in Dam Projects*, edited by A.B. Ota and Anita Agnihotri, 150–78. New Delhi: Prachi Prakashan.

Overholt, C., M. Anderson, K. Cloud and J. Austin. 1995. *Gender Roles in Development Projects: Cases for Planners*. West Hartford: Kumarion Press.

Pandey, Balaji. 1996. 'Impoverishment Risks: A Case Study of Five Villages in Local Mining Areas of Talcher, Orissa.' Paper presented at Workshop on *Involuntary Resettlement and Impoverishment Risks*, March, New Delhi.

Pandey, Balaji. 1997. *Development, Displacement and Rehabilitation in Orissa 1950–1990*. Canada: International Development Research Centre and Bhubaneswar: Institute for Socio-Economic Development.

———. 1998. *Displaced Development: Impact of Open Cast Mining on Women*. New Delhi: FES.

Reddy, U.K.P. 1997. 'Involuntary Resettlement and Marginalisation of Project Affected Persons: A Comprehensive Analysis of Singrauli and Rihand Power Projects.' In *Impoverishment Risks in Resettlement*, edited by H.M. Mathur, 21–42. New Delhi: Sage Publications.

Robinson, W. 2002. 'Remapping Development in Light of Globalization: From a Territorial to a Social Cartography.' *Third World Quarterly* 23(6): 1047–71.

Ravindran, L. and Mahapatra, B. 2009. 'Gender Issues in Displacement: A Study of Irrigation Projects in Southern Orissa.' In *Beyond Relocation: The Imperative of Sustainable Development*, edited by R. Modi, 240–69. New Delhi: Sage.

Razavi, S. and Hassim, S. eds. 2006. *Gender and Social Policy in a Global Context: Uncovering the Gendered Structure of 'the Social'*. Houndmills: Palgrave.

Rothermund, I. 1994. 'Women in a Coal Mining Area.' *Indian Journal of Social Science* 7(3/4): 251–64.

National Thermal Power Corporation (NTPC). 1993. 'National Thermal Power Corporation Limited: Rehabilitation and Resettlement Policy.' In *Rehabilitation Policy and Law in India: A Right to Livelihood*, edited by Walter Fernandes and Vijay Pranjpye, 331–44. Delhi: Indian Social Institute.

Roy, Arundhati. 1999. *The Greater Common Good*. Mumbai: India Book Distributors Ltd.

Sachs, C. 1996. *Gendered Fields: Rural Women, Agriculture and Environment*. Boulder: Westview Press.

Scudder, T. 2005. *The Future of Large Dams: Dealing with Social, Environmental, Institutional and Political Costs*. London: Earthscan.

Sharma, S. 2010. 'The Impact of Mining on Women: Lessons from the Coal Mining Bowen Basin of Queensland, Australia.' *Impact Assessment and Project Appraisal* 28(3): 201–15.

Tulpule, Bagaram. 1996. 'Redefining Development: An Alternative Paradigm.' *Economic and Political Weekly* 9–16 November: 2991–96.

Young, Kate. 1993. *Planning Development with Women: Making a World of Difference*. London: Macmillan.

Zwarteveen, M. and R. Meinzen-Dick. 2001. 'Gender and Property Rights in the Commons: Examples of Water Rights in South Asia.' *Agriculture and Human Values* 18: 11–25.

PART III
Social Perspectives to Inform Mining Policy

Chapter 13

Colonial Legislation in Postcolonial Times

Nesar Ahmad

Cumulative Effects of Coal Mining Projects

Coal India Limited (CIL), a public sector enterprise formed in 1975 by the Government of India, mines about 85 per cent of the coal produced in India. The remaining coal is mined by Singareni Collieries Company Limited (SCCL), another public sector company owned jointly by the central government and Andhra Pradesh government, and some private and public sector companies for their captive use. In Meghalaya coal mining is conducted by private companies. However, the mining policy of the government has changed with the introduction of new economic policy during the early 1990s (Lahiri-Dutt 2007, 29). The *Coal Mines (Nationalisation) Act 1973* was amended in 1993 to allow captive operations for power generation to be carried out by the private sector as well as the public sector. This was in addition to the already existing provision of captive mining for iron and steel companies. Washing operations were privatised and the import of coking coal was placed on the open general license list. In 1996, captive mining was permitted for the cement industry (Ministry of Coal 2005). It is now open to public sector companies to mine coal for commercial purposes, and the government has allowed 100 per cent FDI (Foreign Direct Investment) for captive coal mining (*The Telegraph* 2008). The Ministry of Coal has already allotted as many as 273 mining blocks for captive mining to public and private sector companies across the country (Ministry of Coal Website), which later saw much controversy and court litigations as well on account of alleged practices of corruption.

More than half of these blocks are allotted to the private sector companies. The states where most of the blocks have been allotted for captive mining to the public and private sector companies are Orissa, Jharkhand and Chhattisgarh.

The coal mines are responsible for a major share of involuntary displacement in the country. However, quantitative estimates of displacement only by coal mining are difficult to produce. No official records are available on how many projects have displaced how many people, and one is left with no option but use the indirect estimations. Mining is one of the development activities leading to large-scale displacement. There is no project data available and exact numbers of displaced people are not known. According to an estimate, about 2.5 million people have been displaced due to mining projects in India, which is more than 10 per cent of the total population (21.3 million) displaced by all development projects (Fernandes 1998:251). In a later paper, however, Fernandes (2006: 111) argues that this is an underestimate and revises the total displacement by all development projects in the country to 60 million. The majority of displaced people in the country are from *dalit* or Scheduled Caste (SC) communities and *adivasi*s or Scheduled Tribes (ST), the most deprived and exploited sections of Indian society. The *adivasi*s, for example, comprise only 8 per cent of the total population, but accounted for over 40 per cent of the total displaced until 1990, as noted by Kothari (1996: 1477).

The record of rehabilitation and resettlement of displaced people due to coal mining projects by CIL's subsidiaries is far from satisfactory. There are many studies showing that people displaced by the coal mining projects have not been rehabilitated properly (Ahmad and Lahiri-Dutt, 2006; Ekka 2006; Ahmad, 2003). The coal mining subsidiaries are unfamiliar with the very idea of informing the local people in a systematic manner in advance of commencement of projects, let alone securing their prior consent before displacement. Their participation in rehabilitation programmes is generally very low, and in most of the cases the project officials impose their rehabilitation programmes without offering an opportunity for negotiation or the provision of consent. People in the mining areas suffer for extended durations after ostensible rehabilitation because it is invariably insufficient to meet their needs. People living around the coal mining projects, most of whom are project-affected people, suffer shortages of water, lack of health and education services, and lack of transportation and other services.

This paper provides an overview of the laws and policies governing the involuntary displacement caused by coal mining and the rehabilitation and resettlement of the people displaced by coal mining. It offers two compelling indications that the time has come for these laws to be changed. Since most of the displaced people in India come from the SC and ST communities, which are the most marginalised communities of Indian society, it would be relevant to consider the provisions made by the Indian Constitution for protection of the interests of weaker sections of the Indian community, in particular the SCs and STs; this will enable an assessment of the contradictions existing between these provisions and the coal legislation. The paper also notes the extensive age and often outdated application of these laws and the marginal, often cosmetic changes that have been undertaken so far, before suggesting the need for their comprehensive independent, socially inclusive and open review.

Protection for the SCs and STs by the Indian Constitution

The Indian Constitution provides for special measures to protect the weaker sections of Indian society. Article 46 of the Constitution, for example, directs the state "to promote, with special care, the educational and economic interests of weaker sections, in particular, those of the SCs and STs". Articles 15(4) and 15(5) mandate the state to make special provisions for advancement of any socially and educationally backward classes or for the Scheduled Castes and Scheduled Tribes.

Under article 244(1) of the Indian Constitution the state declares certain areas with a majority of tribal and ethnic populations as Schedule V and Schedule VI Areas in order to protect them from exploitation and preserve their identity. Schedule VI deals with administration of states in the northeastern region of the country, including those of Assam, Meghalaya, Tripura and Mizoram. While Schedule V notifies and declares the Scheduled areas in the remaining parts of the country. So far Scheduled areas have been declared in nine Indian states[1] under the Fifth Schedule.

The Fifth Schedule empowers the Governor of the concerned state to modify, annul or limit the application of any law made by the Parliament or the state legislature in the Scheduled areas. Even though this power has never been used, it is an important provision which can be applied to safeguard the interests of the tribal population. The Governor may also make regulations prohibiting and restricting transfer of land by or among ST members, regulate the allotment of land to members of STs, and regulate business in these areas, such as money lending.

The *Panchayat (Extension to Scheduled Areas) Act 1996*, commonly referred to as PESA, gives special powers to the *Gram Sabhas* and the *Panchayats*[2] in the Scheduled areas. Among other things, the PESA Act mandates that the *Gram Sabha* or the *Panchayat* at the appropriate level shall be consulted before making the acquisition of land in the Scheduled areas for development projects and before resettling and rehabilitating the persons affected by such projects in the Scheduled areas. The PESA also gives rights to the *Gram Sabha* or *Panchayats* to prevent the alienation of tribal lands in the Scheduled areas and restoration of any unlawfully alienated lands to the tribals. It also provides for mandatory prior recommendation by the *Gram Sabha* or the *Panchayat* at the appropriate level in granting prospecting licenses or mining leases for minor minerals in the Scheduled areas. The *Panchayats*, however, is a subject under the state list and states governments are required to conform to the law and regulations in the respective state legislative assembly. Many Indian states in their conformity acts have not mentioned the specific powers of the *Gram Sabha* and *Panchayats* in the Scheduled areas. The Jharkhand *Panchayati Raj Act 2001*, for example, does not mention the power of *Gram Sabha* and *Panchayats* to prevent the alienation of tribal land and restoration of illegally alienated land. Some other states, such as Rajasthan, passed conformity laws that have been criticised for diluting some of

---

1 These nine states are: Andhra Pradesh, Jharkhand, Chhattisgarh, Himachal Pradesh, Madhya Pradesh, Gujarat, Maharashtra, Orissa and Rajasthan.

2 Panchayats are elected local governors in the rural areas of the country at village level (Gram Panchayat), district level (Zila Parishad/Panchayat), and intermediary level (Block Panchayats). Gram Sabha is the assembly of the all the adults of a Gram Panchayat. The Panchayats have constitutional status in India.

the provisions of PESA – but have only recently enacted the rules and regulations implementing PESA, after more than 15 years of procrastination by the Parliament.

Protection of the Land Rights vs. Acquisition of Land

Protection of the land rights of the poor is one of the most important provisions made by the Constitution. Since the land is subject to state control, many Indian states have state-specific Acts which protect the land rights of the tribals and dalits. In many cases these Acts are result of the tribal and agrarian resistance and movements by tribal and other deprived people to protect their rights over land and natural resources. Some of these Acts were passed before Indian Independence (Mohanty 2001). For example, the *Chhotanagpur Tenancy Act 1908*, passed by the colonial government, not only prohibits transfer of tribal land to non-tribal peoples, it also recognises, to some extent, tribal peoples' traditional rights over the land in Jharkhand. A government committee provides a survey of tenancy laws across the country for as many as 25 out of 28 states, and 7 Union Teritories (UTs) that prohibit or restrict the transfer of land belonging to SCs, STs and, in some cases, to Other Backward Castes or OBCs (DoLR 2008, 129). The same committee, however, also states that a 'large number of legislations is no guarantee or indication of efficaciousness of the legislation' (129). Alienation of the land owned by dalits and tribals is documented by many studies (see, e.g., Mohanty 2001; Sharan 2005; and DoLR, 2008).

The Supreme Court of India has ruled in favour of special measures, protecting the land rights of weaker sections of peoples, especially the tribal groups. Its verdict in *Samatha v. State of Andhra Pradesh* – commonly known as the Samatha Judgment – declared that tribal land cannot be transferred to private companies for mining or other purposes as these companies can be treated as legal persons that are non-tribal (*Samatha v. State of Andhra Pradesh* 1997 8 SCC 191). An exception to this was the transfer of tribal land to the government or its instrumentalities for public purposes, which was upheld by the Court.

Regardless of constitutional and judicial protection, however, large-scale development induced displacement of people occurs in the country, resulting from the state's power to acquire any land based on the principle of *eminent domain*, 'which means that the state has the first right to any land which it wishes to use for a public purpose and can forcibly acquire it from a private person if it so desires' (Action Aid nd). This principle, however, contradicts the spirit of the Fifth and Sixth Schedules of the Constitution of India as well as the various Acts which protect the land rights of SC and ST communities by restricting the transfer of lands belonging to these groups to other communities, and also the PESA Act, which provides for protection of land rights of the *adivasis* (STs) in the Scheduled areas. The various legislative enactments seeking to prevent the alienation of tribal land, for example, restrict transfer of land from tribal to non-tribal groups. Similarly, the PESA Act, which conforms with the Fifth and Sixth Schedules of the

Indian Constitution, also requires consultation with the *Gram Sabha* for acquiring land in the Scheduled areas. However, these Acts have been neglected, and in practice land acquisition Acts such as the *Land Acquisition Act 1894* and the *Coal Bearing Areas (Acquisition and Development) Act 1957* are accorded supremacy.

The *Indian Forest Act 1927*, which upholds states' eminence power over forest land – including the land traditionally owned and used by STs and other forest dwellers and common land in Scheduled areas – is another Act which constrains people's rights over and access to forest lands. The *National Forest Policy 1952* and the *Forest Policy 1988* are in conformity with the *Indian Forest Act*. Moreover, the *Forest Conservation Act* of 1980 prohibits the people's access to forest resources, including minor forest produce.

However, the recently passed *Forest Dwelling Scheduled Tribes and Other Traditional Forest Dwellers (Recognition of Rights) Act 2006* – also known as the *Forest Rights Act (FRA) 2006* – provides for legal rights of STs and other forest dwellers who have been residing in such forests for generations but whose rights to those lands were not recorded. The act also puts a restriction on the displacement from the forest land. The act clearly states that "...no member of the scheduled tribe or other forest dwellers family shall be evicted or removed from the forest land under his occupation till the recognition and verification procedure is complete" (section 5, FRA, 2006).

The act also puts a restriction on the displacement from the forest land. The act clearly states that "...no member of the scheduled tribe or other forest dwellers family shall be evicted or removed from the forest land under his occupation till the recognition and verification procedure is complete" (section 5, FRA, 2006).

Both PESA and FRA acts, strengthening the land rights of the advasis (or tribal people) and other forest dwellers are relatively new and have not yet been implemented fully. Therefore the full implications of these very crucial acts are yet to unfold. However, the Supreme Court of India, in a recent order, has asked the government of Odisha, an eastern Indian state, to take the consent of the gram sabhas (general assembly of all the adults of a village) of the villages to be affected by the bauxite mining project to be done jointly by a transnational company, Vedanta and Odisha Mining Corporation Limited, a state government owned company. The FRA, 2006 became the basis of this order. The villagers, mainly Dongaria Kondh and Kutia Kondh tribes, listed as primitive tribes, in all the 12 gram sabhas have rejected the mining proposal and now the Ministry of Environment and Forest will take final decision based on the resolutions of the 12 gram sabhas (for details, see Sharma 2013, Dasgupta, 2013, Bera, 2013).

This intervention by the Supreme Court can become very important precedent and it will now be difficult for the governments to ignore the acts like FRA. In other cases of land acquisition too, the potentially affected people have used these acts to challenge the land acquisition by the governments.

Legal Mechanisms for Land Acquisition

The Coal Bearing Areas (Acquisition and Development) Act 1957 and the Land Acquisition Act 1894

In India, the main land acquisition act, so far, has been the Land Acquisition Act, 1894. This act has been very recently (September, 2013) replaced by a new act passed by the Indian Parliament. Beside this main act, there are as many as 18 sector specific land acquisition acts e.g. the *Coal Bearing Areas (Acquisition and Development) Act 1957* for acquiring land for coal mining, the Land Acquisition (Mines) Act, 1957 for acquiring land for mining etc.

For coal mining by CIL's subsidiaries, land is acquired under the *Coal Bearing Areas (Acquisition and Development) Act 1957*. However, the private and public sector companies that have been allotted coal blocks by the Ministry of Coal, would acquire land under the *Land Acquisition Act 1894* (LAA), which is redundant now. None of these Acts, however, include provisions for rehabilitation and resettlement (R&R) of displaced people; only compensation is monetary reimbursement for land, based on the market rate. The LAA, 1894 was a colonial legacy and accorded no consideration to the basic rights of the people who are evicted from their lands and homes. It gave immense powers to the government to acquire any land for a public purpose. The Act was amended in 1984 to increase the number of agencies that can acquire land. Originally the LAA permitted the acquisition of land only by the government; but now public sector companies could also acquire land directly, and private sector companies could do so through the government. People were not consulted at any stage of acquisition, and their consent was not required. Under the LAA, the people being displaced had no right to object to the acquisition. The Act did not define the term 'public purpose' (Dhagamwar 1997, 113). The Indian government, however, has now passed a new act to replace the LAA, 1894, which we discuss in a later section.

The *Coal Bearing Areas (Acquisition and Development) Act 1957* (CBA) is applicable to situations in which land is acquired for the purpose of coal mining by CIL's subsidiaries; although passed much later than the LAA and after India's independence, it includes provisions that are even more stringent than those of the LAA (Anjum and Manthan 1997, 155). The CBA is highly centralised and powers of taking action under the Act are in the hands of the federal Ministry of Coal and officers of the coal mining subsidiaries of CIL. To acquire land under this instrument, initial and secondary notifications need only be published in the official Gazette. There is no requirement to publish such notices in local newspapers or to notify each affected person. Except for payment of compensation at the market rate, the Act does not provide for any solatium or other payment. The Ministry of Coal, however, has given directives to the coal companies to pay 30 per cent solatium with 12 per cent interest from the date of initial notification. The payment of compensation under the CBA is not essential before taking possession, unlike the LAA, under which 100 per cent payment was required to be paid in advance under normal conditions and 40 per cent of the estimated compensation in emergency conditions. The CBA provides only very limited rights to object to the acquisition, which can be suspended in the case of emergency.

As Anjum and Manthan (1997, 155) suggest, the CBA is even worse than the LAA as far as the interests of the affected population is concerned. No wonder then that the private and public sector power companies seeking to establish captive coal mining projects also want to be given rights to acquire land under the CBA instead of the LAA (*Financial Express* 2008).

Environmental Clearance: Environmental Impact Assessments (EIAs)

Apart from the devastating social impacts of displacing millions, coal mining, particularly open-cut coal mining, has huge environmental costs. It destroys forests, water bodies, and agricultural land, leaving permanent scars on the earth. The environmental destruction caused by coal mines also hurts the poor and marginalised populations living in mining areas and heavily dependent on natural resources for their livelihoods.

With Environment Impact Assessment (EIA) notifications issued in 1994 under the *Environmental (Protection) Act 1986*, the Ministry of Environment and Forests (MoEF) has made it necessary to obtain environmental clearance from the Ministry on the basis of EIA for 29 categories of development activities, including mining. EIA is mandatory for all new and expanding projects with mining lease areas greater than five hectares. Recently this notification has been replaced by the amended EIA Notification of 2006. The requirements for mining projects, however, continue unchanged (Jha-Thakur and Fischer 2008, 445).

Environmental Impact Assessment (EIA) is an important management tool for ensuring optimal use of natural resources for sustainable development. The Ministry has constituted Environmental Appraisal Committees for six sectors including mining. Companies or project authorities submit their documents with the EIA report, which is scrutinised by technical staff of the Ministry prior to being placed before the Environmental Appraisal Committees. If necessary, field visits and on-the-spot assessment are also undertaken. The committee hands down its recommendations, which are then processed by the Ministry for approval or rejection (MoEF website).[3]

A monitoring system has been established whereby MoEF's six regional offices monitor the cleared projects. The main objectives of such a procedure are to ensure that safeguards are in place and that mid-term corrections are undertaken as required (MoEF, GoI, website).

However, the practical situation seems to be very different. The coal mining projects of CIL's subsidiaries and other companies continue to ruin the environment. The manner in which EIA is carried out by the companies is not

---

3   See 'Environmental Impact Assessment Division' on MoEF website: http://envfor. nic.in/division/introduction-8.

known. The international standards of EIA make public consultation a necessary element of the EIA exercise (World Bank 1998; Joyce and MacFarlane 2001). However, consultation with the public is something alien to CIL's subsidiaries and, as a recent study suggests, the private coal mining companies are no different (Lahiri-Dutt et al. 2012). All studies on the impact of mining or mining-induced displacement suggest that the people to be affected by mining operations are not informed in advance, let alone being consulted. EIA includes provisions requiring a public hearing. The public consultation held for Ashoka Expansion OCP in Chatra district in Jharkhand, for example, was conducted as a formality and no consideration was given to the concerns raised by the villagers, according to a local newspaper report (Prabhat Khabar 2007). Though they have no legal standing, the public hearings provide the affected people a forum at which to share their grievances and are enthusiastically attended. The people use public hearings as a platform to voice their concerns and make their resistance known, to avoid which the companies often try to circumvent or manipulate the public hearing process (Lahiri-Dutt et al. 2012, 42). Studies also suggest that the regulatory mechanism to implement EIAs and ensure follow-up action is generally weak and that 'the regulating agencies in India within the OC [open-cut] coal mines are not able to discharge their responsibilities of checking compliance effectively....The regulating agencies act as policing agents, but the power in the regulation process is tilted towards developers' (Jha-Thakur and Fischer 2008, 457).

Right to Resettlement and Rehabilitation

National Rehabilitation and Resettlement Policy, 2007

So far there is no law addressing the rights of the displaced and project-affected population in regard to resettlement and rehabilitation. The Government of India declared its first R&R policy as late as in 1994. A revised draft policy was prepared in 2006 (NPR 2006) to address the admitted failures of the earlier policy and was finalised and formally notified in 2007, with which the *National Rehabilitation and Resettlement Policy 2007* (NRRP 2007) came into effect (GoI 2007). The government has now passed a new act which unifies the land acquisition and resettlement and rehabilitation and provides legal backing to NRRP, 2007.

The policy and the new Act are certainly a step taken in right direction, as they provide for a legal mechanism with which to address the concerns of displaced people via legal rights to R&R. However, they do not live up to the expectations of the displaced and PAPs. The *National R&R Policy* has many weaknesses (ACHR 2007). The policy fails to recognise people's right to reject displacement and upholds the right of the state to acquire private land for the public purpose without informed consent of the people. It deletes the provision of the 2006 draft policy that the emergency provision under Section 17 of the LAA 'should be used rarely' and should be applied only after considering 'full justification' (ACHR

2007). Social impact assessment are only required for the projects displacing more than 400 families, or more than 200 in hilly and desert areas. The policy does not attribute any consultative or other role to the local self-governance bodies such as the *Panchayati Raj* institutions and urban local councils.

The Policy, however, also includes some credible provisions such as those determining that if the land is sold after acquisition, 80 per cent of the net profit earned from the sale is returned to the original land owner. The Policy also provides the option to affected families of accepting up to 20 per cent of their rehabilitation grant –'seven hundred fifty days of minimum agricultural wage or such other higher amount as may be prescribed by the appropriate government' – in forms of share or debentures of the company (Clause 7.14). The state government can raise this proportion to 50 per cent (Clause 7.14). Now, however, the new Act has been passed, incorporating some of the provisions of the Policy.

Right to Fair Compensation, Rehabilitation, Resettlement and Transparency in Land Acquisition Act, 2013

The government has recently (September 2013) passed a new Act in the Parliament which replaces the Land Acquisition Act 1894. The new act unifies land acquisition and R&R and makes R&R legally mandatory. The passing of this law has gone through many interesting phases and has seen various pulls and pushes.

Making of a Law

In 2009 the Indian government passed two bills – the *Land Acquisition (Amendment) Bill* (LAA Bill) and the *Resettlement and Rehabilitation Bill* (R&R Bill) – in the *Lok Sabha*, legislation intended to support the NRRP 2007. The two bills were not passed in the *Rajya Sabha*. In 2011, however, the government introduced a new *Land Acquisition and Rehabilitation & Resettlement Bill 2011* (LARR Bill 2011) (MoRD 2011) in the *Lok Sabha*, abandoning the earlier bills. The bill provided for unifying the land acquisition and R&R and establishing legal provision for rehabilitation and resettlement of the displaced and affected population. The bill was put on the Ministry of Rural Development's website, inviting comments from public. Provisions of the bill were to apply to all other land acquisition acts used by the India government. But when the bill was tabled in the Parliament after receiving comments, the government had added a Schedule IV to it, which provided that the bill would not be applicable in cases of land being acquired under CBA 1957 and 15 other Acts (Schedule IV, LARR Bill 2011). The bill had many other shortcomings as well.

The LARR Bill 2011 was referred to the parliamentary standing committee related to the Ministry of Rural Development, which in its report (Standing Committee on Rural Development 2012) criticised the bill on various grounds. The committee expressed its apprehensions about the government acquiring land for private companies, the minimal role of local government bodies, and the exemption of land acquisition under 16 government laws – including the *Coal Bearing Areas (Acquisition and Development) Act 1957* – from the scope of the proposed law.

Criticism of the LARR Bill 2011 and the subsequently passed Act is concerned with the basis of the model of development, which considers displacement inevitable; the new Bill is based on the logic of the market and is aimed at meeting the growing demand of land for various development projects. This has been the driving motive behind the rehabilitation policy of the government since its inception. The draft of the first Rehabilitation Policy prepared by the Ministry of Rural Development in 1994 stated the need for a rehabilitation policy to meet the growing demand for land in the wake of economic liberalization (MoRD 1994).

The new Act promises higher compensation for the lost land and resettlement and rehabilitation for all displaced and affected people, including landless people. However, problematic areas remain. The definition of 'public purpose' is too broad in the proposed bill. It includes provision to acquire land for private companies and public-private partnership (PPP) projects, but only after obtaining 'informed prior consent' of 80 per cent of the land losers and those working on government assigned land in case of land being acquired for a private company and that of 70 per cent in case of PPP projects. In the original bill the provision was to take consent of all the affected population, including the land less. Such provision, however, is not intended for land being acquired for government projects. There is also no restriction or limit on private companies' purchase of land on their own. The new Act also has no provision of approval of local self governance institutions like PRIs or ULBs. However, it provides for taking approval/consent of local institutions of self governance only in the VI Scheduled areas. The NRRP 2007 and the new Act introduced in the parliament take displacement as given and favour the development projects, including that of the private companies, at the cost of the rights of the displaced and project-affected populations (see MoRD 2013a and MoRD 2013b for details of the new Act.). In case of land being sold after the acquisition, the new acts have provision to share only 40 per cent of the appreciated value with the original owners.

LARR bill 2011 was subject to much criticism by the people's organizations, state governments, and bodies of industries. As mentioned above, it was referred to the Parliamentary Standing Committee which, after many consultations, suggested various changes. The government, however, decided to proceed with the bill, with an alternative title and further dilution of its provisions in order to placate the industry (Development Debate, 2012). It has now been passed as the *Right to Fair Compensation, Rehabilitation, Resettlement and Transparency in Land Acquisition Act 2013* by the Indian Parliament. What is important from the perspective of coal mining industry and mining industry in general, is that the new Act provides exemption from its provisions in the cases of the land acquired under 13 union acts, including the Coal Bearing Areas (Development and Acquisition) Act, 1957, the Land Acquisition Mines Act, 1885, Damodar Valley Corporation Act, 1948. So the CIL and its subsidiaries would continue the acquire land under

CBA, 1957. However, within one year, the provisions of the new law which relates to compensation, rehabilitation and resettlement will be applied to the 13 Acts by a notification of the central government.

The claims by the government to resolve the ongoing land conflicts in the India's rural and tribal areas with the passing of the new Act can, at best, be termed as a wishful thinking. According to media reports about 85 to 90 per cent of the land is acquired under the acts which are out of the purview of the new Act passed by the parliament and the misery of the people being displaced by those acts are only partially to be addressed by the new Act (only after one year) and the opposition to land acquisition efforts by the government is, therefore, likely to continue (Yadav, 2013).

CIL's Rehabilitation and Resettlement Policy 2012

Coal India Limited has its own R&R policy, which was announced in 1994 under the pressure of the World Bank. This policy has been revised many times. The latest revision occurred in 2012 (CIL 2012) after the Indian government had drafted the *Land Acquisition and Rehabilitation Bill 2011*. Prior to this, CIL R&R Policy 2008 was in effect, which was a revised version of earlier R&R policy and which followed the NRRP 2007. The Earlier versions of CIL R&R Policy have been criticised for their many shortcomings (see, e.g., Anjum and Manthan 1997). Despite revisions, the basic principles of CIL policy remain same.

However, in the R&R Policy 2012 there are some marked differences. One, for example, is the language. The R&R Policy 2008 clearly states: 'In light of growing difficulties many subsidiaries face in land acquisition highest priority will be given to avoiding or minimizing disturbance of the local population' (Para 6). Clearly CIL's objective has been to dissipate and evade public discontent to ensure that the acquisition of land proceeds smoothly.

The 2012 R&R Policy uses the language of participation, consultation and transparency. One of the objectives of the policy is '[t]o ensure a humane, participatory, informed consultative and transparent process for land acquisition...' (CIL 2012: 2).

Another marked difference from the R&R Policy 2008 is in offering employment with the subsidiaries of CIL. The most attractive provision of CIL's policy so far has been permanent employment for those people losing their lands – one position for every two or three acres of land. One must recall, however, that the record of CIL awarding employment to the affected families has been extremely poor: according to one study members of only 30 per cent of the affected families were given paid work (Fernandes 1998, 251). But now with technological advancements in the coal mining industry, CIL's dependence on highly mechanised open-cut mines – which contribute 80 per cent of total coal production in the country – and the policies adopted by CIL in the reform era, it is difficult for CIL to offer employment to unskilled, displaced and project-affected people. During the last two decades, the focus in CIL has been on the retrenchment of workers in the name of rationalisation of manpower; the industry has reduced its paid workforce from 670 thousand in 1985 to 443 thousand in 2006 (Ministry of Coal 2007, 97).

The focus in CIL's R&R Policy 2008 was on exploring options other than giving employment to displaced people. The 2008 Policy clearly stated that 'mine jobs would be offered only in exceptional circumstances with the approval of the subsidiaries' Board' (Para 15). The emphasis was instead on self-employment and additional monetary compensation.

To its credit, the 2012 Policy permits the option of providing permanent employment to the dispossessed in addition to compensation for lost land, which is to be paid according to 'the provisions of the concerned Act or the State Govt. notification' (CIL 2012, 7). The Policy states that 'For every two acres of land one employment [position] *can be considered*' (CIL 2012, 7: emphasis added). Although the 2012 Policy appears to be non-committal on permanent employment, the strict exclusion of the provision of employment in the 2008 Policy has been omitted. The 2012 Policy offers the option to people owning less than two acres of land to club their lands to the extent of two acres and nominate any one person for employment with the subsidiary company: a practice which existed in the earlier policy. The 2012 Policy includes provision for 'Lumpsum Monetary Compensation' for those dispossessed who are not eligible for employment. They are entitled to 'Rs. 500,000 (Rs. 5 lakhs) for each acre of land on pro-rata basis.' The Policy provides for persons eligible for employment to opt for monetary compensation if they do not wish to take up employment (CIL 2012, 8); such compensation can be paid in monthly or annual installments.

The Policy requires that community facilities are to be provided at the rehabilitation sites. The subsidiaries are supposed to provide a school and playground, roads with street lights, sewer drainage, a well for the supply of drinking water, a community centre, a place of worship, a dispensary, and grazing land for cattle(CIL 2012, 10).

CIL R&R Policy 2012, however, perpetuates many shortcomings. The Policy has no provision for an independent social impact assessment, which is mandatory in the National R&R Policy 2007 and also in the LARR 2011 and in its revised version. The only requirement is a non-government organization (NGO) survey of the project-affected villages that lists the projected-affected persons (PAPs).

Although the 2012 Policy states that it will make the acquisition process 'humane, participatory and informed and transparent', nowhere does it specify how and in what ways this will be achieved. No consideration is given to when and how the 'Resettlement and Rehabilitation Committee' will be constituted, apart from reference to the District Collector. The post-implementation social audit is to be conducted 'in consultation with the authorities.' In the final analysis, in spite of using the language of participation and consultation, this policy is likely to be implemented in a top-down manner that ignores the needs and views of the displaced and affected population.

In view of *Right to Fair Compensation and Rehabilitation and Resettlement Act 2013* now enacted and its provisions related to the compensation and rehabilitation and resettlement to be applied to the 13 other land acquisition acts including CBA, 1957, it would be interesting to see how the R&R policy of the CIL will be revised to incorporate the provisions of the new Act.

Are CIL's R&R Programmes Working?

Research into CIL's R&R programmes demonstrates that the earlier-mentioned facilities are rarely provided at rehabilitation sites (see, e.g., Ahmad 2003). Maintenance and operation of these facilities are also an important issue. Generally the subsidiaries construct a school building but they assume that teachers will be provided by the government. In Jharkhand, this never used to happen. In some cases – for example, in Benti village, displaced by the famous Piparwar open-cut project in Chatra district – local youth would teach in those school buildings and charge fees to students. Recently para-teachers have been appointed under Sarva Siksha Abhiyan (SAA) in those schools. The same is true for maintaining and operating tubewells, if they are ever provided. The lack of drinking water facilities and health services are a common problem at all the rehabilitation sites and project-affected villages (Ahmad 2003, 38).

The World Bank had funded 25 coal mining projects of CIL under the Coal Sector Rehabilitation Project (CSRP). Additionally, a loan for the Coal Sector Environmental and Social Mitigation Project (CSESMP) was also approved by the World Bank to mitigate the environmental and social impacts of the 25 open-cut mines funded under CSRP (Herbert and Lahiri-Dutt 2004, 2403). Contracted employment to the PAPs is one major component of the CSESMP (Roy 2006, 142).

CIL also has community development (CD) programmes in the vicinity of the coal mines, within a radius of eight km. The CD programmes focus on infrastructure development, social services – education, health, drinking water, skill-development and capacity building (CIL nd). In addition, CIL adopted an Indigenous People Development Plan (IPDP) in 1995, which provides that the subsidiaries will outline a plan of CD assistance to the affected population within one km of the boundary of the mines. IPDP has three components: community assets (infrastructure); community activities; and training and capacity building (Roy 2006, 145).

CIL also makes provision for involving NGOs in implementation of the RAPs and IPDP. CIL selected 11 NGOs, mainly to assist mine-level staff in planning and implementing mine-specific IPDPs and RAPs. These NGOs assisted CIL's subsidiaries in planning and implementing IPDP and RAPs in all 25 OCP areas (Roy 2006, 146). According to Roy (2006, 148), only 25 per cent of entitled PAPs in 14 OCP areas could be brought under non-farm self-employment income restoration plans during 1998 and 1999. By mid-2002 – the year CSESMP closed – 1,724 entitled PAPs completed skills training, of which only 58 per cent have been able to obtain an income (Roy 2006, 148). Roy also observes that in those 14 OCP areas, 42 per cent of entitled PAPs are able to earn some income due to secondary economic opportunities afforded by the mining industry (Roy 2006, 148). In some areas CIL subsidiaries have also tried to involve PAPs in land-based income generation activities. Plenty of land is available in mining areas as overburden (OB) dump sites, reclaimed land after mining is over and if reclamation occurs, and vacant mines. A pilot project on land-based income-generation activities in one SECL mining area has involved PAPs in a cooperative, and has apparently achieved promising results (Roy 2006, 150).

Studies conducted in both Parej OCP and Ashoka and Piparwar OCP areas suggest that existing policies are not greatly improving the living standards of the displaced and PAPs. Ranchi-based organisation Xavier Institute of Social Service (XISS) was selected for Parej East OCP by the CCL and operated there during the period 1996–2002 (Ekka 2006, 153). Ekka (2006, 153) presents the findings of a survey conducted by XISS in 2004 at the request of CIL and the World Bank to obtain a clearer picture of the impacts of IPDP and RAP on PAPs in Parej. The study found that these efforts have had only marginal impacts (Ekka 2006, 165):

> On examining the six major RAP components, it can be inferred that the CCL has been able to mitigate adverse social impacts only marginally. Most PAPs and PAFs either relocated or resettled and tried to reestablish livelihoods on their own. The training for individual and group economic activities had partial success for lack of backward and forward linkages.

The situation in the Piparwar and Ashoka OCP areas in the Chatra district is also unchanged. These are older projects, initiated in 1985, for which compensation, rehabilitation and resettlement have been completed. Studies conducted in Benti village – which is one of the most affected villages of the two OCPs – suggest that CD programmes are not able to minimise social impacts (Ahmad 2003; Lahiri-Dutt and Ahmad 2011). People are still struggling for their livelihood and basic facilities such as a potable water supply, health services, and education. The people in this village, living in hamlets only one to two km from open-cut mines, are still dependent for on a natural water source – Jobhia jharna – which is one-and-a-half kilometers from the village; their children attend SSA schools, and those who can afford are sending them to private schools; health facilities are minimum five to ten kms away (Lahiri-Dutt and Ahmad 2011).

CIL's R&R Policy 2008 cites the need to consider corporate social responsibility (CSR), perhaps for the first time, in and around the villages where land is being acquired. It states that one to 2.5 per cent of the total earnings of CIL shall be provided for CSR and the priority should be given to displaced persons. It also provides for social assets, infrastructure, and skills development and vocational training.

Conclusion: Need for Change

Coal mining is one of the major causes of displacement of people from their homes and lands and takes away their means and livelihood. The CBA 1957 and now redundant LAA 1894 have given immense powers to the state and the public sector coal mining companies. These Acts are at odds with some of the Constitutional protections provided to the weaker groups of society like STs and SCs. It took post-colonial India many years to develop its first national rehabilitation policy, which was published in 1994. CIL also inaugurated an R&R policy in 1994, after 19 long years of its formulation and more than two decades of nationalisation of coal mining.

CIL and its subsidiaries have not shown any respect for their own policies as far as rehabilitation and resettlement is concerned. The World Bank (WB) has not displayed any capacity to supervise CIL's subsidiaries, in spite of its guidelines and directives. The WB Inspection Panel – an internal assessment body that ostensibly holds the Bank accountable for violations of its policies and procedures – have identified a number of violations of the Bank's guidelines and operational directives, to which management have turned a blind eye. It has also identified CIL and CCL's seemingly willful disregard of its guidelines and operational directives in Parej East OCP (Herbert and Lahiri-Dutt 2004, 2408). There is a need to establish mechanisms that hold the management of CIL and its subsidiaries accountable in the event that the NRRP and CIL's R&R Policy are not followed properly. CIL's R&R Policy includes provision for a committee to review the grievances of the PAPs, but it does not specify its powers.

The problem, however, is that the new Act and CIL R&R Policy 2012 consider displacement as inevitable; there is a need to search for options that avoid this outcome, such as reviving the use of underground instead of open-cut mining. Although open-cut mining is not a viable option for extracting deeply embedded coal deposits, it is heartening that the Planning Commission in the eleventh Five-Year Plan (2007–12) underlined the importance of underground mines and called for 'promoting underground mining operations for extraction of deep seated deposits [of coal]' (Planning Commission 2008, 380). Underground mines require less land compared to open-cut mines and are environmentally less destructive. A renewed emphasis on underground mining with technological advancement, increased productivity and environmental responsibility can decrease the extent of displacement. This, however, is a long-term strategy. Meanwhile, dignified means of rehabilitation and resettlement of displaced and project-affected people must be instituted that are founded on respect for people's rights. With the passage of the newly passed *Right to Fair Compensation and Rehabilitation and Resettlement Act 2013*, its provisions related to compensation and rehabilitation and resettlement to be applied in cases of land being acquired under CBA 1957 within one year, for the first time there will be legal mandate to provide R&R to the displaced population. The options to make PAPs stakeholders in mining companies would provide them with a sense of ownership in the project, especially in the absence of employment opportunities in mining. Since most OCPs have a short life of 25 years or less, reclaiming the land after mining is completed and returning it to its original owners or their associations/cooperatives for income generation activities could also be considered. The land-based income generation activities in one SECL OCP area involve cooperatives of the PAPs and appear to be successful. Such examples could be replicated in other areas.

The most important issue, however, is obtaining the consent of the displaced and affected people before the acquisition of land. The institutional mechanisms such as the newly passed *Right to Fair Compensation, Rehabilitation, Resettlement and Transparency in Land Acquisition Act 2013* are limited by market-directed goals and aim to supply enough land to meet the ever-increasing demand of mining in a fast-growing economy without respecting the constitutional protections. The currently used CBA 1957 and other land acquisition Acts do not require the consent of land users before acquiring their land. The newly passed *Right to Fair Compensation, Rehabilitation, Resettlement and Transparency in Land Acquisition Act 2013* has limited provision of taking consent of the affected population in some cases of land acquisition. This, however, would not be applicable to the cases of land acquired under CBA, 1957.

Land must not be acquired without securing prior and informed consent of communities. This means that communities should retain the crucial right to refuse proposed projects. Acts such as PESA and FRA can be amended to require this, not just in the Scheduled areas but throughout the country. Also the provisions of the newly passed *Right to Fair Compensation, Rehabilitation, Resettlement and Transparency in Land Acquisition* Act 2013 should be applicable in all cases of land acquisition and the provision to give exemption to land acquisition under 13 other union government acts should be removed.

References

Action Aid. nd. 'Tribal Land Rights: Myth or Reality.' available on http://www.indlaw.com/ActionAid/?Guid=f7ef1f2e-db9c-4327-b648-b43949e9bff4, last accessed February 18, 2010.

ACHR (Asian Centre for Human Rights), 2007, 'India's failed National Rehabilitation and Resettlement Policy, 2007',available on http://www.achrweb.org/Review/2007/198-07.html, last accessed January 10, 2010.

Ahmad, Nesar and Lahiri-Dutt, Kuntala. 2006. 'Engendering Mining Communities: Examining the Missing Gender Concerns in Coal Mining Displacement and Rehabilitation in India.', *Gender Technology and Development* 10(3), Bangkok.

Ahmad, Nesar. 2003. *Women, Mining and Displacement: Report of a Pilot Study Conducted in Jharkhand*, Indian Social Institute, New Delhi.

Anjum and Manthan. 1997. 'A Review of the Policy of Coal India' In Walter Fernandes and Pranjpey, Vijay (ed.), 1997, *Rehabilitation Policy and Law in*

India: A Right to Livelihood, Indian Social Institute, New Delhi and Econet, Pune.

Bera, Syantan, 2013, 'Niyamgiri Answers' available on http://www.downtoearth. org.in/content/niyamgiri-answers, last accessed on October 1, 2013.

CIL. 2012. Rehabilitation and Resentment Policy of Coal India Limited 2012, available on http://www.coalindia.in/Documents/Policies/CIL_ RR_2012_100412.pdf, last accessed on January 18, 2013.

CIL. 2008. 'Resettlement and Rehabilitation Policy.' available on http://www. coalindia.in/Documents/Policies/PolicyOnCILRR17062008.pdf, last accessed on January 18, 2013.

CIL. Undated. 'Policy for Community and Peripheral Development' available at http://www.coalindia.nic.in/policy per cent20on per cent20community.htm last accessed on June 4, 2011.

Committee on State Agrarian Relations and Unfinished Task in Land Reforms (Department of Land Resources, Ministry of Rural Development). 'Report' (Chapter four), available on http://dolr.nic.in accessed on February 10, 2010.

Development Debate, 2012, 'Government ignores Parliamentary Committee's Recommendations: Makes Acquisition Bill further pro-industry' available at http://www.developmentdebate.in/2012/08/government-ignores-parliamentary.html, last accessed on October 1, 2013.

DoLR (Department of Land Resources), 2008, Report of the Committee on State Agrarian Relations and Unfinished Task in Land Reforms (Department of Land Resources, Ministry of Rural Development), available on http://dolr.nic. in/dolr/reports.asp, last accessed on February 10, 2010.

Dasgupta, Kumkum. 2013. 'Vedanta's India Mining scheme thwarted by local objections' available on http://www.theguardian.com/global-development/ poverty-matters/2013/aug/21/india-dongria-kondh-vedanta-resources-mining, last accessed on October 1, 2013.

Dhagamwar, Vasudha. 1997. 'The Land Acquisition Act: High Time for Changes' In Walter Fernandes, and Vijay Pranjpey (ed.), 1997, *Rehabilitation Policy and Law in India: A Right to Livelihood*, Indian Social Institute, New Delhi and Econet, Pune.

Ekka, Alex. 2006. 'Mitigating Adverse Social Impacts of Coal Mining: Lessons from the Parej East Open Cast Project', *Social Change* 36(1), March 2006, New Delhi.

Fernandes, Walter. 1998. 'Development induced Displacement in Eastern India' In Dubey, S.C. (ed.) *Antiquity to Modernity in Tribal India* Vol. 1, Inter-India Publication, New Delhi.

Fernandes, Walter. 2006. 'Liberalization and Development-induced Displacement', *Social Change* 36(1), March 2006, New Delhi.

Financial Express, 2008. 'Power Companies Demand amendment in Coal Bearings Act' *Financial Express*, dated August 26, 2008, internet edition, available on http://www.financialexpress.com/news/Power-cos-demand-amendment-in-Coal-Bearings-Act/353257/, last accessed January 8, 2013.

GoI, 2007. National Rehabilitation and Resettlement Policy 2007, available on http://www.dolr.nic.in/, last accessed on January 2, 2010.

Herbert, Tony and Lahiri-Dutt, Kuntala. 2004. 'Coal Sector Loans and Displacement of Indigenous Populations: Lessons from Jharkhand', *Economic and Political Weekly*, June 5, 2004, Mumbai.

Jha-Thakur, Urmila and Fischer, Thomas. 2008. 'Are open-cast coal mines casting a shadow on the Indian environment', *International Development Planning Review* 30(4), 2008.

Joyce, Susan A and MacFarlane, Magnus. 2001. 'Social Impact Assesment in the Mining Industry: Current Situation and Future Directions,' IIED, available on http://pubs.iied.org/G01023.html, last accessed on June 4, 2012.

Kothari, Smitu. 1996. 'Whose Nation? Displaced as Victims of Development', *Economic and Political Weekly*, June 15, 1996, Mumbai.

Lahiri-Dutt, Kuntala. 2007. 'Coal Mining Industry at the Crossroads: Coal Policy in Liberalizing India'. Paper presented at *1st International Conference on Managing Social and Environmental Consequences of Coal Mining in India*, organized on November 19–21, 2007 in New Delhi, also available on http://crawford.anu.edu.au/pdf/staff/rmap/lahiridutt/JA9_KLD_Coal_liberalising.pdf, last accessed on January 8, 2013.

Ministry of Coal. 2007. Annual Report, 2006–07, Ministry of Coal, New Delhi.

Ministry of Coal. Website. www.coal.nic.in, last accessed on 17 March, 2012.

Lahiri-Dutt, Kunlata, Krishnan, Radhika, and Ahmad, Nesar. 2012, 'Land Acquisition and Dispossession Private Coal Companies in Jharkhand' in *Economic and Political Weekly*, February 11, 2012.

Lahiri-Dutt, Kuntala, and Ahmad, Nesar, 2011. 'Considering Gender in SIA' In Frank Vanclay and Esteves, A.M. (eds) (2011) *New Directions in Social Impact Assessment: Conceptual and Methodological Advances*, Cheltenham (UK): Edward Elgar. ISBN 978 1 84980 117 1.

Ministry of Coal. 2009. 'Details of Coal Block which stand allotted' available at http://coal.nic.in/alloblock050209.pdf, last accessed on January 8, 2013.

Ministry of Coal. 2005. 'Eligibility to Coal Mining' available on: http://coal.nic.in/eligibility_to_coal_mining.htm, last accessed on January 8, 2013.

Ministry of Coal. 2007. Annual Report 2006–07, New Delhi.

Ministry of Coal. 'Captive Coal Mining Blocks' available on http://www.coal.nic.in/welcome.html, last accessed on October 1, 2013.

MoEF, 2006. 'Environment Impact Assessment Notification 2006', available on http://moef.nic.in/divisions/iass/notif/notif.htm, last accessed on October 1, 2013.

MoEF website. Environment Impact Assessment Division. http://envfor.nic.in/division/introduction-8.

Mohanty, BB. 2001. 'Land Distribution among Schedule Castes and Schedule Tribes', *Economic and Political Weekly*, October 6, 2001, Mumbai.

MoRD. 2013a. Frequently Asked Questions on Right to Fair Compensation and Transparency in Land Acquisition and Rehabilitation and Resettlement Act,

2013, available on http://rural.nic.in/sites/downloads/general/LARR_FAQs_post_Parliament.pdf, last accessed on October 1, 2013.

MoRD. 2013b. Right to Fair Compensation and Transparency in Land Acquisition and Rehabilitation and Resettlement Act, 2013: An Overview, available on http://rural.nic.in/sites/downloads/general/RTTFC_in_LARR_2013.pdf, last accessed on October 1, 2013.

MoRD. 2011. The Draft National Land Acquisition and Rehabilitation and Resettlement Bill, 2011, available on http://rural.nic.in/sites/downloads/general/LS per cent20Version per cent20of per cent20LARR per cent20 per cent20Bill.pdf, last accessed January 8, 2013.

MoRD (Ministry of Rural Development). 1994. Draft National Rehabilitation Policy.

Prabhat Khabar (*Hindi Daily*). 2007. News about the public hearing on Ashoka Expansion OCP in Chatra district in Jharkhand, Prabhat Khabar (Hindi Daily), Ranchi, dated March 14, 2007.

Roy, M. P. 2006. 'Approaches to Reconstructing Livelihoods: CIL Experience with Self Employment Scheme' in *Social Change* 36(1), March 2006, New Delhi.

Sharan, Ramesh. 2005. 'Alienation and Restoration of Tribal Land in Jharkhand: Central Issues and Possible Strategies', *Economic and Political Weekly*, October 8, 2005, Mumbai.

Sharma, Kalpana. 2013. 'Naturally Rich' available on http://www.thehindu.com/opinion/columns/Kalpana_Sharma/naturally-rich/article4985247.ece, last accessed on October 1, 2013.

Standing Committee on Rural Development. 2012. 'The Land Acquisition, Rehabilitation and Resettlement Bill, 2011, 31st Report' MoRD, available at: http://dolr.nic.in/dolr/downloads/pdfs/Land per cent20Acquisition, per cent20Rehabilitation per cent20and per cent20Resettlement per cent20Bill per cent202011 per cent20-per cent20SC(RD)'s per cent2031st per cent20Report.pdf, last accessed on June 2, 2012.

The Telegraph. 2008. 'FDI freedom for captive coal' Telegraph, Kolkata, dated April 8, 2008, also available on http://www.telegraphindia.com/1080408/jsp/business/story_9107780.jsp, last accessed on January 8, 2013.

The Hindu. 2012. 'GOM approves land acquisition bill' available on http://www.thehindu.com/news/national/gom-approves-land-acquisition-bill/article4003181.ece, last accessed on January 8, 2013.

World Bank 1998. Environmental Assessment of Mining Projects, available at http://www.elaw.org/system/files/22.pdf, last accessed on June 4, 2012.

Yadav, Anumeha. 2013. 'Why the Land Wars won't end', available on http://www.thehindu.com/opinion/op-ed/why-the-land-wars-wont-end/article5182473.ece, last accessed on October 1, 2013.

Chapter 14

On the States' Ownership and Taxation Rights over Minerals in India

Amarendra Das

Introduction

The basic problem in the management of mineral resources in India is that the global and local communities unevenly share its benefits and costs. Inadequate transfer of benefits to the local communities generally makes mineral-rich areas the least developed regions. In India, states such as Chattisgarh, Jharkhand, Madhya Pradesh and Odisha persist as less-developed regions in spite of being endowed with rich mineral resources. Although these states have formulated legislation to mobilise additional revenue, national law has inhibited this. Due to revenue constraints, states fail to undertake special development programmes in the mineral hinterlands.

The causal effect of constitutional design and the structure of government on the management of natural resources have been widely discussed in the resource curse literature.[1] Devolution of power between different tiers of governments within a country – such as central, provincial and local – has serious bearing on the management of mineral resources and, hence, on the economic growth of the country and its regional economies. Inept devolution of power between different tiers of government not only causes sub-optimal use of resources, but also constrains the growth process.

This chapter appraises the constitutional provisions for the ownership and regulation of minerals in India under the law and economics framework. It is argued that in order to ensure intra and inter-generational equity, ownership of minerals should be vested with state governments and regulatory powers vested with the central government. However, for the development of mining areas, state governments should be provided with adequate elbow-room to mobilise revenue secured from mineral extraction. Therefore, the present system of uniform royalty rates determined by the central government should be abolished and states should be enabled to determine freely their royalty rates and other levies.

The rest of the chapter is organised as follows: the second section describes the constitutional framework for the ownership and regulation of minerals in India. The third section explains the significance of minerals in state economies. The

---

1   For example, see Andersen and Aslaksen (2008) and Brunnschweiler and Bulte (2008).

fourth section narrates the initiatives taken by state governments to raise additional revenue, and scrutinises their validity in accordance with the relevant legal and economic perspectives. The fifth section critically appraises the alternative of transferring the rights to local governments.

Ownership and Regulation of Minerals in India

India has a quasi-federal parliamentary and democratic form of government with a powerful central administration. The jurisdictions of central and state governments are clearly demarcated in the constitution of the country in three lists: central, state, and concurrent.[2] The authority of the central government is reflected in its overriding control of the subjects enumerated in the concurrent list. The Constitution of India confers mine ownership rights to state governments. However, the central government retains regulatory authority over all major minerals defined in Schedule A[3] of the *Industrial Policy Resolution 1956* (IPR). Regulatory power over the minor minerals enlisted under Schedule B has been retained by state governments. The jurisdictions of central and state governments over minerals have been clearly defined in the *Mines and Minerals (Regulation and Development) Act 1957* (or MMRD). Central government regulates the major minerals by controlling exploration, extraction and trade of minerals, and determining the royalty rates. Thus, states' right over minerals are subject to the central government's regulatory control.

Although provisions have been made to ensure the benefits of minerals, the cost of mining is disproportionately borne by the local communities. Environmental and social costs in the mining areas are enormous. Environmental degradation occurs in many forms: deforestation; pollution of ground and surface water; air and noise pollution; loss of biodiversity and waterways. The social costs of mining are effected in many ugly ways: displacement of local communities – mostly *adivasis* – and the psychological and emotional harm caused by this; loss of agricultural land; health problems due to pollution and the loss of livelihood sources due to deforestation. In order to minimise the deleterious effects of mining on the environment and social fabric, various institutional arrangements have been initiated by the central and state governments. However, the incumbent institutional mechanism has failed to safeguard the environment and the interests of local communities.

---

2   The subjects listed in the concurrent list gives power to both centre and state governments to act upon. However, the decision of the central government remains binding upon the state governments.

3   Minerals namely iron ore, coal and lignite, mineral oils, mining of iron ore, manganese ore, chrome-ore, gypsum, sulphur, gold and diamond, mining and processing of copper, lead, zinc, tin, molybdenum and wolfram and minerals specified in the Schedule to the Atomic Energy (Control of Production and Use) Order, 1953 (MMSME 2007).

Significance of Minerals in the State Economies

Although mining constitutes a negligible share of India's Gross Domestic Product (GDP), amounting to around three per cent, it plays an important role for a number of state economies. The contribution of mining activities to the state economies can be measured by the share of mining and quarrying (M&Q) activities in the Net State Domestic Products (NSDP) and the share of revenues generated from minerals as a proportion of total revenues of states. Data provided by the Reserve Bank of India(RBI)–on components of NSDP at factor cost – shows that in 2004–05, M&Q constituted 10.82 per cent of the NSDP of Chhattisgarh. Other major states where M&Q constitutes a major share in the NSDP are Jharkhand, Meghalaya, Assam, Odisha, Madhya Pradesh and Goa, with 10.25 per cent, 8.63 per cent, 8.27 per cent, 7.12 per cent, 4.55 per cent and 4.26 per cent share respectively. Table 14.1 depicts the share of the mining and quarrying sector in the NSDP of major mineral producing states of India. It is observed that the mining production of Odisha has declined substantially in recent years. Starting from 2008–09 the share of the M&Q sector in NSDP has declined from 6.76 per cent in 2007–08 to 3.54 per cent in 2011–12. This is due to the revocation of mining licenses of companies that flouted environmental and mining laws.

Table 14.1   Contribution of Minerals to State Economies: Recent Trend

States	2004–05	2005–06	2006–07	2007–08	2008–09	2009–10	2010–11	2011–12
Andhra Pradesh	2.60	2.56	3.42	3.19	2.40	2.37	2.37	2.24
Arunachal Pradesh	2.04	2.42	2.21	2.24	2.00	2.34	1.75	1.81
Assam	8.27	8.09	7.88	7.91	5.74	5.24	4.91	4.72
Chhattisgarh	10.82	10.87	10.11	9.48	9.41	10.21	9.48	9.09
Goa	4.26	4.04	5.18	3.98	2.68	5.22	4.46	3.98
Gujarat	3.40	2.96	2.74	2.47	2.14	1.92	1.69	
Jharkhand	10.25	10.82	9.81	7.93	10.15	10.06	9.84	9.54
Madhya Pradesh	4.55	4.30	4.33	4.49	4.01	3.51	2.91	
Meghalaya	8.63	8.05	7.75	8.23	5.42	5.32	5.34	5.35
Odisha	7.12	7.17	7.68	6.76	6.53	5.90	4.38	3.54
Rajasthan	2.01	1.99	2.58	2.59	2.10	2.32	2.18	

Source: Estimated from the RBI data on components of net state domestic product at factor cost by industry of origin (at constant prices, base 2004-05)http://rbi.org.in/scripts/ AnnualPublications.aspx?head=Handbook%20of%20Statistics%20on%20Indian%20 Economy Data accessed from RBI website on 09-02-2013

Mineral-rich states derive a sizeable amount of revenue from mining activities. The major source among them is from royalties collected from the mining firms. Besides royalties, states collect dead rent from lessees who have not been operating their mines and thus not paying any royalties. In addition to royalties and dead rent, states also receive some revenue from the initial application fee payable by a concession seeker; an annual fee payable by the Reconnaissance Permit (RP) and Prospecting Licence (PL) holder on the basis of the area held; surface rent; sales tax or value added tax (VAT); local area tax, e.g., Panchayat tax; and stamp duty. Some states, for example, Odisha and West Bengal, have imposed a cess as well as a surcharge on minerals to mobilise additional revenue for special purposes. However, revenues from all these sources are meagre, even in comparison with the modest returns from royalties and dead rent. There are no systematic data on the total revenues collected by states from mining sources. Because royalties are the major source of revenue from mining, the contribution of royalties from mining to the total revenue receipts (TRR) of the major states is provided here. Table 14.2 depicts the total amount of royalty collected by major mineral-producing Indian states and the share of royalties as a percentage of TRR for the year 2009–10. In absolute nominal value the mineral royalties generated by Odisha government from major and minor minerals including coal is highest compared to other mineral rich districts. However, in percentage terms the share of mineral royalties in the total revenue receipts of the Chhattisgarh state government stands at the top with an 8.1 per cent share. The other mineral rich states where the share of mineral royalties exceeds 5 per cent of TRR are Goa, with 7 per cent, Jharkhand with 6.9 per cent, and Odisha with 6.3 per cent.

Loss of Revenue

With the help of two indicators in Table 14.2 – share of royalties in TRR and share of M&Q in NSDP – the list of states where mining constitutes a major economic activity are: Jharkhand, Chhattisgarh, Odisha, and Goa. In spite of a huge mineral base, these states have remained the most laggard states, with low economic growth and low per capita income. Although states own the rights over minerals and collect extraction revenues on various accounts, the central government regulates their exploration and extraction activities. Due to several reasons, state governments have not been able to realise the potential benefits from minerals. The most crucial reasons are revenue loss due to adoption of wrong criteria for collecting royalties, and irregularity in the revision of royalty rates. Similarly, due to poor monitoring systems, states have foregone enormous revenue. Although different sources provide different estimates for the revenue loss to the states, presented here are indicative figures revealing the nature of this failure of governance. Three major mineral-producing states with numerous reports of illegality are Goa, Karnataka and Odisha. The justice MB Shah Commission report, which was tabled in Parliament in September 2012, states that Goa's iron ore mining scam is

Table 14.2   Share of mineral royalties: total revenue receipts of major mineral-producing states

| State | Royalty from Minerals | | | | Total revenue receipts@ | Mineral royalty as a percentage of TRR |
	Major non-coal minerals*	Minor Minerals#	coal royalty$	Total=Major+Minor+ Coal		
Andhra Pradesh	370.4	826.5	546.2	1743.1	64678.35	2.7
Chhattisgarh	474.4	94.3	898.4	1467.2	18153.66	8.1
Goa	285.9	1.4	-	287.3	4100.28	7.0
Gujarat	192.9	222.0	-	414.9	41672.36	1.0
Jharkhand	202.3	30.1	1142.1	1374.5	19840.77	6.9
Karnataka	427.1	64.3	-	491.4	49155.7	1.0
Madhya Pradesh	351.5	213.4	931.3	1496.1	41394.7	3.6
Odisha	894.4	202.6	872.4	1969.5	31445.3	6.3
Rajasthan	987.3	502.8		1490.1	35385.01	4.2

Source: *=IBM, 2011; #=IBM, 2011; $ =http://www.coal.nic.in/AnnualPlan2010-11.pdf

@=*http://rbi.org.in/scripts/AnnualPublications.aspx?head=State+Finances+%3a+A+Study+of+Budgets*

Note: TRR: Total revenue receipts

worth nearly 35,000 crore.[4] Similarly, the loss of revenue reported in Odisha and Karnataka are 300,000 crore[5] and 50,000 crore[6] respectively. Governance failure has also been reported in the government-managed Coal India Limited (CIL).

With a view to increasing the supply of coal in the country, the Ministry of Coal (MoC) de-reserved 48 coal blocks of CIL for the private sector. However, the de-reservation of CIL blocks has not yielded the desired results. Captive Coal mining is a mechanism intended to encourage private sector participation in coal mining on account of the limitations of CIL to increase production to meet the growing demand for coal, and to ensure a supply of coal to the core infrastructure sectors, namely, electricity generation, and steel and cement production. The delay in introducing transparency and objectivity to the process of allocating coal blocks caused a huge loss to the public exchequer. The allocation of 194 coal blocks with an aggregate 44,440 million tonnes to different government and private allottees have been estimated at INR 18,559.34 crore as on 31 March 2011 for opencast mines or opencast reserves of mixed mines. The report by the government of India Controller and Auditor General (CAG) pointed out that the government could have availed itself of a proportion of this financial benefit by expediting a decision on competitive bidding for the allocation of coal blocks. The CAG has also pointed out that the Coal Controller's Organisation did not conduct a physical inspection of allocated coal blocks in order to ascertain progress and production in comparison with the progress and production reported by the allottees as per the MMDR Act 1957. The correctness of the data furnished by the allottees, therefore remains unverified. Such governance lapses have caused enormous revenue losses to the mineral-rich states.

Usually royalties are collected according to three criteria: (i) quantity based or, more commonly, tonnage-based; (ii) *ad valorem*, that is, percentage of the value of the product; and (iii) profit-based, that is, percentage of profit. Quantity-based royalty is collected on the basis of rupees per tonne of specific mineral. Although this system is easy to implement, it does not reflect the monetary value of minerals. The value-based royalty system, *ad valorem*, is fixed on the basis of actual value of the mineral. A certain percentage of the value of mineral is collected as royalties. This is considered to be the superior method. In the profit-based system, royalties are collected as a percentage of profit, i.e., revenue minus cost. This system is believed to be non-distortionary in terms of investment decisions. However, the estimation of royalties in the second and third systems involves difficulties due to the lack of accurate data regarding the cost and sales price of the minerals, and it also involves cumbersome administrative processes.

---

4   http://www.hindustantimes.com/India-news/Goa/Goa-mining-scam-worth-Rs-34-935-crore-Justice-Shah-Commission/Article1-926249.aspx. Accessed on 10th February 2013.

5   http://www.odishaminingscam.com/press-release/. Accessed on 10th February 2013.

6   http://article.wn.com/view/2012/08/09/Karnataka_mining_scam_may_have_cost_exchequer_Rs_50000_crore/. Accessed on 10th February 2013.

Although the regime has been moving towards *ad valorem* rates in India, as many as 22 minerals still attract tonnage-based rates. The transfer to *ad valorem* rates has been constrained due to the unavailability of accurate data on prices of minerals. The result has been lump-sum revenue loss for states. This is evident in the estimations of the High Level Committee on National Mineral Policy 2006. From these estimations the revenue loss only from iron ore is presented, which accounts for more than 50 per cent of the value of major non-energy minerals produced. It is estimated that if the *ad valorem* rates were fixed at 7.5 per cent for all grades of iron ore, the revenue from the production of 142 million tonnes would have been about Rs 1600 crore, assuming Rs 1,500 per tonnes as the average sale price. On the same assumption, the royalty revenue from iron ore would have been Rs 260 crore instead of Rs 42 crore in Chhattisgarh, Rs 180 crore instead of Rs 23 crore in Jharkhand, Rs 456 crore instead of Rs 72 crore in Odisha, and Rs 418 crore instead of Rs 79.75 crore in Karnataka (GoI 2006). Therefore, the total revenue foregone due to the continuance of faulty mechanisms for collecting royalties is Rs 957 crore: a colossal sum equivalent to nine-and-a-half billion Australian dollars.

The MMRD Act of 1957 empowers the central government to determine rates of royalties for different minerals, a determination that is uniformly applicable to all states. In order to minimise uncertainty, royalty rates are revised once in every three years. The central government establishes a study group comprising representatives of state governments and the mining industry to revise the royalty rate. Past experience shows that royalty rates are not revised in a timely fashion, and this causes huge revenue loss to state exchequers.

Due to the faulty pricing mechanism in coal pricing, state governments are losing a huge amount of revenue. The new Coal Distribution Policy announced by the Ministry of Coal in 2007 envisaged distribution of coal to small and medium consumers in an efficient manner. However, no monitoring mechanism has been put in place by CIL for verification of the end-use of coal.

States' Initiatives for Additional Revenue

As discussed earlier, the ownership right over minerals is vested with states. Hence, the right to collect tax and raise revenue remains in the State List. Entry 23 of the State List relates to the regulation of mines and minerals development. However, it is expressly subject to the provisions of the Union List with respect to regulation and development under the control of the Union. Entry 54 of the Union List provides for regulation of mines and mineral development to the extent to which such regulation and development under the control of the Union is declared by parliament that Entry 23 of List II has not been made subject to any specific Entry of List I. Parliament has enacted the *Mines and Minerals (Regulation and Development) Act 1957* (MMRD Act) to provide for the regulation of mines and the development of minerals under the control of the Union in the public interest.

Under these circumstances, states are not allowed to impose any other tax on minerals. Nonetheless, mineral-rich states have formulated a number of legislative instruments that mobilise additional revenue from mining companies in the form of cess. However, states have hardly succeeded in this attempt. For example, the Government of Odisha has enacted three separate forms of legislation, in 1962, 1992 and 2004, for collecting additional revenue from mining companies and, in all cases, the court has struck them down. As a means of realizing resources, the Government of Odisha imposed a tax on mineral-bearing lands through the *Odisha Cess Act 1962*. This Act was challenged in the Odisha High Court and subsequently in the Supreme Court, and finally was determined as outside the powers of the government to enforce by the Supreme Court in 1991. Subsequently, in order to promote employment and implement production programmes in rural areas, the state enacted the *Odisha Rural Employment, Education and Production Act 1992* (OREEP Act). But this Act was also declared ultra vires by the Supreme Court in 1992. Similarly, in the year 2004 the Government of Odisha enacted the *Odisha Rural Infrastructure Socioeconomic Development Tax Act 2004*, which envisaged the augmentation of revenue by imposing tax on mineral-bearing land. However, the mining companies protested against this move and challenged the Act in the Odisha High Court. The court held that the state legislature was not competent to enforce the Act and rejected the rules formulated under it. Most importantly, the bench directed the state to refund any amount of tax collected under the Act to the mine owners.

The state government had cited augmentation of additional resources for development of infrastructure, promotion of education, employment and socio-economic programmes in under-developed rural mining areas as the objective behind imposition of the tax on mineral-bearing land. In the 2004 case between West Bengal and Kesoram Industries Limited, (SCI, 2004) the government had proceeded with imposition of the new tax in the light of the Supreme Court judgment upholding the validity of levy of cess on mineral-bearing land. A constitutional bench of the Supreme Court examined the validity of the levy of cess on coal and mineral-bearing land and tea plantations. In its 2004 judgment the Supreme Court held that the power to tax or levy in order to augment revenue shall continue to be exercisable by the state legislature in spite of regulation or control having been assumed by another legislature, that is, the Union.

In this context it is imperative to scrutinise the legitimacy of constraints imposed on states to obtain additional revenue from mineral extraction by imposing specific levies.

Validity of Uniform Royalty Rates across States

The rationale behind the uniform royalty rate, as provided by the central government, is to provide equal scope for all states to develop their mineral extraction industries. It is evident from the arguments provided by the Government of India, Department of Mines before the Sarakaria Commission, as follows:

> Levy of various cesses, surcharges at different rates has an impact on the development of minerals. This affects uniformity and introduces uncertainty in mineral prices, some of whom may have to compete not only in the national market but also in the international market. Cesses are usually levied as a percentage of royalty and therefore when levied at different rates in different States they distort the uniformity in royalty rates. Hence States need to exercise restraint on imposition of such levies, so as not to affect uniformity or competitiveness (Sarakaria 1995: 429).

The arguments of the central government, however, stand on weak theoretical foundations for following reasons:

- Competitiveness of mining firm is largely driven by the productivity of mining which is determined by the quality of minerals, cost of production and adoption of modern technology.
- Uncertainty in the prices of minerals is not substantially influenced by the royalty rate, but by market forces: demand and supply factors.
- So far as the development of minerals is concerned, apart from price, many other factors play more important roles to attract private investment. For example, improving the quality of infrastructure and institutions attracts private investment. Investment requirements for the development of infrastructure are not uniform across states.

Keeping royalty rates uniform across states does not serve any purpose, but rather prevents states from realising their revenue potential. Given the right of states to the ownership of land, they should be free to determine the royalty rates. Leaving states to determine their own royalty rates will in no way detract from the regulatory power of central government. Central government should regulate the exploration and extraction of minerals by providing clearance for mining projects. In this context, the following analysis of important judicial interpretations of the MMRD Act 1957 and the rights of central and state governments will demonstrate the extent of the states' right to levy tax on mineral extraction.

The Preamble to the *Central Act 67* of 1957 mentions that it is: 'An Act to provide for the development and regulation of mines and minerals under the control of the Union' and that taxation and the levying of fees are not subjects dealt with by the Act (SCI, 2004). The constitutional Bench of the Supreme Court (SCI, 2004) pointed out that:

a power to regulate, develop or control would not include within its ken a power to levy tax or fee except when it is only regulatory. Power to tax or levy for augmenting revenue shall continue to be exercisable by the Legislature in whom it vests i.e. the State Legislature in spite of regulation or control having been assumed by another legislature i.e. the Union .... In the garb of exercising the power to regulate, any fee or levy which has no connection with the cost or expenses of administering the regulation, cannot be imposed; only such levy can be justified as can be treated as part of regulatory measure (SCI 2004).

In *Union of India vs Harbhajan Singh Dhillon* (1971) (as cited in SCI, 2004), the court ruled that the power to legislate in respect of a matter does not carry with it a power to impose a tax under the constitutional scheme. Power to tax must be explicit otherwise there will be no power to tax: an implied power to tax does not exist.

The court further noted that there is a difference between 'power to regulate and develop' and 'power to tax':

The primary purpose of taxation is to collect revenue. Power to tax may be exercised for the purpose of regulating an industry, commerce or any other activity; the purpose of levying such tax, an impost to be more correct, is the exercise of sovereign power for the purpose of effectuating regulation though incidentally the levy may contribute to the revenue (SCI 2004).

The government has general authority to raise revenue and to choose the methods of doing so. Nevertheless, cases that deal with matters of taxation as the *primary purpose* are to be regarded as:

If revenue is the primary purpose, the imposition is a tax. Only those cases where regulation is the primary purpose can be specially referred to the police power. If the primary purpose of the legislative body in imposing the charge is to regulate, in such case, the charge is not a tax even if it produces revenue for the public (Cooley 1924: 98–99 as cited in SCI 2004).

It is inappropriate to regard the imposition of cess in all cases as increasing the price uncertainty of minerals and impeding the development of mineral extraction projects. States should have enough autonomy and rights to mobilise revenue in their own sphere. The Supreme Court in *Automobile Transport (Rajasthan) Ltd. v. The State of Rajasthan & Ors.*, (1963) observed that:

if it be held that every law made by the Legislature of a State which has repercussion on tariffs, licensing, marketing regulations, price-control etc., must have the previous sanction of the President, then the Constitution in so far as it gives plenary power to the States and State Legislatures in the fields allocated to them would be meaningless. A reasonable tax or fee levied by State legislation cannot therefore be construed as trenching upon the Union's power and freedom to regulate and control mines and minerals (SCI 2004).

It should be kept in mind that the subject of tax is different from the measure of the levy. A financial levy must have a mode of assessment, but this mode of assessment does not determine the character of a tax. To be a tax on land, the levy must have some direct and definitive relationship with the land. So long as it is a tax on land, it is open for the purpose of levying tax to adopt any one of the well-known modes of determining the value of the land, such as annual or capital value of the land or its productivity. The methodology adopted, having an indirect relationship with the land, would not alter the nature of the tax as being one on land. The nature of the machinery used for assessment is often complicated and is not of much assistance except insofar as it may throw light on the general character of the tax. The annual value is not necessarily income, but only a standard by which income may be measured. Merely because the same standard or mechanism of assessment has been adopted in legislation covered by an entry under the Union List and also by legislation covered by an entry in the State List, the latter legislation cannot be said to have encroached upon the field meant for the former.

The Supreme Court in case between the government of West Bengal and Kesoram Industries Ltd. and Ors (SCI, 2004) held that merely because a tax on land or a building is imposed by reference to its income or yield, it does not cease to be a tax on land or a building. The income or yield is taken merely as a measure of the tax and does not alter the nature or character of the levy. It still remains a tax. No one can say that a tax under a particular entry must be levied only in a particular manner. The legislature is free to adopt such method of levy as it chooses; so long as the essential character of the levy does not extend beyond the boundary of the particular entry, the manner of levying the tax would not have any vitiating effect.

Is it Possible to Transfer the Rights to Local Governments?

The question that arises is: Should ownership and taxation rights over minerals be transferred to local governments? For pertinent reasons, this would be an inefficient solution. In order to ensure inter and intra-generational equity, conserve the environment and minimise transaction costs in the free flow of mineral resources, ownership rights should be vested with state rather than local governments.[7] Local governments might face technical and financial constraints to regulate and develop mineral resources efficiently. On the other hand, the state will gain from economies of scale in regulation. However, there is a strong reason to transfer

---

7   See Das and Joe (2008).

more resources to local governments for the development of economic and social infrastructure in mining regions.

Conclusion

One can see that the present scheme for the devolution of power between central and state governments over mineral resources is inappropriate. Although the central government should retain regulatory power, state governments should be given enough maneuverability to reap the benefits of their natural mineral endowments. The present system of uniform royalty rates determined by the central government should be removed and States should be freed to determine their royalty rates and other levies. For ensuring the efficient use of the revenues generated from minerals, efforts should be taken to increase transparency in the management of mineral extraction.

References

Andersen, Jørgen Juel and Silje Aslaksen. 2008. 'Constitutions and the Resource Curse.' *Journal of Development Economics* 87(2): 227–46.

Brunnschweiler, Christa N. and Erwin H. Bulte. 2008. 'The Resource Curse Revisited: A Tale of Paradoxes and Red Herrings'. *Journal of Environmental Economics and Management* 55(3): 248–64.

Cooley, Thomas McIntyre. 1924. 'A Treatise on the Law of Taxation including the law of Special Assessment.' 4th Edition. Chicago: Callaghan and Company.

Das, Amarendra and William Joe. 2008. 'Community Participation in Monitoring Coal Production.' *Economic and Political Weekly* XLIII(6): 77–79.

GoI (Government of India). 2006. 'National Mineral Policy: Report of the High Level Committee.' New Delhi: Planning Commission.

MMSME (Ministry of Micro, Small and Medium Enterprises). 2007. 'Industrial Policy Resolution.' Government of India. Available online at http://www. laghu-udyog.com/policies/iip.htm#Indus2 (accessed on 15 April 2007).

Reserve Bank of India (RBI). 2008. Budget documents of the respective state governments. Available online athttps://59.160.162.25/businessobjects/ enterprise115/desktoplaunch1/InfoView/main/main.do?objId=6169 (accessed on 9 July 2008).

Sarakaria, R.S. 1995. 'Sarakaria Commission Reports: India, Centre State Relations Commission.' New Delhi: Government of India.

SCI (Supreme Court of India). 2004. 'CASE NO.: Appeal (civil) 1532–1533 of 1993'. Available online at http://judis.nic.in/supremecourt/qrydisp. aspx?filename=25825 (accessed on 4 July 2008).

Chapter 15

Key Policy Issues for the Indian Coal-Mining Industry[1]

Ananth Chikkatur and Ambuj Sagar

Background

Electric power and services that derive from it are critical for almost all modern economic activities, as well as for broader human development. The availability of electricity in India, however, lags far behind industrialised – and even many industrialising – countries: the per-capita electricity consumption in 2010 for the nation was only 620 kWh, which was about one-fifth that of China and just over one-thirteenth that of the Organization for Economic Cooperation and Development (OECD) average (World Development Indicators, World Bank). India has also long suffered from an insufficient supply of electricity – in November 2011, the average shortage of power to meet peak demand – April to November 2011 – was estimated to be 9 per cent (CEA 2012). The quality of power supply in the country is also very poor, with unstable voltages and frequent power failures; and this lack of an adequate and reliable supply of power is often considered as a critical constraint to industrial development.

Recently, with growing commodity prices there has been an increasing focus on energy security, which has generally devolved to a greater emphasis on using domestic resources, especially coal. Coal use currently accounts for more than 50 per cent of total primary commercial energy consumption in the country and about 70 per cent of total electricity generation. About 80 per cent of coal produced in India is used for power generation. Hence, the power and coal sectors are interlinked in India, and coal is likely to remain a crucial energy source for at least the next 30–40 years – especially since India has significant domestic coal resources relative to other fossil fuels, and substantial extant infrastructure for coal-based electricity production and distribution.

1 We acknowledge financial support for some of this work through the Harvard Kennedy School's project on Energy Technology Innovation Policy, from the David & Lucile Packard Foundation; a gift from Shell Exploration and Production; and general support grants from BP Alternative Energy and Carbon Mitigation Initiative. AC notes that the views expressed in this chapter are his alone and that they do not reflect the views of ICF International.

There are, however, a number of daunting issues that need to be addressed as India attempts to increase coal supply to meet its growing energy needs, with mitigation of the environmental and social impacts of coal mining being the most critical. Sustainable development of the Indian coal sector will require improvements in the capacity to maintain the increased production of coal in the country in an environmentally and socially acceptable manner. It is from this point of view that this chapter briefly reviews the key challenges in the Indian coal sector and proposes policy approaches to move it towards a more sustainable path.

Key Challenges

The coal sector faces two main challenges: first, that of meeting the high demand for coal, particularly for power generation; and second, that of dealing with ongoing and past socio-environmental issues.

The high demand for coal

Production of coal increased nearly six-fold since Indian coal companies were nationalised in the early 1970s, with an annual production of 533 tonnes (Mt) of raw coal and 38 Mt of lignite in 2010–11 (MOSPI 2012). Most of the production has occurred in the state-owned collieries of Coal India Limited (CIL) and Singareni Collieries Company Limited (SCCL), which between them account for about 95 per cent of current coal production. Much of the coal produced was of the non-coking type, as coking-coal reserves are limited; non-coking coal is primarily used for power generation. Furthermore, open-cut mining has dominated India's coal production since the 1970s as underground mining became more expensive.

Coal demand is expected to continue to increase, driven primarily by the power-generation sector. According to the Planning Commission (2006), long-term scenarios indicate that annual coal consumption by the power sector will range between 1 to 2 billion tonnes (BT) by 2031–32, with the total coal demand varying anywhere between 1.5 and 2.5 BT. Coal demand of 2.5 BT occurs in a scenario in which coal is the dominant fuel of choice; a demand of 1.5 BT occurs in a scenario in which nuclear, hydroelectricity, gas and renewable resources are promoted aggressively and demand side management, coal use efficiency, and transport efficiency are all increased (Planning Commission 2006). The Integrated Energy Policy Committee determined that the annual coal and lignite demand will be about 2 BT by 2030 – of which the power sector alone will demand 1.4 BT of coal and lignite (Ministry of Coal 2006). Other energy resources exhibit a range of practical problems that preclude them making a dominant contribution to the near-to-mid term growth in the power sector: some are uneconomic – as in the case of naphtha and liquefied natural gas; or have insecure supplies – diesel and imported natural gas; or are simply too complex and expensive to build – such as nuclear and hydroelectricity (Chikkatur and Sagar 2007).

The high demand for coal is already beyond domestic supply capacity, leading to increased imports. Coal imports are likely to continue to increase significantly over the next 20–25 years in order to meet projected demand. About 11 to 45 per cent of total coal demand, that is, coal imports of 70 to 450 million tonnes of oil equivalent (Mtoe) (Planning Commission 2006), are likely to be met by coal imports. This is a significant deviation from the past, when imported coal comprised only a small fraction of coal consumption. Imports were primarily for coking coal that is used in the steel industry; however, the power sector has recently been importing greater quantities of coal to mitigate domestic coal shortages. At the same time, the cost of imported coal has been increasing in the international market, adding further pressure to demand for domestic coal.

Increasingly demand for coal is for consistent, high quality. However, the quality of coal in India has been worsening over the decades because of increased open-cut mining combined with disincentives inherent in the grading structure used to determine coal quality. Improving coal quality is an important issue, as better and consistent quality improves the performance of coal power plants. Increased use of beneficiated coal – i.e., mineral ore that has been separated from valueless associated material ('gangue') for further processing or direct use – is limited by institutional and financial constraints. Coal used to be sold in grades with wide bandwidths, a process that functioned as a strong disincentive to improve coal quality. Coal washeries were not considered legitimate coal end-users and hence were not part of the transportation linkage process. However, some of these issues are expected to be resolved with the provision of contractual fuel supply agreements between CIL and consumers, as per the *New Coal Distribution Policy*, enacted in 2007. The success of this distribution policy is yet to be evaluated.

Constraints on expanding production   Increasing coal demand in India requires dramatic expansion and acceleration of coal exploration, production and processing in the country. Although the Indian coal sector does possess significant exploration and resource assessment capacity, this capacity is increasingly under strain. The Geological Survey of India (GSI) and the Central Mine Planning and Design Institute limited (CMPDI) – a subsidiary of CIL – are the main exploration agencies in India (see Chikkatur and Sagar 2007). The key limiting factors for increasing exploration capacity are the limited domestic technological capacity and low availability of suitable human resources. Based on research conducted by the authors in 2007, virtually no new geologists and geophysicists have been hired for coal exploration since 1990. Further, there has been very little investment in upgrading drilling machines and associated technologies, adapting and deploying new exploration technologies and carrying out indigenous exploration in research and development (R&D).

Expanding production from existing and new mines has been constrained in part by lack of investment in underground mining and inability to resolve socio-environmental problems associated with open-cut mining (see next section). Given that open-cut mining causes a high level of environmental damage, many analysts have called for greater investment and planning in underground coal production in the country (Chand 2005; Ministry of Coal 2006). However, increasing underground mining projects requires significant new investment in manufacturing and human resources.

Socio-environmental issues

The human and social impacts resulting from coal mining have been significant by any measure; issues relating to displacement and rehabilitation and resettlement (R&R) have been a major and persistent problem for the Indian coal industry. Most of the tenancy land acquired for mining in the country has been transferred through application of the *Land Acquisition Act 1894* or the *Coal Bearing Areas (Acquisition and Development) Act 1957*, which has led to a large pool of involuntarily displaced persons (Banerjee 2004).

Displacement of people due to coal mining in India is a particularly significant phenomenon that occurs for several reasons: (i) coal is the largest mining industry in India and it requires the largest amount of land for mining; (ii) coal reserves are located in heavily forested parts of the country inhabited primarily by tribal groups and socially weak communities; (iii) unlike other development projects, the location of mining is fixed by geological factors and there is no flexibility in this.

Despite the injurious social impact of displacement in coal mining, there is very little detailed official data concerning the number of displaced people and their status. Partly this is because coal mining has a very long history in India and displacement was not considered an important metric until recently. Also, displacement data prior to nationalisation of coal are difficult to obtain.

Nonetheless, there are some estimates of displacement, primarily undertaken by Walter Fernandes and colleagues at the Indian Social Institute. In total, displacement from all development projects between 1951 and 1991 in India has been estimated to be about 21 million people (Fernandes 1995 as cited in Bala 2006); more recent estimates for the period 1951–2000 suggest that the number could be as high as 60 million (Fernandes 2007). Only 29 per cent of displaced people are estimated to have been rehabilitated, leaving almost 13.2 million uprooted people (Roy 1994 as cited in Saxena 2006). Scheduled tribes bear the disproportionate brunt of displacement: while they constitute only about 8 per cent of the country's population, they comprised almost 40 per cent of displaced people until 1990 and nearly 50 per cent by 1995 (Guha 2005).

Mining of all minerals has been the second-largest cause of displacement – recent estimates put the number of those displaced by mining at 5 million (Fernandes 2007). It is also estimated that tribal peoples and dalits account for somewhere between 25 per cent (Fernandes 2007) and 40 per cent (Sethi 2006). Note that dams for irrigation and hydropower account for nearly 77 per cent of the total development-related displaced population, and mining – for all minerals – accounts for about 12 per cent (Sethi 2006). Moreover, displacement not just due to mining has resulted in greater impoverishment. In India it is estimated that up to 75 per cent of displaced people are worse off as a result of their displacement (Fernandes et al. 1989 as quoted in Cernea 2003).

Even with R&R, there are multiple dimensions of human, social and economic impacts on displaced people: the breakdown of family and community structures; greater class and caste conflicts, with women, elderly and children particularly vulnerable; loss of livelihoods and worsening of family and individual economic situations due to disruption (Verma 2004). Even if not displaced, local communities surrounding the projects often suffer from local environmental impacts of mining activities, including water scarcity; water pollution and air pollution, with human health and agricultural impacts; and deforestation, with a concomitant loss of livelihood.

The environmental impacts of coal mining are also considerable: (i) release of total suspended particulates (TSPs) and respirable particulate matter of 10 micrometres or less (PM_{10}) from the fugitive dust during open-cut mining and transportation operations (Chaulya 2004; Singh 2006); (ii) damage to water resources from coal mining, washing and associated activities through water-use, runoff, acid mine drainage and damage to aquifers from open-cut mines (Singh 2006); (iii) land degradation and deforestation, especially because of the large areas required for extractive processes as well as for the overburden in open-cut mining; (iv) the consequences of deforestation, with concomitant effects on biodiversity and wildlife corridors (*Telegraph* 2005). More than 0.8 million hectares of land in the country are currently retained for mining (Jain 2003).

Governance and Institutional Problems

A number of governance and institutional issues also play a critical role in the coal-mining sector. These are briefly outlined below.

Government and public-sector institutions' dominance

The Ministry of Coal determines policies and provides guidelines for all matters regarding exploration, production, supply, distribution and sale of coal and lignite. The Ministry is in administrative control of major coal-producing companies including CIL, SCCL, and Neyveli Lignite Corporation (NLC). It also oversees the Coal Controller's Organisation, which grants permission for opening new seams and mines, collects and publishes data on the coal sector, collects excise duties, and monitors progress in captive mining. In addition, various other ministries play contributing roles in the coal sector, including the Ministries of Power, Mines, Environment and Forests, Labour and Finance. The Planning Commission, which sets the long-term vision and priorities for the government, provides overall policy guidance and specific growth targets for CIL.

Given the government dominance of the coal sector, the actions and policies of any single government agency can affect the trajectory of the entire sector; conversely, any significant changes require bringing on board a number of key organisations. Institutional reforms in the coal sector that could greatly advance the sector's sustainable development are difficult because the government bureaucracy has resisted these changes, as have many other groups. The coal sector is politically powerful and the heavy presence of labour unions makes legislative changes – such as amendments to existing laws – difficult to enact. Therefore, institutional reforms have often been placed within the confines of existing legislation – leading to convoluted policies.

Corruption and lack of transparency

Problems with governance and corruption constrain the growth and productive development of the sector. For example, as the recent CAG (2012) report indicated, the process of allocating coal blocks has suffered from lack of transparency and susceptibility to political interference – this led to significant windfall gains for those who were allocated the blocks and also hindered competition. A system less exposed to corruption, such as a market-based auction process, would benefit the public interest.

As another example, it is commonly – although not always publicly – accepted that illegal mining is a problem; yet, the extent of the problem is not fully known. Issues such as lack of accurate data collection on depleted reserves – not merely cumulative production – and the general lack of transparency and independent oversight in coal mining are problematic. Moreover, coal mining companies are reluctant to reveal such information, because of the fear of retribution and the loss of private gains. The governance and corruption issue is generally uncomfortable for many individuals and groups, as potential and real conflicts of interest exist at all levels of administration and management.

Poor R&R policy design and implementation

Displacement issues, especially R&R, have been persistent problems for the Indian coal industry, in large part because of the lack of coherent public policy in this area and lack of unwavering commitment by the industry. Although CIL has had a R&R policy in place since 1994, and has recently updated it, the government's R&R policies have all been criticised, both for their formulation processes and because of the perception that they support industry over the concerns of the people, particularly in matters of land acquisition. In addition, the land acquisition process has recently come under intense scrutiny with the promulgation of the new Special Economic Zone policies.

At the same time, the R&R activities of firms often leave much to be desired. CIL, for instance, has had an ambitious R&R policy since 1994, but subsequent experience with a major World Bank-funded project intended explicitly to assist in environmental and social mitigation during expansion of 25 coal mines revealed serious shortcomings in implementation: Central Coalfield Limited (CCL, a CIL subsidiary) failed in meeting the goal of successful rehabilitation or even CIL's own policies, resulting in a lack of adequate compensation for lost assets and loss of income (Herbert and Lahiri-Dutt 2004; World Bank 2002). On the environmental side, CCL did not plan for reclamation of the mined land, despite it being included as part of the project plans (Herbert and Lahiri-Dutt 2004). To make things worse, it appears that many of the practices condoned by CCL contravened what one would expect of good-faith efforts (Herbert and Lahiri-Dutt 2004). A wide gap between stated policies and implementation has given rise to an atmosphere of mistrust, which is difficult to overcome and hinders future rehabilitation efforts.

The lack of desire or ability to address and amicably resolve the appalling past social and environmental record of the sector could become a key limiting factor for further growth. People who have observed the fate of project affected-people over the past several decades no longer want to transfer their land for mining projects, as they do not see any benefit in it. Moreover, the public sector units cannot blame all of the R&R problems on the private firms that held the mines prior to nationalisation. Forcible expulsion and attempts to cheat people of fair compensation will serve to worsen the public's relations with – and perception of – government and industry. This issue is relevant not only for the coal sector, but also for all future industrial and infrastructure projects, as competition for different land-use activities will only increase over time.

Towards Better Policies

Lack of transparency in decision making and robust governance are the largest barriers to the sustainable development of the Indian coal-mining sector. These challenges are widespread across different government sectors, and many of these issues can be resolved only in the political domain. Here, we focus on the policy-making and implementation aspects in the coal sector and suggest significant steps that could align policy-making with the sustainable development imperative for the coal-mining sector.

Moving beyond traditional notions of energy security

Currently, energy policy in the country is driven by dealing with energy security, and energy security is narrowly defined as increasing and protecting fuel supply. For example, the *Integrated Energy Policy* explains that energy security 'at its broadest level, is primarily about ensuring the continuous availability of commercial energy at competitive prices to support its economic growth' (Planning Commission 2006). However, the Policy also discusses energy security of the poor and provision of lifeline energy needs of its households with safe, clean and convenient forms of energy. While this issues are mentioned, it is clear that policymakers are mostly focused on expanding fossil fuel supply rather than managing demand or placing increased supply in the context of other social changes. Hence, it is critical for Indian policymakers to move beyond a narrow definition of energy security and begin dealing with the more politically difficult task of bringing about social and environmental security.

Managing coal demand

The present and future role of coal in the Indian power sector, and visa-versa, cannot be understated. The strong influence of the power sector on coal demand implies that better power policies can help to reduce the demand for coal, especially in the short term; thereby providing time and breathing room for the coal sector to devise and implement appropriate policies and institutional changes. The current overemphasis on coal production forces the government into a default state of panic that exerts pressure for production at all costs, leading to the problems discussed earlier. Although a detailed assessment of technology policies for coal power has been presented elsewhere (Chikkatur and Sagar 2007), a key policy element of a broader energy strategy – greater energy efficiency – is discussed briefly below.

There is significant potential for improving the efficiency of all elements within the country's existing power system. On the generation side, the efficiency of existing power plants based on sub-critical pulverized coal (PC) technology has great potential for improvement. The average net efficiency of the overall sector is 29 per cent, with the best units (500 MW) achieving 33 per cent (Chikkatur 2005). Nearly all existing power plants can improve their efficiency by at least 1–2 percentage points, and improvement by one percentage point will reduce coal use, and corresponding air pollution and CO_2 emissions, by three per cent (Deo Sharma 2004). Current losses in the Indian transmission and distribution (T&D) system are very high: reducing these losses to a more manageable – though still high – 10 per cent will release power equivalent to about 10–12,000 MW of capacity (CEA 2007), thereby reducing coal demand. Existing efforts to upgrade the T&D system by modernising the existing infrastructure and introducing new technologies must be accelerated through steps such as expanding high-voltage lines, improving integration among regional grids, and improving monitoring and metering of distribution networks.

Improving the efficiency of end-use applications is also important. Each kW saved in end-use is equivalent to almost 1.8 kW of generation, once auxiliary consumption at the power plant and T&D losses are taken into account. There is great potential for end-use energy-efficiency gains in the country. For example, it is estimated that the deployment of energy-efficient lighting, more efficient refrigerators in households, and more efficient motors in industry could save as much as 10 per cent of national power generation (Shrestha et al. 1998).

Improved data collection and assessment   Robust data and data analysis are critical for devising appropriate policies. In the coal sector, although some data – such as production, number of operating mines, type of mining technology, manpower productivity – are readily available from the CCO, critical data – such as the total economically mineable resources, depleted reserves, number of displaced people, abandoned mines – are missing. A good understanding of domestic coal reserves, including information about the economic feasibility and technical extractability of coal resources, is essential for better energy planning and policies, which is in turn necessary to ensure that social and environmental conditions are fully considered. There are significant problems with the methodologies used for assessing Indian coal resources (Chand 2005; Chikkatur 2005; Ministry of Coal 2006).

Detailed data and analysis from mining operations, including depleted reserves, is important in assessing mining efficiency, management and whether available reserves are being used to the fullest extent. Data regarding collection, collation and time trends of environmental and social factors related to coal mining is critical for better R&R policies – and often, such data is missing or not easily available. Furthermore, there is very little information about abandoned mines, the reasons for abandonment and the amount of remaining reserves in abandoned mines – therefore, we do not have an understanding of possible reclamation potential or barriers to reclamation.

A consensus roadmap for coal exploration and extraction

A systematic effort to develop a technology roadmap for the coal exploration and extraction sector is necessary in order to incorporate the latest technological developments and enable proper consideration of options for mitigating environmental and social impacts resulting from exploration and extraction, while enhancing the effectiveness of these activities. Particular effort is needed to increase underground mining, an area where India has lagged behind even as other countries, especially China, have made progress in recent decades. A 2007 Expert Committee report has provided some significant proposals for increasing underground mining (Ministry of Coal 2007). Other associated options such as underground coal gasification and coal bed methane extraction must also be part of a strategic roadmap, as these technologies have potentially significant environmental benefits and costs relative to conventional options. The development of a roadmap will also support overall planning in the coal sector by assisting in the timely introduction of technologies that could improve mine and worker safety while reducing impacts on the broader environment.

Responsive and participatory decision-making

There are a large number of stakeholders in the coal sector. Reconciling their interests and resolving their conflicts will require the development of a transparent, inclusive, and responsive process that is seen as fair by all involved. In effect, it is important to adopt a philosophy of 'planning with people' that empowers all stakeholders and allows them to participate in decision-making, rather than 'planning of people' or 'planning for people' (Burton 1976 as cited in Sidaway 2005).

Therefore, it would be useful to begin by developing a vision for the coal production sector as part of a larger visioning exercise for the overall coal power sector. This would require building consensus across various stakeholders regarding the need for coal production in the country and how benefits could be shared appropriately while limiting the social and environmental costs borne by a minority of them. For example, at a 2008 workshop at Ranchi, where both industry and non-government organisation (NGO) representatives were present, a common consensus-based vision was arrived at: 'Coal, a national asset, must be assessed, extracted and used in a scientific and viable manner with due responsibility for working conditions as well as ecological and social sustainability to meet the human and economic development needs in the country' (Chikkatur and Sagar 2009a).

While agreement on a vision is necessary, the vision must also then allow for the prioritisation of the pivotal issues that need to be addressed, filling in missing data and analysis gaps and developing appropriate policies. Nonetheless, a common vision is the first critical step.

Environmental policies

The Environmental Impact Assessment (EIA) process conducted by the Ministry of Environment and Forests (MoEF) is a crucial step of the project approval process. The EIA process is complex and the cost–benefit analysis is often laborious and difficult; yet, the EIA is a very important tool that provides relevant data and information for the public, for project proponents, and for the regulatory authorities (MoEF). An EIA should be performed by reputed and accredited agencies with stakeholder input, and involve input from local people.

However, there have been numerous documented problems with EIA preparation. For example: some faulty EIAs have been cut-and-pasted from other EIAs (see, e.g., Dutta 2007); very little time is provided for accurate data collection – often rapid EIAs without full seasonal data are used for expedience (Dubey 2006); and the project proponent directly funds the agency preparing EIAs, rather than providing funds to an independent authority which then chooses a preparation agency. Furthermore, EIAs are prepared after nearly all of the project-related activity is already complete – thereby placing enormous pressure on regulatory authorities, particularly the MoEF, to approve projects; this pressure also creates the potential for corruption. Project-affected people and NGOs feel that the EIA process is often subverted and that important issues are sidelined or ignored.

On the other hand, many in the mining industry feel that the EIA process is too slow and impedes the timely progress of exploration and mine development activities. Some of them regard the clearance decision-making process as arbitrary and opaque. Moreover, coal mining has to take place in areas where coal is located, therefore, mining companies do not much have a choice in terms of where they can mine. In contrast, they consider agriculture and other livelihoods can be relocated to other areas. However, local people who have lived on the to-be-mined areas have historical, cultural, and economic attachment to their lands. These lands can also have significant biodiversity and can be homes to a wide variety of flora and fauna, which cannot be transplanted.

These different perspectives lead to a contentious and unproductive situation in which the focus is more on the shortcomings of the EIA process rather than on working collaboratively with the MoEF in developing a process that takes into account these concerns while paying full attention to environmental issues. A streamlined and more transparent process which removes uncertainties from the approval process would be highly desirable for the mining industry; local communities and NGOs, on the other hand, would very much support a public hearing process that is seen as thorough and responsive.

In effect, the EIA and the public hearing processes can be a venue for consensus-building between government agencies, industry, and the citizenry to protect the local environment as well as the rights of project-affected people. Assessing environmental impacts could happen first, so that the viability of a project can be assessed on techno-economic grounds and on environmental bases simultaneously. The EIA process should be taken seriously by all stakeholders and subverting this process through fraudulent EIAs and improper data collection will only hinder reconciliation and stakeholder consensus.

R&R Policy

Cernea (2006) has noted that risks to the environment and risks to investors are given attention, but that social risks are often ignored: 'some risks, to some stakeholders, are considered while other risks to other stakeholders – like the risk-set imposed on displaced populations – are "beyond the horizon" in the mindsets, attitudes, and management discourse.' However, this attitude has to be changed to ensure that R&R policies in India are successful.

The goal of R&R efforts should be to ensure that project-affected persons (PAPs) are as well or better-off than before displacement. Thus, R&R efforts should be seen as development projects and the policy goal should be to enable the dispossessed to re-establish themselves productively and improve their livelihoods and lives. It is only through such an approach that the PAPs will see themselves as sharing in the development benefits that arise from mining, as opposed to the jaundiced perspective of historical practices from which they have not benefited during national development, despite having given up their lands. Distribution and equity issues must receive explicit consideration during the R&R process, with special consideration being afforded the most vulnerable groups, which also have the least bargaining capacity. Direct compensation mechanisms are necessary, but need to be improved. However, compensation cannot be the only mechanism used to redress the grievances of displacement. Mechanisms such as safety nets to protect the vulnerable and provide appropriate institutional and financial capacity should be instituted to ensure resettlement with welfare improvement.

As the World Bank states, 'resettlement activities should be conceived and executed as sustainable development programs, providing sufficient investment resources to give the persons displaced by the project the opportunity to share in the project benefits' (World Bank 2001, as cited in Cernea 2003). The Extractive Industries Review (World Bank 2003) notes that it is particularly critical for local communities to receive benefits from extractive industries projects. Some crucial steps to ensure this are:

- Engaging in consent processes with communities and groups directly affected by the projects.
- Revenue sharing with local communities.
- Systematic monitoring using poverty indicators.
- Incorporation of public health components in projects.
- Helping to build the capacity of affected communities.
- Establishing independent grievance mechanisms.

The review also suggests: instituting integrated environmental and social impact assessments instead of disaggregating the two; planning for closure, especially by ensuring funds for appropriate end-of-life activities and helping those who may lose jobs; plans for emergency prevention and response; and addressing the legacies of the past.

The potential role of management in the extractive industries in promoting sustainable development in mining projects cannot be overstated. Cernea (2006) suggests that the mark of far-sighted management is that it engages in the recognition, rather than denial, of risks. Thus, mining firms must anticipate, identify and mitigate all reasonable project risks, including those of displaced people. This involves paying particular attention to landlessness, joblessness, homelessness, marginalisation, food insecurity, loss of access to common property resources, increased morbidity and community disarticulation. Reducing and reversing these risks require explicit strategies as well as adequate funds (Cernea 1997).

Unfortunately, both the new *National Policy on Resettlement and Rehabilitation* (NPRR 2007) (MoRD 2007) and CIL's subsequent R&R policy (CIL 2008) do not go far enough. More importantly, it is less policy content as their implementation that matter. Though the 2008 policy is not discussed in this chapter, some preliminary comparisons between CIL's 1994 and 2008 policies are presented here. One major change with the new CIL policy is that it focuses on project-affected families (PAFs), rather than PAPs – and compensation is provided on a family basis. The Policy also covers considerable detail and the discretion for subsidiaries is more limited. Overall, the 2008 policy attempts to move people away from an expectation of mine jobs. It allows 'a provision of monetary compensation additional to the value of land to those PAFs who forego any claim – perceived or otherwise on mine jobs' (CIL 2008, 9). It states that 'as a policy, mine jobs will be offered only in exceptional circumstances with approval of subsidiaries' boards and PAPs getting jobs under such circumstances would not get additional monetary compensation in lieu of employment' (CIL 2008, 9). In essence, CIL is trying to deal with the reality that mining requires skilled labour, and that a large unskilled labour population is unproductive. However, many of the PAFs – and PAPs – may resist, as they continue to see employment as critical to their economic wellbeing. Hence, CIL will need to demonstrate that other compensation options are in their interest and not just in CIL's interests – this task is made more difficult as CIL and its subsidiaries need first to address their past R&R legacies.

Dealing with past legacies

It is of crucial importance in the Indian context that past legacies are redressed, particularly given our history and poor experience with displacement and the almost singular focus on employment in mining projects as the primary compensation mechanism. For example, the draft rehabilitation policy prepared by the National Advisory Council had only a 10-year retrospective provision (Sethi 2006), although this seems to have been omitted in the NRRP 2007. Employment as a compensatory mechanism for land acquisition did serve as an attractive incentive for PAPs to transfer their land for coal projects in the past. The PAPs appeared satisfied with the prevailing norm of one job per two or three acres of land acquired. However, this situation is no longer tenable as coal mining is becoming increasingly mechanised and coal companies are shifting towards contract mining, requiring fewer numbers of people in new projects. At the same time, people's demand for employment is increasing as unemployment in the area continues to soar. Thus, it is critical for the coal sector, as well as for the local people, to determine alternative R&R approaches beyond mere employment.

Exploring innovative approaches

Innovative approaches that promote environmental and social protection while allowing for the desired increases in coal extraction must be explored. In this context, it would be useful to learn from international experiences and policy experiments in these areas. In the United States, for example, the *Surface Mining Law* specifies that a coal mining permit requires posting of a reclamation bond, which ensures the availability of funds to reclaim a mine site if the operator does not complete the reclamation plan presented in the permit. The Working Group for Coal and Lignite for the Eleventh Plan proposed the notion of 'green credits', which would allow coal-mining companies to conduct afforestation in advance and be given green credits for acres of new forest created. The utility and feasibility of such ideas need to be widely discussed and debated and, if appropriate, considered for implementation. Other ideas include: accreditation and benchmarking of EIA preparation agencies; involving NGOs and local people in the EIA preparation process; building up environmental data through satellite imagery; public dissemination of environmental information; leasing of land for mining, rather than acquisition; and providing a share of the revenues from mining directly to PAPs or *Panchayat*s.

As the Indian coal sector stands poised for greater growth, the need for more comprehensive and humane planning and implementation of coal mining projects and a transition towards sustainable development of the sector is crucial. While this will require progress on a number of fronts, perhaps the most critical element from a sustainable development perspective will be the willingness of the various stakeholders and decision-makers to work together to reduce and manage the conflicts between the environment, the rights of local communities, and the demands of the coal sector. In the end, a sustainable coal sector will not only require reduction of the environmental and social impacts of coal mining, but will also necessitate equitable sharing of the benefits of coal mining activities.

References

Bala, M. 2006. 'Indian National Policy on Resettlement and Rehabilitation and the Marginalization of Women.' *Social Science Research Network*. Available online at http://papers.ssrn.com/sol3/papers.cfm?abstract_id=942913 (accessed on 5 March 2008).

Banerjee, S. P. 2004. 'Social Dimensions of Mining Sector.' *IE(I) Journal – MN* 85: 5–10.

Bureau of Energy Efficiency (BEE). 2008. *Verified Savings Report for the Year 2007–08*. Delhi, India: Bureau of Energy Efficiency, Government of India. See: http://beeindia.in/content.php?page=miscellaneous/energy_savings_achieved.php

Bureau of Energy Efficiency (BEE). 2010. *Verified Savings Report for the Year 2009–10*. Delhi, India: Bureau of Energy Efficiency, Government of India. See: http://beeindia.in/content.php?page=miscellaneous/energy_savings_achieved.php

Central Electricity Authority (CEA). 2004. *Draft Report on National Electricity Plan (Volume I)*. Delhi, India: Central Electricity Authority, Government of India.

Central Electricity Authority (CEA). 2007. *Report of the Working Group on Power for 11th Plan*. CEA, Government of India. See: http://cea.nic.in/planning/WG%2021.3.07%20pdf/03%20Contents.pdf (accessed May 2008).

Central Electricity Authority (CEA). 2012. *Monthly Review of Power Sector Reports (Executive Summary)*. CEA, Government of India. See: http://cea.nic.in/executive_summary.html

Cernea, M. M. 1997. 'The Risks and Reconstruction Model for Resettling Displaced Populations.' *World Development* 25(10): 1569–87.

Cernea, M. M. 2003. 'For a New Economics of Resettlement: A Sociological Critique of the Compensation Principle.' *International Social Science Journal* 55(1): 37–45.

Cernea, M. M. 2006. 'Resettlement Management: Denying or Confronting Risks.' In *Managing Resettlement in India: Approaches, Issues, and Experiences*, edited by H. M. Mathur, 19–44. New Delhi, India: Oxford University Press.

Cernea, M. M. 2007. 'Financing for Development: Benefit-Sharing Mechanisms in Population Resettlement.' *Economic and Political Weekly* 24 March 2007: 1033–46.

Chand, S. K. 2005. 'Can Domestic Coal Continue to Remain King?' *TERI Newswire* 1–15 April 2005. Available online at http://www.teriin.org (accessed on 15 October 2005).

Chaulya, S. K. 2004. 'Assessment and Management of Air Quality for an Opencast Coal Mining Area.' *Journal of Environmental Management* 70(1): 1–14.

Chikkatur, A. 2005. 'Making the Best Use of India's Coal Resources.' *Economic and Political Weekly* 40(52): 5457–61.

Chikkatur, A. P. and A. D. Sagar 2007. *Cleaner Power in India: Towards a Clean-Coal-Technology Roadmap.* Discussion Paper 2007–06, Belfer Center for Science and International Affairs, Harvard University, Cambridge, MA, December 2007.

Chikkatur, A. P. and A. D. Sagar 2009a. *Proceedings of the Second Research Seminar on Coal Mining Technologies and Socio-Environmental Issues.* Ranchi, January 10–12, 2008 (Unpublished).

Chikkatur, A. P. and A. D. Sagar. 2009b. 'Rethinking India's Coal-power Technology Trajectory.' *Economic and Political Weekly* 14 November 2009.

Coal India Limited (CIL). 2008. *Resettlement and Rehabilitation Policy of Coal India Ltd.* Coal India Limited (CIL), Calcutta Available online at: http://coalindia.nic.in/policy%20on%20CIL%20R&R%20%2017062008.pdf (accessed July 2009).

Comptroller and Auditor General (CAG). 2012. *Performance Audit on Allocation of Coal Blocks and Augmentation of Coal Production.* Report of the Comptroller and Auditor General of India, 21 May 2012. See: www.cag.gov.

Deo Sharma, S. C. 2004. 'Coal-fired Power Plant Heat Rate and Efficiency Improvement in India.' *Workshop on Near-Term Options to Reduce CO2 Emissions from the Electric Power Generation Sector in APEC Economies*, February 2004, Asia Pacific Economic Cooperation (APEC), Queensland, Australia. Available online at http://www.iea.org/dbtw-wpd/Textbase/work/2004/zets/apec/presentations/sharma.pdf (accessed on March 2006).

Dubey, S. 2006. *EIA: Foundations of Failure.* Available online at http://www.indiatogether.org/2006/mar/env-eiafail.htm (accessed on 5 May 2008).

Dutta, R. 2007. *First Report on National Level Environment Impact Assessment Response Centre*. Available online at http://www.ruffordsmallgrants.org/files/ Rufford%20Final.doc (accessed July 28, 2009).

Fernandes, N., Das J. C., and Rao S. 1989. ,Development and Rehabilitation: An Estimate of Extent and Processes.' In *Development, Displacement, and Rehabilitation*, edited by W. Fernandes and E. Ganguli-Thukral, New Delhi: Indian Social Institute.

Fernandes, W. 1995. 'Rehabilitation of Displaced Persons: A Critique of the Draft National Policy.' *People's Action* 3(2): 3–10.

Fernandes, W. 2007. 'Mines, Mining, and Displacement in India.' In *Proceedings of the 1st International Conference on Managing the Social and Environmental Consequences of Coal Mining in India*, edited by G. Singh, D. Laurence and K. Lahiri-Dutt, 19–21. New Delhi: Indian School of Mines University, November 2007.

Guha, A., 2005. 'Resettlement and Rehabilitation: First National Policy.' *Economic and Political Weekly* 40: 4978–4802.

Herbert, T., Lahiri-Dutt, K. 2004. 'Coal Sector Loans and Displacement of Indigenous Populations.' *Economic and Political Weekly* 39: 2403–9.

Jain, K.K. 2003. *Evolution of Environmental Management in Mining Industry in India*. ENVIS center on environmental problems of mining (June 2003). See: http://www.geocities.com/envis_ism011/table_kkj_pdf.html (accessed April 2007).

Kanbur, R. 2003. 'Development Economics and the Compensation Principle.' *International Social Science Journal* 55(1): 27–35.

Ministry of Coal. 2006. *Report (Part-I) of the Expert Committee on Road Map for Coal Sector Reforms*. Delhi, India: Ministry of Coal, Government of India. Available online at http://www.coal.nic.in/expertreport.pdf (accessed on 15 October 2008).

Ministry of Coal. 2007. *Report (Part-II) of the Expert Committee on Road Map for Coal Sector Reforms*. Delhi, India: Ministry of Coal, Government of India.

Ministry of Rural Development (MoRD). 2007. *National Policy on Resettlement and Rehabilitation*. Delhi, India: Ministry of Rural Development, Government of India.

Ministry of Statistics and Programme Implementation (MOSPI). 2006. *Energy Statistics 2012*. Government of India. Available online at: http://mospi.nic. in/mospi_new/upload/Energy_Statistics_2012_28mar.pdf (accessed on 10 January 2013).

Planning Commission. 2006. *Integrated Energy Policy: Report of the Expert Committee*. Planning Commission, Government of India. Available online at http://planningcommission.nic.in/reports/genrep/rep _intengy.pdf (accessed on 1 April 2008).

Roy, D. 1994. 'Large Projects for Whose Benefit?' *Economic and Political Weekly* 29(50): 3129.

Saxena, N. C. 2006. 'The Resettlement and Rehabilitation Policy of India.' In *Managing Resettlement in India: Approaches, Issues, and Experiences*, edited by H. M. Mathur, 99–123. New York: Oxford University Press.

Sethi, N. 2006. 'India's Rehabilitation Policy under Scanner Again.' *Down to Earth*, Centre of Science and Environment, New Delhi, 20 June 2006.

Shrestha, R. M., Natarajan, B., Chakaravarti, K. K., Shrestha, R. 1998. 'Environmental and Power Generation Implications of Efficient Electrical Appliances for India.' *Energy* 23: 1065–72.

Sidaway R. 2005. *Resolving Environmental Disputes: From Conflict to Consensus*. London: Earthscan.

Singh, G. 2006. 'Environmental Issues with Best Management Practice of Coal Mining in India. *Responsible Mining – A Multi-stakeholder Perspective.*' TERI, New Delhi, February 2006.

The Telegraph. 2005. 'Alarm for Wildlife Habitat.' *Telegraph* 15 October 2005.

Verma, M. K. 2004. 'Development-induced Displacement: A Socio-economic Study of Thermal Power Projects.' *Man in India* 84(3/4): 209–45.

World Bank. 2001. *Operational Policies 4.12: Involuntary Resettlement*. Washington, D.C.: World Bank.

World Bank. 2002. *The Inspection Panel Investigation Report – India: Coal Sector Environmental Social Mitigation Project.* World Bank, Washington, D.C. 25 November.

World Bank. 2003. *Striking a Better Balance: The Extractive Industries Review*. Washington, D.C.: World Bank.

Index